PREHISTORIC LIFE

PREHISTORIC LIFE

Gemuendina, a specialized early Devonian chordate. The long fin-rays suggest the skate, a group to which it is only distantly related. Photograph of a reconstruction, after Broili.

PREHISTORIC LIFE

PERCY E. RAYMOND
LATE PROFESSOR OF PALEONTOLOGY
HARVARD UNIVERSITY

CAMBRIDGE : HARVARD UNIVERSITY PRESS : 1967

COPYRIGHT, 1939, 1947
BY THE PRESIDENT AND FELLOWS OF HARVARD COLLEGE

DISTRIBUTED IN GREAT BRITAIN BY
OXFORD UNIVERSITY PRESS, LONDON

Fifth Printing

PRINTED IN THE UNITED STATES OF AMERICA

CONTENTS

	Introduction	vii
I.	Fossils	3
II.	Collecting Fossils	12
III.	The First Animals and Plants	19
IV.	Pre-Cambrian Life	29
V.	The Ordovician Fauna	40
VI.	Petrified Butterflies	47
VII.	The Aquatic Arachnids	55
VIII.	The Radiates	62
IX.	The Beginnings of the Chordates	73
X.	The Origin of the Vertebrates	84
XI.	The Rise of the Air-Breathing Vertebrates	96
XII.	Amphibia, the First Tetrapods	108
XIII.	The First Reptiles	119
XIV.	The Terrible Lizards	128
XV.	More about Dinosaurs	144
XVI.	Marine Reptiles	153
XVII.	Flying Reptiles	164
XVIII.	From Scales to Feathers	173
XIX.	Squids, Devilfish, and Chambered Shells	189
XX.	Insects: The First Aviators	200
XXI.	Archaic Mammals	209
XXII.	The Mammals Inherit the Earth	220
XXIII.	Beasts of Prey	228
XXIV.	Vegetarians Seize Their Opportunity	238
XXV.	Some Genealogies	253
XXVI.	Mammals of Yesterday	264
XXVII.	The Ancestry of Man	274
XXVIII.	The Importance of Plants	292
XXIX.	Retrospect and Prospect	300
	Index	313

INTRODUCTION

We have been told, perhaps too often, that "the proper study of mankind is man." But what of man's ancestors, man's background? Facts mean little unless they can be interpreted in terms of contemporaneous and previous history. Morse invented the telegraph, Bell the telephone, and Marconi the "wireless." Each is heralded as a wonder-worker, and deservedly so. But would any one of them have accomplished anything had it not been for the less spectacular labors of his predecessors in the study of that field of physics now commonly labeled Electricity and Magnetism? No great advance in knowledge has been sudden. An inspiration occurs only to one who knows what has previously been discovered.

In other words, progress depends upon information accumulated in the past. Until recently it has been impossible for any animal, even man, to preserve records that might benefit future generations, but before the invention of writing, the wise men of prehistoric tribes handed on traditions. Even earlier than the wise men, animal instincts were communicated from generation to generation by mothers, with or without the assistance of the less responsible fathers. It is impossible to say when instincts, as teachable qualities, arose; but back in the distant past there were countless animals, orphaned before birth, whose environment furnished their only education. To them, life was purely an experiment. Various paths of evolutionary change were open to them. Their opportunities were, of course, largely controlled by their ancestry. Fish-roe could not produce oysters, nor would oyster-spat develop into fish. "Like begets like." But, although there are at present thousands of kinds of fish, and many kinds of oysters, it is possible to trace genealogies back to a time when there was only one kind of fish and one kind of oyster. Indeed, it is possible to go still further, to a time when there was neither fish nor oyster, only ancestral forms, more or less like them.

It is the purpose of this book to trace the history of life from the time of its first appearance on the earth to the present. That history has been one of constant change, for better or for worse. Practically all Nature's changes are evolutionary in that they are orderly, sequential. The inorganic, as well as the organic, evolves. Erosion changes a plateau into mountains and then into plains. The present surface of the earth is the result of evolutionary processes. As will be shown, physical and organic evolution have gone hand in hand. It is difficult to say which has had the greater control, the innate qualities of living matter or the environment. Organisms are plastic, environments rigid. Man cannot change the climates of the tropics or of the polar regions, but he can live in both. So far as possible, he adapts his habits to life under abnormal conditions, but no one will deny the fact that the environment changes him.

INTRODUCTION

Anyone familiar with the geology of the earth's surface will realize that the higgledy-piggledy distribution of the rocks precludes the inference that it is the result of design. If it had been, there would have been symmetry, such as actually exists beneath the crust. Geologists have so far unraveled the history of the rocks as to show that their present constitution, attitude, and distribution result from processes which have acted differently and at different times on various regions throughout a period of some two billion years. There have been rhythmic sequences of mountain building, vulcanism, emergence, and submergence, and the evolution of life seems to have been controlled by them.

Under these various environmental controls, the progress of life has not been one grand march from *Amoeba* to man. Their plasticity has permitted animals and plants to follow as many paths as the environment opened to them. Some led to pleasant places, and the groups which followed them have survived; others were ways which ended in the swamps of despair and extinction. Man's ancestors took the hardest trail of all; not, in fact, a single trail, but a series of bypaths along which they slunk for millions of years before they emerged upon the highroad.

Unless one has a sense of chronology, this history will be entirely meaningless. If he does not know it already, the reader should at once commit to memory the "geological timetable" on page ix. The estimates in years are those now generally accepted as reasonably accurate, the basis being the rate of decay of uranium through the radioactive series of derivatives to the final product, lead.

This book is the result of a constantly changing series of lectures which I have given at Harvard for the past seventeen years under the title of Paleontology 1. When I came to Harvard in 1912, I knew nothing about paleontology, although I had held responsible positions as paleontologist previously, and was recognized as an "authority" in my particular field. When I found that I was expected to teach the subject, I set out to learn something about it. As my education progressed, I found that I could best interest students by telling them of the things which most interested me. If this book has any value, it lies in the fact that I have tried to present the general results of the work of hundreds of paleontologists, not specific details.

Theoretically, education should proceed from the general to the special. That, however, is not the tradition in paleontology. Most paleontologists started as collectors of fossils. It naturally follows that taxonomy, the science of identification, differentiation, and naming of specimens, should be their first interest. Many never progressed beyond this stage; hence there has been a tendency to train students along narrow lines. It is easy to become a specialist in some particular field in a few years. This tendency has been fostered recently by the fact that men who know particular subjects are sure of getting jobs. Educational short cuts to technical positions do not appeal to the writer. His own experience, which, as has been explained, was an education in reverse, convinces him that the broader the background, the better the specialist.

INTRODUCTION

Acknowledgements of assistance rendered are expected, and are due. First and foremost I would place the twenty-five unfortunate classes on whom I have experimented. I have told them much that was not so, and made them learn theories which I have later abandoned. Questions which they asked led to ideas which I have since exploited as my own.

No one person, except the editor, has read the whole book. My daughter, Ruth Elspeth Raymond, has been very helpful with suggestions and criticisms. Professor Frank Carpenter rewrote the chapter on insects; Mr. Henry Crosby Stetson, Dr. Theodore E. White, Mr. William E. Schevill, Mr. Henry Seton, Dr. Fred B Phleger, Professor Alfred S. Romer, and Mr. William C. Darrah have read particular chapters and offered criticisms and suggestions, not all of which have been accepted. I realize that many of my views are heterodox, and assume full responsibility. To Dr. Thomas Barbour I am indebted for release from curatorial duties over a trying period during which the book was rewritten, and for encouragement which led to its completion.

Mr. Edward A. Schmitz made all the drawings not otherwise credited. He deserves my thanks, as does Professor Kirtley F. Mather, who loaned nine illustrations from an as yet unpublished book. The contributors of the photographs are acknowledged under each cut.

Paleontologists are so active that any book depending upon their discoveries is out of date before publication. All but one of the chapters were written before the end of the year 1936. Some notes and comments were inserted in 1937 and 1938. But no one who is himself engaged in original investigation can possibly read the total yearly output of his fellows. To rectify sins of omission, I hope for the coöperation of such teachers as may use this book.

P. E. R.

CAMBRIDGE, MASSACHUSETTS
September 1, 1938

GEOLOGICAL TIMETABLE

Cenozoic Era
- Recent
  - Began about 15,000 years ago.
- Pleistocene
  - Began about 1,500,000 years ago.
- Tertiary
  - Pliocene
    - Began about 15,000,000 years ago.
  - Miocene
    - Began about 30,000,000 years ago.
  - Oligocene
    - Began about 40,000,000 years ago.
  - Eocene
    - Began about 50,000,000 years ago.
  - Paleocene
    - Began about 60,000,000 years ago.

Mesozoic Era
- Upper Cretaceous
  - Began about 105,000,000 years ago.
- Lower Cretaceous
  - Began about 120,000,000 years ago.
- Jurassic
  - Began about 150,000,000 years ago.
- Triassic
  - Began about 180,000,000 years ago.

Paleozoic Era
- Permian
  - Began about 225,000,000 years ago.
- Pennsylvanian (Upper Carboniferous)
  - Began about 270,000,000 years ago.
- Mississippian (Lower Carboniferous)
  - Began about 300,000,000 years ago.
- Devonian
  - Began about 345,000,000 years ago.
- Silurian
  - Began about 375,000,000 years ago.
- Ordovician
  - Began about 435,000,000 years ago.
- Cambrian
  - Began about 540,000,000 years ago.

Proterozoic Era
  Began about 1,000,000,000 years ago.

Archaeozoic Era
  Began about 1,500,000,000 years ago.

PREHISTORIC LIFE

PREHISTORIC LIFE

I
FOSSILS

> The *adventitious Fossils*, which are but the *Exuviae* of *Animals* have been erroneously thought a sort of *peculiar Stones*. Cotton Mather, *The Christian Philosopher*

Much has been written in recent years about the early history of the earth in so far as it can be deduced from astronomical and physical data. The evolution of the world would have been futile, however, had it not been for the introduction of life. As to how life originated, geology unfortunately gives little information, but that the earth has supported life for countless millions of years is clearly shown by the remains of animals and plants entombed within the sedimentary rocks during their accumulation and preserved to the present time. These remains serve a twofold purpose. Not only do they give a clue to the history of life upon the globe, but, when properly studied and interpreted, they reveal much of its physical history. What prehistoric implements are to the archaeologist, or the inscriptions incised by ancient peoples upon enduring rock are to the historian, such are fossils to the geologist. Fortunately, their study is not nearly so difficult as that of artifacts or inscriptions, nor does it require so technical a training; yet it produces results of the same order of accuracy.

To study fossils it is necessary to have some knowledge of living animals and plants, for fossils are either more or less perfectly preserved remains of organisms, or evidences of their former existence. At best they are less complete than recent specimens, so that, unless something is known of the modern fauna and flora, one is totally unable to interpret the fragments found in the rocks. Let us, then, review briefly some of the fundamentals of biology.

It should be borne in mind that all matter is either organic or inorganic. If it is organic, in the true sense, then it is the result of life processes. Either it is, or has been, alive, or it was produced by something living. Chemists, unfortunately, use the term organic in another sense. Because carbon is one of the principal constituents of living organisms, they have designated as "organic chemistry" that branch of their subject which deals primarily with the compounds of carbon. In this book the term is used only in relation to what are called organisms, the chief of whose characteristics is "life." What life is, no man can say, but we know that an organism has a power different from that of any inorganic substance in that it can take in foreign chemical compounds, absorb them into itself, and still remain structurally the same as before. In chemical reactions, on the other hand, if elements already in combination take to themselves a new element or combination of elements and unite with them, a product with properties decidedly different from the original is formed. Furthermore, in addition to absorbing and excreting external matter (food) without losing its identity, an organism has the power of growth from within, instead of by accretion from without, and of reproduction, that is, of making new individuals like itself.

There are two great groups of organisms, plants and animals. If one compares the higher animals with the herbage they eat, it seems absurd to ask what the difference is between them; yet when one studies some of the tiny and simple living forms found in fresh-water pools, it is not easy to decide to which group they belong. They may have the green coloring matter of a plant but the mobility of an animal. Indeed, there are some groups of these small beings whose classification is still uncertain, for they are called plants by the botanists and animals by the zoölogists. Ordinarily we think of animals as being capable of locomotion, whereas plants are fixed in one spot; yet in reality many animals are fixed just as immovably as plants. We also think of animals as having arms and legs and a head; but many animals possess no more of these organs than a plant. There is, in fact, no one criterion which can be relied upon to distinguish the two groups. Nevertheless there are several ways in which they generally differ. These may be tabulated.

PLANTS (in general)	ANIMALS (in general)
Take inorganic matter as food and convert it into organic.	Take organic food.
Contain green coloring matter (chlorophyll).	Lack chlorophyll.
Contain cellulose.	Lack cellulose.
Lack power of locomotion.	Have power of locomotion at some stage in life.
Use carbon dioxide and some oxygen.	Use oxygen.
Excrete oxygen and carbon dioxide.	Excrete carbon dioxide.
Eat by absorption through the surface.	Eat by a mouth and digestive canal.
Have indefinite growth.	Have definite growth.
Lack a nervous system.	Have a nervous system.

There are, however, exceptions to all of these general "rules."

Life depends for existence on sunlight. Since light is a kind of kinetic or active energy, it would be quickly exhausted by the individual receiving it if it could not be transformed into potential or stored energy. Plants effect this transformation by the aid of chlorophyll, which enables them to make starch from inorganic compounds; animals obtain their energy by eating plants or other animals. It is barely possible that some of the earliest organisms on the globe were free-moving chlorophyll-possessors which both stored energy in the form of starch and made use of it in movement; such organisms did double duty. Since an active existence requires concentrated food, there was an opportunity for division of labor and hence for the origin of two great groups, one which attended to the storing of energy, and a second which captured and used what the other had produced. Plants settle down to life on a single spot, manufacturing food; animals use plants or other animals as food and expend the energy they have acquired in motion. Plants, living on inorganic matter, have an assured food supply, are quickly settled in life, and merely vegetate. Animals, forced to hunt for their prepared food and, if necessary, to fight for it, have more points of contact with their environment, and are apt to change in a greater variety of ways.

As most people know, the simplest animals are microscopic creatures, found commonly in fresh water, which look like bits of transparent jelly. Some of them have no particular shape, or rather a constantly changing one. They possess no heads or bodies, arms or legs, eyes or mouths, or even any digestive tract: in fact, they are organless, except for a spot of slightly denser material called a nucleus. A well-known example of such a creature is the *Amoeba*, familiar to everyone who has looked through a microscope.

It may be convenient, if one lives in water and doesn't want to go anywhere in particular, to have a jellylike consistency, but most animals have some sort of device for stiffening the body. This strengthening matter, the skeleton, may be either external or internal. An external skeleton serves secondarily as a means of protection. The most common material of skeletons is calcium, either in the form of carbonate or phosphate or as a combination of the two, but it may be silica or hornlike chitin. Plants as well as animals have a stiffening material, the woody tissue, which, like chitin, is a carbohydrate, called cellulose. Silica, too, is present in some plants, giving their cutting power to grasses and sedges. Many of the aquatic plants (algae) secrete large amounts of calcium carbonate. Most organisms, in short, possess "hard" parts, bones, shells, spicules, or woody material. Were it not for these relatively indestructible portions there would be little record of the history of ancient life. "Soft" parts, the protoplasmic tissues, flesh, and cartilage, decompose very rapidly through the action of bacteria. Hard parts, although they decay, do so slowly; consequently there is a chance that they may be preserved as fossils.

What is the process of becoming a fossil? It is merely preservation, either by the checking of decomposition or by the replacement of the hard parts by some relatively durable substance. Anything unfavorable to the life of bacteria impedes decay. Very dry air, a low temperature, sea or bog water, burial in mud or volcanic ash, an encrustation of pitch, gum, or calcium carbonate, all have a more or less preservative effect, so that decomposition is either retarded or entirely prevented. When bacteria are entirely excluded not only the hard but the soft parts as well may be preserved as in a modern refrigerating plant. The most famous instances of cold storage are those of the remains of mammoths and rhinoceroses occasionally found in the frozen gravels and ice of Siberia. Another case of remarkable preservation is that of insects in amber (Fig. 1). While it was a sticky gum exuded from a species of pine, numerous insects were trapped in it, to be preserved as it hardened. Although amber of many ages is known, the most abundant insect-bearing material is found in the Oligocene strata on the Prussian shores of the Baltic.

Suppose an organism to be buried in some substance which retards decay. Conceivably the sand, clay, or calcareous ooze which surrounds it may become sufficiently compacted while the organism retains its original form to hold its shape. Then subsequent decay of the object will leave a hollow mold (Fig. 3). This may be preserved as such, in which case it would itself be a fossil, or percolating waters may eventually fill it with calcite, silica, pyrite, mud, or even sand. Examples of this sort, with the original outer form preserved as a cast, commonly reproduce only the skeleton, for the soft parts decay rapidly even when partially protected. Many plants, whose woody tissue decays less rapidly than flesh, are preserved in this way, sand-casts of tree trunks, branches, and roots being common in late Paleozoic strata. Many mollusks and other animals with hard external skeletons are similarly preserved.

There is another type of replacement in which decay of the skeleton occurs in the presence of mineralized waters, so that for each particle removed the water gives up a bit of its mineral matter, producing a delicate replica of the whole original structure (Fig. 4). The conditions favorable to this process have, however, rarely obtained. A few localities have furnished most such specimens, a noteworthy one being a thin layer of coal in the Carboniferous of England. The so-called "coal balls" found in it are really calcareous concretions in which a part of the vegetation which formed the coal has been replaced by calcium carbonate. Those who have visited the Petrified Forest of Arizona or other areas of "bad lands" in the western states have probably noticed the great amount of petrified wood which, although entirely changed to stone, still shows the characteristic rings of growth, knots, and other features of modern trees. This type of preservation is much more commonly found in the replacement of plants than of animals.

Some fossils are so preserved that they retain indications of the shape of the internal organs of animals, even though no tissues actually remain. Thus there are certain creatures which commonly ingest large quantities of mud with their food. The mud-filled alimentary tracts of a few such organisms have been recovered, showing the shape of stomach and intestines (Fig. 3, at right). Coprolites, the fossil excrement of animals, also show something of the shape of the alimentary canal but are particularly interesting because many of them contain undigested remains of food. Impressions of skin are sometimes found, and, more rarely, impressions of other soft, perishable tissues. Even jellyfish, whose bodies consist mostly of water, and other equally delicate organisms have been found at a few places where the rock is of fine grain. The two most famous localities for such fossils are the outcrop of Mid-Cambrian strata above Burgess Pass, near Field, British Columbia, discovered and

Fig. 1. Contrasts in states of preservation. At left, a natural cast in sandstone of the interior of the shell of a brachiopod. Coarse-grained though the matrix is, the heart-shaped muscle scars are conspicuous. At right, a slab of sandstone, showing the trail of a dinosaur; halfway along the slab it stepped into the water. Between the two, an insect in amber. Photograph at right through the courtesy of W. E. Corbin.

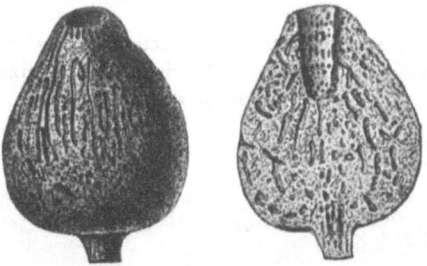

Fig. 2. A Cretaceous sponge, seen from the side and in median-section, showing form, canals, and cloacal cavity into which the excurrent canals empty. Many sponges are preserved in this way, for particles of sediment became entangled in the spicular mesh, permitting the retention of the original form. From J. F. Pictet, *Traité de paléontologie*.

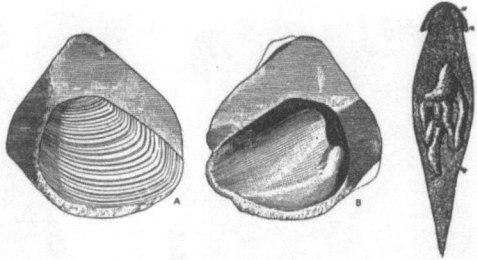

Fig. 3. At left, a flint nodule from the English chalk, which, on being split open, revealed the cavity formerly occupied by a clam shell. From such a mold an artificial cast can be made. At right, an unusually well-preserved amphibian from the Pennsylvanian with mud-filled cast of the stomach and alimentary canal. From R. L. Moodie.

Fig. 4. At left, a magnified transverse section of a calcified stem of *Lepidodendron*, showing cell structure. At right, a carbonized Pennsylvanian fern. Photograph at the left by courtesy of W. C. Darrah.

Fig. 5. Dinosaur bones partially uncovered in quarry. Photograph by courtesy of Dr. Barnum Brown.

explored by Dr. C. D. Walcott (Figs. 7, 10), and the quarries in the lithographic limestone of Upper Jurassic age at Solenhofen, Bavaria. These latter have been worked for many years and have furnished some of the most remarkable and important of fossils, running the gamut of the animal kingdom from jellyfish to flying reptiles and toothed birds. These localities will be repeatedly mentioned in the succeeding chapters.

Skeletons which consist essentially of compounds of carbon — that is, those of plants (Fig. 4), fish, and invertebrates with chitinous coverings, such as insects, crustaceans, and similar animals — are commonly preserved as a black, filmy residue, showing the form more or less perfectly but much or even entirely flattened. The change in this case seems to be a chemical one, involving the loss of the volatile constituents and reducing the composition to a state approximating that of coal. The bark of plants is especially resistant to decay; hence it is not unusual to find rhizomes and stems of trees represented by carbonized remains on the surface of a cast of the interior woody tissue. Some fossil fish are similarly preserved, the blackened scales surrounding a mass of matrix, as if the skin had been stuffed by a clever taxidermist.

Artificial structures made by organisms are occasionally found. Some burrowing animals lined their habitations with bits of shells or sand which they cemented together, whereas others made nests which are occasionally preserved in the rocks. The most abundant artificial structures are the various implements of prehistoric man.

To recapitulate, the states of preservation in which fossils are found may be tabulated.

Actual remains of organisms
> with the soft tissues preserved
> with only the skeleton preserved
> with the skeleton partially changed by addition of mineral matter
> with tissues carbonized

Natural molds of organisms

Replacements of the hard parts of organisms
> Skeleton petrified
>> without loss of structural details
>> with loss of structural details
>
> Skeleton replaced by natural casts
>> mechanically, by infiltration of sand or mud into natural molds (often called external casts)
>> chemically, by deposition of mineral matter in the same
>
> Natural casts of cavities in the hard parts (commonly called internal casts)

Impressions of organisms

Tracks, trails, and burrows

Artificial structures made by organisms

Coprolites

The paleontologist must be familiar with all these various states of preservation, for the same sort of animal may appear to him in various guises. It will have one appearance as a mold, another as an internal cast (Fig. 1), and, as we have shown, there are sundry other possibilities. Furthermore, the shells or bones may have been crushed or distorted by pressure during their long retention in the rocks. Generally they are decolorized, and the original substance may have been replaced by sand, mud, silica, opal, carbonate of lime, barite, pyrite, marcasite, collophane, or, more rarely, salts of copper or even of silver.

The term "fossil," which referred originally to anything dug up from the earth, has been restricted to the various classes of organic remains listed above. It is difficult to frame a brief definition which is both inclusive and exclusive, but it may be said that fossils are the remains of organisms or the direct evidences of their former existence, preserved by natural causes in the earth's crust. This definition, although generally acceptable, has the fault of being rather too inclusive, since it makes no reference to the time of burial. As a matter of convenience many paleontologists, including the author, arbitrarily exclude from the category of fossils all things which have been buried since the beginning of historic times. Such a course helps the paleontologist to avoid duplicating the work of the archaeologist, botanist, and systematic zoölogist, and yet leaves a sufficiently indefinite line of separation to enable each paleontologist to decide for himself how nearly he shall approach modern times. Care should be taken not to confuse the terms "fossil" and "petrifaction." To petrify is to turn to stone, and it is evident from what has been said that not all fossils are petrified. On the other hand, not everything which is petrified is a fossil, for to be a fossil an object must be organic in origin. So far as possible the adjective "fossilized" should be avoided. It is at best a nearly meaningless word, usually employed as a synonym for "petrified," with which it is not synonymous; it is therefore much misused.

Because the prime necessity for the preservation of an organism is that it shall be protected from decay by some covering, it follows that aquatic animals and plants are much more apt to be preserved than the inhabitants of the land. Consequently, marine invertebrates and fish are much more common as fossils than terrestrial organisms are. A land animal stands little chance of becoming a fossil unless it happens to die in a bog or by a stream, although the preservation of bones in caves under the protection of a stalagmitic or earthy cover is common.

The task of the paleontologist is to reconstruct from such materials as he finds in sedimentary rocks the animal and vegetable life of the period during which the strata were in the process of accumulation. Too often he must rely upon incomplete, distorted, and broken objects which are not easily interpreted. The solution of his difficulties can come only through comparison: first, with other materials from the same strata, in the endeavor to bring together scattered parts belonging to the same sort of animal or plant; and second, with such modern organisms as appear to be related. A wide knowledge of the comparative anatomy of modern organisms is therefore a necessary part of the equipment of the paleontologist. Fragmentary material must be pieced together to build up a whole skeleton. Then from scars of muscles, shapes of bones or shells, and such other features as may be preserved, it may be possible to arrive at a rough approximation of the form of the animal as it appeared in life. Such restorations, however, are always to be labeled as tentative, for the entire anatomy and osteology of few extinct animals is satisfactorily determined. Some things, such as coloration, length of fur on mammals, amount of fat, and the like, must be inferred almost entirely from our knowledge of living creatures. We cannot prove, for instance, that any fossil camel had a hump. On the other hand, some soft parts, although boneless themselves, may be recorded; for example, the proboscis of an elephant is registered by recognizable modifications of the nasal bones of the skull.

Knowledge of paleontology obviously progresses not in a single, direct line but by irregular steps along a wide frontier. New materials are constantly coming to light, revealing not only new kinds of animals and plants, never before seen by man, but new facts about extinct organisms long imperfectly known. Each so-called species of extinct creature is really an artificial creation, built up by man on the basis of remains found in the rocks. Each represents the sum total of the best information available at the time, but every scientist admits that his conception of a species is liable to change with the discovery of new material or new methods of study. It frequently happens that what is called a single species by the original describer will be seen as four or five or ten species by some later worker with a wider knowledge of the subject.

If one is to be able to speak of any particular kind of animal or plant, it must have a name; so to each kind, or species, a name is given by the person who first publishes a description of it. Many experiments were made before a definite system of nomenclature was finally reached, about the year 1758. It was natural to try not only to assist the memory by applying a descriptive name but also to indicate the relationship of the particular animal or plant to other organisms. Men naturally like to get their knowledge into as orderly, usable, and easily remembered form as possible, and so with the naming there became involved the idea of classification. It is obvious that certain modern animals are more closely related to each other than they are to others. Anyone would say at first glance that a cat and a tiger are more closely related to each other than either is to a cow. Yet the cow is more like a cat than it is like a fish. Thus the classification of plants or animals is built up about degrees of likeness or unlikeness. The name given to the particular kind is intended to furnish a knowledge of at least one degree of relationship, as well as to serve as a convenience in mentioning it. The earlier writers, who were unnecessarily descriptive, gave names a whole sentence long. Linnaeus, the great Swedish naturalist, about 1758 set the fashion now followed, of cutting the name down to two words, the first or generic name indicating relationship, and the second or specific name suggesting, ideally, some outstanding characteristic of the organism described. It is usual to compare the generic name with the family name among people and the specific name with the Christian name. The generic name is given to a group of species which are found to be very closely related in the structure of their bones, teeth, muscles, et cetera, but each of the various kinds within a group has a specific name. Thus the genus of the cats is *Felis*; the lion is *Felis leo*, the tiger *Felis tigris*, and the house cat *Felis catus*. The generic name shows their evident close relationship, and the specific name indicates which particular kind of cat is meant.

It will be noted that all of these names are in Latin or latinized Greek. This seems to many a great drawback and may make the study of natural history repellent. One can hardly blame a beginner for finding such names as *Strongylocentrotus droebachiensis* and *Sphaerocoryphe pseudohemicranium* rather clumsy at first, but few scientific names are as bad as that. The Latin names are, as a matter of fact, a tremendous advantage, for they are in general use all over the world. Although one may not be able to read any other part of a Russian or Japanese book on natural history, one can at least understand what animals and plants the author is enumerating. If one had to know the common names of the organisms in all languages, dialects, and localities of the world, it would be impossible to get any idea of the fauna and flora of the globe. Everyone is familiar with the fact that in various sections even of the United States common names of animals and plants are differently applied.

There are different ideas about the relationships of various organisms to one another; hence one should not expect to find textbooks in agreement about classification. In a general way, the species is the unit in the system, just as the inch is the unit in the widely used British system of mensuration. Although it is true that no definite number of species is required to make up a genus, the next higher unit, there is nevertheless a feeling that if a genus has a very large number of species it is capable of subdivision; hence subgeneric names may be employed to designate groups of closely related species. Genera are brought together in families; large families may be split by the erection of groups called subfamilies. Families are gathered under orders, or in some cases, into superfamilies under the orders. The orders are considered as subdivisions of classes or, in large classes, of subclasses. The classes are major groups under the phyla, the phylum being the largest unit ordinarily used, although the term "kingdom" is still in use for the two great divisions of organisms, animal and vegetable.

Although the species is the unit, it can be subdivided, just as the inch can be divided into a certain number of barleycorns and lines. Many systematic zoölogists recognize subspecies, and even give names to varieties, so that the Latin name may be a trinomial or even a polynomial, thus reverting to the condition which obtained before the time of Linnaeus. Fortunately, few paleontologists have got beyond the trinomial; most of us retain the binomial system. After all, a name is a purely artificial thing, employed for convenience. If the endeavor to make it descriptive, either of characteristics or relationship, causes it to become so long as to be cumbersome, it ceases to fulfill its original purpose.

A surprising number of students each year ask for an example of the use of these terms; one is therefore inserted here.

> Kingdom *Animalia*. Includes all animals
> Phylum *Chordata*. All animals with a notochord
> Class *Mammalia*. Animals which suckle the young
> Subclass *Eutheria*. Mammals with placenta
> Order *Primates*. Mammals with flat nails
> Suborder *Anthropoidea*. Tailless, semi-erect or erect primates
> Family *Hominidae*. Erect, large-brained anthropoids
> Genus *Homo*. Anthropoids with the modern type of brain
> Species *Homo sapiens*. Men with chins, "even as you and I"

Many an innocent youth will in later life express himself in print. He should remember that generic names begin with a capital letter, specific names with a small one. Both should be printed in italics.

II

COLLECTING FOSSILS

Fossiles give joy to *Galen's* soul,
He digs for knowledge, like a Mole;
In shells so learn'd, that all agree
No fish that swims knows more than he!

John Gay, "To a Lady"

Where are fossils found? Many people know them only as exhibited in museums and do not encounter them in the ordinary routine of their lives. In my own case, I began as a schoolboy a collection of the rocks and minerals to be found within ten or fifteen miles of my birthplace in southwestern Connecticut and was naturally led to read such books on geology as were in my home or in the local library. Since these books devoted a great deal of space to fossils, it became my highest ambition to add some of them to my collection, and even though the books explicitly stated that fossils did not occur in granites and gneisses, the rocks of the surrounding country, I spent many a fruitless day searching for them. The books did say that they were to be found in limestone, but I hunted through the limestone quarries north of my usual haunts with no better success. As a result of years of such efforts I became firmly convinced that fossils were scarce objects, an impression which I have since found to prevail not only among those who live upon the ancient metamorphic rocks of New England but also among those who dwell in regions where the strata are really highly fossiliferous. It is extraordinary how unobservant many people are. Some years ago I was collecting in northern Alabama, and during the morning I picked up a few arrowheads. We had occasion to stop with a local cotton-grower for lunch. I asked him if Indian implements were common thereabouts. He replied that he had never seen any. While waiting for the meal to be prepared, I found a couple of dozen very nice points in his driveway and garden. In fact, I became so absorbed in the search that I unobservantly walked into his beehive, and got properly stung.

As a matter of fact, fossils occur almost everywhere. There are several localities for them even among the "everlasting hills" of New England. They may be expected in any region of unmetamorphosed sedimentary strata, and if one looks at a geological map of North America, he finds that such rocks cover a much greater area than those of igneous or metamorphic origin. There is, it is true, a huge area of the latter extending from Hudson's Bay to the Labrador coast, and there are two long narrow strips of them, one down the eastern and one down the western side of the continent. Much greater areas, however, are underlain by fossiliferous sedimentary strata. One such is the coastal plain from Massachusetts southward to Mexico; another covers the vast region south of the Great Lakes and the St. Lawrence, extending to the Gulf of Mexico and westward practically to the Pacific coast; and still another, an extension of the last, includes nearly all of Canada west of eastern Manitoba, northward to the Arctic and northwestward through Alaska. In fact, three-quarters of the earth's surface has beneath its mantle of soil stratified rocks that are more or less fossiliferous. They are present beneath the surface of the ocean as well, although not easily reached. Mr. Henry C. Stetson has recently collected Cretaceous fossils from strata in place on the wall of a canyon on the southern side of Georges bank in latitude 40°24′30″ N. and longitude 68°07′30″ W., at depths of from 1600 to 1950 feet below the surface. Fossils are not, therefore, rare, but occur in inexhaustible numbers. Consequently they should be better known and understood than they actually are.

Not all unmetamorphosed sedimentary rocks contain fossils, of course, nor are they entirely absent from metamorphosed sediments. In general it may be stated that they are more apt to be present in limestone than in any other kind of rock, and least apt to be found in fine-grained shale or coarse-grained sandstone, particularly if the rocks are red in color. There are, however, exceptions to every rule, and any sedimentary rock deserves a search.

Hunting for, and collecting, invertebrate fossils requires no particular preparation or equipment, although the person who knows what he is looking for usually finds the best specimens, and a certain amount of skill is required if it is necessary to break them from the rocks. The paleontologist is often asked how deep one has to dig in order to find fossils. As a matter of fact, one seldom does any digging unless a particular layer is found to be so productive that it is worth while to remove other layers to get at it. If sedimentary rocks had been left by nature in the position in which they were laid down, with the older buried beneath the younger, it would be necessary to dig deeply to obtain representatives of the more ancient faunas, but there are few strata, except those deposited in the deep oceans, which have not been subjected at some time to folding, faulting, and uplift, so that rocks of all ages appear at the surface at one place or another. To find their fossils it is only necessary to visit the exposures in cliffs, along the sides of ravines, in stream beds, quarries, and excavations along railroads and highways. In some cases one must break the fossils from the rocks which contain them, but the best specimens are those which through the action of frost, rain, alternations of heat and cold, or other natural causes have been freed, or as it is called, weathered, from the matrix. Such processes work most rapidly upon the relatively soft strata of the central states from Ohio westward to Minnesota and southward to Texas and Alabama. Hence the "interior" region has become especially famous for its fine fossils. Collecting under such conditions does not differ greatly from picking up shells along the seashore, for really wonderfully preserved material can be readily obtained.

Although invertebrate fossils are common and fully as worthy of study as any, the ones which attract most attention are the larger and more showy bones of the vertebrates. These are seldom found or collected except by trained men who go to the regions where they are known to occur. Occasional specimens are found by people not professionally engaged in their study, but these are apt to be ruined by the zealous collector, who is always overanxious to secure his specimen and is ignorant of the methods by which it might be preserved. It is perhaps worth while to describe briefly three of the most famous regions for vertebrate fossils, the chalk of western Kansas, the Tertiary deposits east of the Rocky Mountain region, and the dinosaur beds of the United States and Canada.

The chalk deposits of western Kansas were first explored in the years between 1870 and 1875. At that time the roving bands of Indians that infested the country made it necessary for expeditions to be accompanied by detachments of troops from the near-by army posts. Indians were not, however, the only or the worst difficulties in the way of the searcher for fossils in that region. It was then almost completely unsettled, as large parts are even today. Since the chalk beds are bare of vegetation except for a few shrubs, the howling winds blow the calcareous dust into eyes, nose, and mouth, and cause painful inflammations. Furthermore, in an expanse of country a hundred miles long and forty wide, there are only a half dozen springs of fresh water, and to make camp the early collectors often had to hunt for hours for moist ground in which the borings of crayfishes indicated the presence of water a short distance beneath the surface. When such a place was found and a well dug, a supply of alkaline water was procured, but, although liquid, it was exceedingly disagreeable to the taste and had extremely unpleasant effects.

The prize fossils of this region consist of fish ten to fifteen feet long, aquatic reptiles fifteen to fifty feet long, flying reptiles, and toothed birds. Although the rock is soft, it is not possible to dig out the individual bones of the skeleton separately, for all are crushed, distorted, and flattened to such an extent that, if they were separately removed, most of the thinner ones would be broken and destroyed, the larger ones hopelessly mixed up, and both so moved from their natural association as to lose much of their significance. The material must therefore be taken out in slabs. The finest specimens are those which, when found, are entirely covered by rock excepting for the tip of some extremity such as the snout, tail, fin, or paddle. Since few of the skeletons are complete, when one sees a bone, or bones, projecting on an outcrop, the first thing to be done is to determine what the animal is and how much of the specimen still remains. No bones are removed until the rock has been cleared from the entire surface. Then if the exploratory work shows that all or a considerable part of a skeleton is present, and not, as too often proves to be the case, merely one or two odd bones, enough of it is uncovered to learn its extent. When it has been outlined, and as much of the cover has been taken away as seems safe, a trench is dug around the specimen, the depth varying with the size of the fossil and the thickness of the layer in which it is embedded. If the animal is small, covering an area of no more than five by seven or eight feet, an attempt is made to get it out in one slab, the exposed surface having first been covered with wet paper, or cheesecloth soaked in poisoned gum arabic or glue, to make the bones firm and keep them in place. Often it is necessary to follow this coating with burlap soaked in plaster of Paris. After the upper surface has been secured against breakage, the layer in which the fossil is preserved is pried up by means of wedges, picks, and bars, and tilted up with levers; a frame is built around it, the specimen is turned over, and the box is completed. Then it may be hoisted with an improvised derrick onto a wagon or truck and started for the railroad. When specimens are too large to be moved as one slab, they are cut into the requisite number of pieces, the division being made where as few bones as possible will be affected, care being taken to preserve all parts of broken bones. In some cases an almost entire skeleton has been found which has been exposed so long that it is badly weathered and seems hardly worth collecting. Such specimens have been saved by coating the upper surface with plaster and cleaning the matrix from the lower side, a process which reveals the unweathered sides of the bones, the plaster taking the place of the original rock.

The fresh-water Tertiary deposits of the high plains east of the Rockies, and of the intermontane valleys within and west of them, provide an extensive field for search in beds ranging in age from Paleocene to Pleistocene. Almost all the states in the vicinity of the Rocky Mountains and westward to the Pacific coast have furnished one or many good localities for fossil mammals, but the Oligocene and Miocene beds of South Dakota, western Kansas, western Nebraska, and eastern Colorado and Wyoming have been especially prolific. Modern expeditions to these regions maintain a central movable camp from which one or several trained collectors go out to spend their days in prospecting along the sides of the bluffs and dry "draws" of the "bad lands." Although living conditions are usually not so bad in the Tertiaries as in the Cretaceous chalk, the temperature may be extremely high, water bad or absent, and surfaces glaringly white in the burning sun. Most mammals are found as individual teeth, bones, skulls, or parts of jaws, but once in a while, perhaps two or three times in a season, an entire or nearly entire skeleton is found. In reading descriptions of new species one frequently finds that the paleontologist has written somewhat as follows: "This splendid skeleton was found by Mr. Smith in the lower part of the Oligocene near Pole Creek, Nebraska. It lacks only the skull, backbone, pelvis, manus, and pes, but the remainder of the specimen is in a marvelous state of preservation, except for some crushing and distortion of the limb bones." When two or three bones are gathered together, the paleontologist waxes enthusiastic.

The process of excavating a single bone or a few bones of the skeleton of a mammal is relatively simple. If only a skull or a few large, solid bones are present, they are dug out with pick and awl, care being taken not to remove all the matrix, since the extra rock will make shipment less perilous. To extract a whole skeleton or fragile bones of any sort, it is necessary, as has been said, to proceed with great care, cutting a trench around the specimen, saturating bones and matrix with gum arabic or shellac, bandaging them with strips of cheesecloth soaked in poisoned gum or paste, and finally reinforcing all with splints held by burlap soaked in plaster of Paris.

At a few localities bones or skeletons have proved to be so abundant as to make collecting a process of quarrying. Such were the deposits at Manhattan, Kansas, in the early days and at Agate Springs, Nebraska, more recently. At these places a particular layer was found to be made up almost entirely of bones, so that blocks were taken out for removal to museum laboratories where the bones could be cleaned from the matrix. In these operations, horses, plows, and scrapers sufficed to remove the overburden; the layer was mapped out, split into blocks of convenient size, the sections numbered, marked to show how they fitted into adjacent blocks, and a careful plan drawn so that bones and skeletons broken up in the process of removal might be reunited when removed from the matrix. At such places the collector becomes a super-quarryman, engaged in a tedious but interesting task.

The chief localities for remains of dinosaurs in North America are in the third region, along the flanks of the Rocky Mountains and the plains to the eastward, from Alberta to Colorado, the richest finds of Jurassic dinosaurs having been made in Colorado, Utah, and Wyoming, and of Cretaceous species in eastern Wyoming, central Montana, and along the Red Deer River in Alberta, Canada.

These great American reptiles are of comparatively recent discovery, the first having been brought to the attention of paleontologists and the scientific public in 1877, although hunters and travelers had seen the bones and had even brought back sections of them as pieces of petrified wood before that. In 1877 three observers, one a school teacher, another a professor in the School of Mines at Golden, Colorado, and the third a section foreman on the Union Pacific railroad, found dinosaur bones. As the late Samuel W. Williston once said, their discovery was not nearly so remarkable as that the dinosaurs had remained so long unnoticed. "The beds containing them had been studied for years by the geologists of the Hayden and King surveys, yet in some areas acres were literally strewn with bones and fragments of bones, and at what has since been known as the Bone Cabin quarry in central Wyoming, a Mexican sheep herder had built the foundations of a cabin by cording up the huge limb bones of dinosaurs."

One of the discoverers, Professor Arthur Lakes of Golden, was hunting for fossil leaves in the basal Cretaceous sandstone near Morrison, Colorado, in March, 1877, when he came upon an enormous vertebra partly protruding from the rock. Since he had heard of Professor O. C. Marsh of Yale in connection with discoveries of toothed birds in the chalk of Kansas, he communicated with him, with the result that the great bone, rock and all, was expressed to New Haven at an expense of ten cents per pound for transportation. Professor Marsh promptly described, from this single vertebra, a new dinosaur which he predicted would be found to have a length of 115 feet. Professor Lakes was at once set to work to collect from the beds which had furnished the bone, and the name Morrison is still retained for the Upper Jurassic formation in which these great reptiles are found. A Mr. O. Lucas, an amateur botanist teaching school at Garden Park, Colorado, was another lucky discoverer, who, also in March of 1877, stumbled on some bones in a little ravine not far from Canyon City. He had heard of Professor E. D. Cope of Philadelphia, who later became Marsh's rival, and who was one of the most brilliant naturalists this country has ever produced. Cope described a new dinosaur from the fragment sent him, but one of Marsh's collectors beat him to the locality, which later produced some remarkable specimens. The third discoverer was the famous Bill Reed, long noted as a professional hunter, section hand, preparator, and, at the time of his death a few years ago, curator of paleontology in the University of Wyoming. His discovery was made in the vicinity of Como, Wyoming, also in 1877.

In the wild scramble to dig out bones which the rivals, Marsh and Cope, described and supplied with names, all sorts of valuable and important skeletons and bones were hacked to pieces and lost, but later, as the collectors learned the value of time, patience, gum arabic, shellac, glue, and plaster, better and better specimens were obtained.

Hunting for dinosaurs is the same process as hunting for any other vertebrate fossil. The areas and formations which can be expected to yield specimens are well known, and because of the large size of the bones one would think that by this time all possible specimens must have been discovered. But anyone who has been in the "bad lands" realizes how innumerable are the little dry draws or ravines that intersect the hills, how great the exposures, and above all, how rapid erosion is when it does rain. Instead of being exhausted, each year sees the dinosaur country producing new and more nearly perfect specimens, for the storehouse appears to be constantly renewing itself. As with other vertebrates, it is much more common to find individual bones than entire or even partial skeletons. When a good skeleton is found, the pleasure of prospecting is over, and the collector sits down to one, two, or even three seasons of hard work. It can readily be imagined that, if one is lucky enough to detect the tip of the tail of a seventy-foot dinosaur projecting from a cliff of hard sandstone, it will take some digging to get to the skull at the other end of the vertebral column — if it chances to be there. The process settles down to a regular quarrying operation, except that, for fear of shattering the specimen, one has to use much less dynamite than he would like to. In most cases the overburden, down to a respectful distance from the specimen, may be removed by drilling and blasting. Then begins a slow process of exploration to locate and trace the skeleton. As each bone is uncovered it must be hardened with gum or shellac, pasted, and wrapped, as already described (Fig. 5). After the initial removal of the overlying rock, the greater part of the work is prosecuted with a shoemaker's awl and a whisk broom, seemingly inadequate instruments for an attack upon a dinosaur. As might be expected, various ingenious shifts have been employed in getting out the great skulls and bones of the dinosaurs. A single vertebra of *Diplodocus* or *Brontosaurus* in its petrified state may weigh from 120 to 150 pounds, and a limb bone 600 to 800 pounds. Among the hardest things to handle are the huge skulls of ceratopsians, which have in some cases been brought back in one piece. The largest single fossil so far known to have been handled was a skull of *Triceratops*, which when boxed weighed 6,850 pounds. When one thinks of loading a box weighing nearly three and a half tons, and dragging it seventy miles over a roadless tract of canyons and gullies, he realizes that the collector needs something more than luck to accomplish his task.

After the specimens have been wrapped, crated, and shipped to the laboratory, the most interesting but at the same time the slowest and most monotonous process in the business begins. The specimens in their packings are mounted on circular revolving tables, where the box, bandages, et cetera, are removed. In modern work this is easy, as the gum and plaster readily soak off, especially when the precaution has been taken to put a layer or two of paper next to the bone. In the old days of shellac and plaster, however, the taking off of the bandages was no simple matter. I have seen the whole outer layer of a bone come off with the bandage, which meant infinite labor on the part of the preparator in removing the fragments from the bandage and sticking them back on the bone. Even with the best collecting, it takes three men two or three years in the laboratory to clean up one dinosaurian skeleton, and another year to mount it. Oftentimes it is found that the bones are hard on the outside but powdery within, which means that a plaster composition, plus an iron rod as a support, must be run into the bone. Or the bone may be complete but so shattered that it has to be taken to pieces a little at a time, and the fragments cemented back into place.

In mounting a specimen, the modern principle is to give it a lifelike posture, with as little as possible of the support showing. In the case of big dinosaurs, temporary mounts of parts of the backbone are made, plaster casts of the undersurface of the vertebrate secured, and a steel support cast from the mold so produced. The various parts of the axial skeleton are adapted and fastened to one another, then attached to a couple of upright rods, usually pipe of suitable diameter. The legs are held in place by fitting half-oval pieces of iron to the inner surfaces and bolting or otherwise fastening the bones to them. Museum technique is constantly being changed and improved as zealous preparators strive to excel one another. Possibly few paleontologists realize how much they owe to the skill of the men who actually handle the bones and assemble them in the only positions which their shapes and articulations show to be possible.

Fig. 6. At left, four sorts of Foraminifera. The three at the left have calcareous shells; the next is of the arenaceous type. From Joseph A. Cushman. At right, a Carboniferous radiolarian. From D. Rüst.

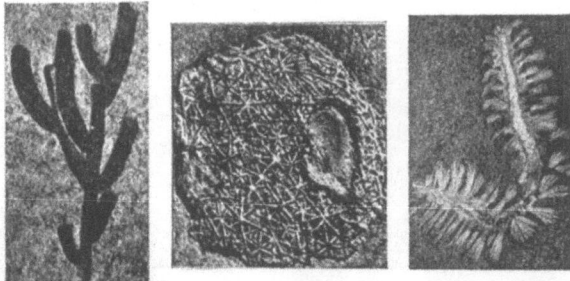

Fig. 7. Three Mid-Cambrian fossils from the Walcott quarry at Burgess Pass, near Field, British Columbia. At left, a branching sponge, one of the oldest known colonial animals. In middle, a sponge showing large siliceous spicules. At right, an annelid worm with long bristles on the parapodia. From C. D. Walcott.

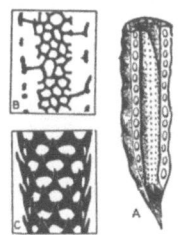

Fig. 8. An archaeocyathid sponge. A, a vertical section, showing central cavity, inner and outer walls, and pores. From V. Okulitch. B, C, the spicular mesh. After R. and W. R. Bedford. B and C are enlarged, A reduced in size.

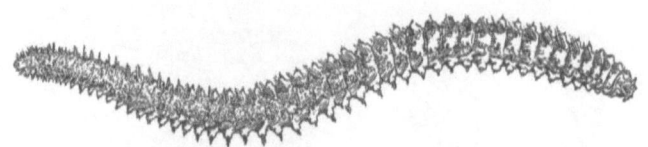

Fig. 9. At left, a modern annelid worm, *Amphinome*. From an original drawing. At right, an Eocene encrusting bryozoan, to illustrate the small size of the apertures of the individual zoöecia. Six times natural size. From Canu and Bassler.

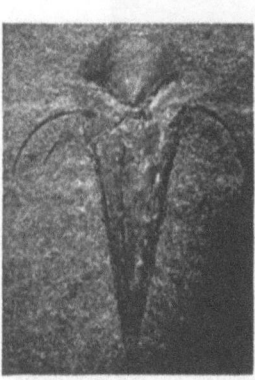
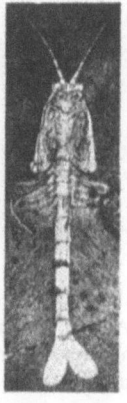

Fig. 10. Three Mid-Cambrian fossils. At left, a swimming gastropod, *Hyolithes*, with operculum and foot-supports. In middle, a crustacean, *Waptia*, superficially similar to the modern *Apus*. From C. D. Walcott. At right, a crustacean, *Sidneyia*, with eurypteroidal form. Photograph by courtesy of Charles Resser.

III

THE FIRST ANIMALS AND PLANTS

For according to the proverb, the beginning is half the whole business.

Plato

The oldest aggregation of animals and plants which is adequately known is found in rocks of Cambrian age. This, the first authentic record of life of the globe, is of more than usual interest. What sorts of organisms lived during those ancient times, hundreds of millions of years ago? Were they such as exist today, or sufficiently like them to fit into a classification based on modern animals and plants? Do they meet the expectation of believers in the doctrine of evolution in being exceedingly simple?

Before the actual record is examined, it is necessary to review in outline the classification of modern animals. For our purposes the animal kingdom may be divided into thirteen great groups, or phyla, although this may not be in agreement with the best modern practice. Their characteristics will be summarized, beginning with the simplest.

The Protozoa are animals which, although unicellular, have the power of performing all the important functions of life; namely, feeding, digestion, assimilation, and reproduction. The last is accomplished in two ways: asexually, by fission, the parent dividing its substance to produce individuals like itself, or sexually, by the fusion of two animals and subsequent division. Nearly all Protozoa are exceedingly minute. Only two subdivisions include members which secrete shells. The Foraminifera produce in rare instances skeletons as much as an inch or two in diameter, but this is exceptional, for most of them are less than a millimeter across. These animals have two types of shells, one formed by the agglutinization of foreign objects such as particles of silt (Fig. 6), the other a secretion of chitin, of carbonate of lime (Fig. 6, at left), or of silica. The agglutinated or "arenaceous" forms appear to be primitive. Another group, the Radiolaria (Fig. 6, at right), build tiny siliceous shells which have so many openings that they form only the most sketchy framework of a skeleton.

The Porifera or sponges are in some respects comparable to the Protozoa, for a single specimen is made up of many cells which, although mutually dependent, are individually practically like protozoans. This structure suggests that sponges originated in aggregations of single-celled animals which did not completely separate at the time of reproduction but remained together to form a composite individual.

Digestion takes place in some of the individual cells, without benefit of a real mouth, stomach, or other definitely formed organs (Fig. 13 C). Most sponges have, however, a common skeleton, which, like that of the Foraminifera and Radiolaria, is internal. This framework may be horny and flexible, as in the common bath sponge, or it may be made up of flinty or calcareous needlelike rods, called spicules (Fig. 7).

The coelenterates include several sorts of animals, many of them attached to foreign bodies and plantlike in their growth. Most of them are colonial; that is, the young remain attached to the parents from which they budded. Two of the more important groups will be discussed in some detail on later pages. They are the simplest animals which are provided with a mouth and digestive cavity, but they lack an anal opening, so that waste must be ejected through the mouth. In the single cavity are carried on the digestive, reproductive, and excretory functions. Most of them show more or less perfect radial symmetry, with the parts repeated in multiples of four or six, the mouth being surrounded by a ring of delicate, fingerlike tentacles. The most important classes are the corals with calcareous skeletons (Fig. 12 D), graptolites (Fig. 15) and other hydroids with chitinous support, and the jellyfish with none.

Next in order are the four great phyla which used to be called by the inclusive name of "Vermes," or worms. The greatest achievement of this group was the production by its highest representatives, the segmented Annulata (Fig. 9), of a complete digestive tract within a body cavity, a head with eyes, an adequate nervous system, and an excretory apparatus. Most of the "worms" are bilaterally symmetrical, and they are not colonial, but separate individuals with powers of locomotion. Not all of them share these characteristics, however, for although some are primitive, many are highly specialized parasites. Since all lack skeletons, they are not well known as fossils. Two of the groups, the flat worms and the wheel worms, are practically unknown in the rocks, and few representatives of the round worms have been found. The fourth group, the segmented or annulate worms, have left many more traces, such as impressions, burrows, jaws, or tubes.

The echinoderms, like the coelenterates, have radial symmetry, but they are not colonial, and the parts are repeated in multiples of five. Their skeletons, made up of calcareous plates set edge to edge, are really internal, although near the surface. This phylum contains eight groups, cystoids, edrioasteroids, crinoids, blastoids, sea urchins, starfish, brittle stars, and sea cucumbers, all except the last well represented as fossils; some of them will be discussed later.

The bryozoans are minute colonial animals which, although abundant nowadays, are seldom noticed. Superficially they resemble coelenterates, having saclike bodies with the mouth surrounded by tentacles, but they are more highly organized, having, among other things, a complete gut within the body cavity and no radial partitions (Figs. 9, 12 C). They secrete lacelike or more solid calcareous skeletons, which are

THE FIRST ANIMALS AND PLANTS

common as fossils. They are important to the geologist but of no great evolutionary significance, and will not be further discussed.

The brachiopods, unlike their near relatives, the bryozoans, are not colonial but exist as separate individuals. They are perhaps the most frequently met of the fossils of the Paleozoic rocks. Although they are still present in the oceans they are relatively rare and unimportant; hence little attention is paid to them. Like the more familiar clam and oyster, the brachiopod has two shells, but they are placed above and below the body instead of at the sides. Two somewhat different kinds are known. Those known as the Inarticulata (Fig. 11) have thin chitinous shells, reinforced with

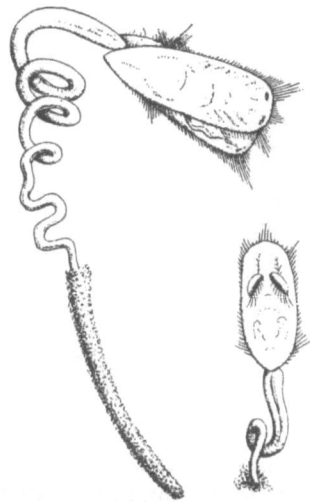

Fig. 11. The modern inarticulate brachiopod, *Lingula*, showing shells and pedicle. From E. S. Morse.

more or less phosphate and carbonate of lime. The others, the Articulata (Fig. 12, A, B), secrete calcareous shells with interlocking processes on the hinge.

The phylum Mollusca includes three familiar groups, the snails or gastropods, the bivalves or pelecypods, and the cephalopods; two less important classes need not detain us. Most of them have calcareous shells; the gastropods have a spirally enrolled, unsymmetrical conch, the pelecypods two similar valves joined together at the hinge by a ligament, and the cephalopods straight, curved, or spirally enrolled, bilaterally symmetrical, chambered shells. The student should, however, be warned that there are numerous exceptions to all three of these general statements.

The Arthropoda is the largest and most highly diversified of the phyla. The members are characterized by bilateral symmetry, a fundamentally chitinous exoskeleton which is periodically shed and renewed, and a segmented body with segmented appendages (Fig. 10). Exceptionally, as among the barnacles, a calcareous shell is produced. The classes are: the Crustacea (lobsters and the like), with the first pair of appendages flexible tactile organs; the Arachnida (spiders, scorpions, et cetera), with claws as the first appendages; the Diplopoda (thousand-legged worms), with two pairs of appendages on each segment back of the head; the Chilopoda (centipedes), with poison fangs just back of the head, and the Insecta, the greatest group in the whole animal kingdom, with three pairs of walking legs and, typically, two pairs of wings.

The last phylum, the Chordata, includes those animals with an internal axial support known as a notochord. In most of the groups the notochord is replaced at

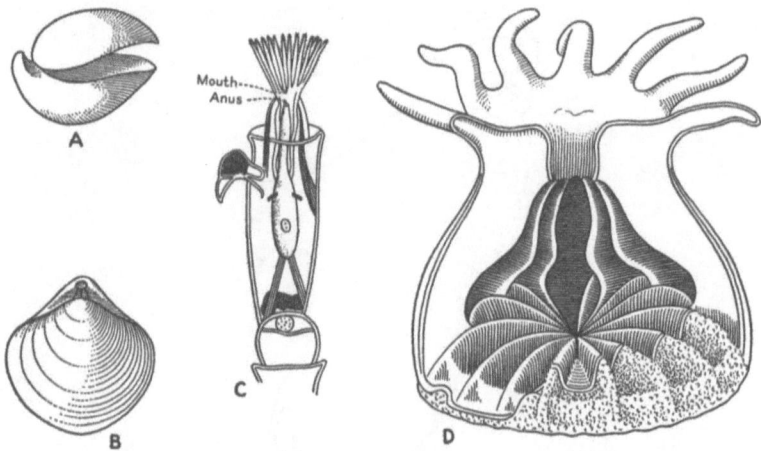

FIG. 12. A, B, lateral and dorsal views of an articulate brachiopod. C, a bryozoan, drawn as though transparent, showing tentacles about the mouth, digestive organs, and muscles within the body cavity. After Parker and Haswell. D, a coral polyp, the upper portion cut vertically to show tentacles, mouth, oesophagus, and body cavity with radial mesenteries. Beneath, dotted, is the beginning of the skeleton. Redrawn from a wall chart by Paul Pfurtscheller.

an early period in life by the vertebrae of the backbone; hence the best known of these animals are those called Vertebrata. In the latter category are the Pisces (fish), Amphibia (frogs and toads), Reptilia (lizards and snakes), Mammalia, and Aves (birds).

Ever since the days of the great Franco-Bohemian paleontologist, Joachim Barrande, it has been customary to refer to the animals whose remains are found in the Cambrian rocks as the first or "primordial" fauna. But the ten thousand or more feet of Cambrian strata were not formed in a moment; their deposition is generally supposed to have occupied from ninety-five to a hundred million years. It is therefore absurd to consider all the faunas which succeeded one another during that long time as contemporaneous. All geologists recognize three distinctly different faunas in the Cambrian, one in the lower beds, another in the middle, and the third in the upper part of the strata. It would be just as logical to group together all the animals of the Upper Cretaceous, Cenozoic, and Recent as representing the state of evolution of life in the late Mesozoic as it is to say that the total Cambrian fauna is the congeries which was in existence at the beginning of that period.

The Lower Cambrian fauna of the world, so far as it is at present described, consists of about 455 species, distributed among the phyla as follows: Protozoa, none; Porifera, 18.5 per cent; Coelenterata (all jellyfish) and Echinodermata (cystids and edrioasteroids), taken together, 2 per cent; "Vermes" (tubes, trails, and burrows), 4.25 per cent; Brachiopoda, 27.5 per cent; Mollusca (all gastropods, and all but two bilaterally symmetrical forms), 11.5 per cent; and Arthropoda (84 per cent trilobites, 16 per cent other Crustacea), 36.25 per cent. These statistics probably give a more accurate idea of the shell-bearing members of the oldest Paleozoic fauna than those based upon the Cambrian as a whole. The proportion of sponges and free-swimming gastropods is relatively much higher, of arthropods considerably lower, and of brachiopods somewhat lower, than in the lists published by other writers.

It is evident that the oldest Cambrian fauna is diversified and not so simple, perhaps, as the evolutionist would hope to find it. Instead of being composed chiefly of protozoans, it contains no representatives of that phylum, but members of seven higher groups are present, a fact which shows that the greater part of the major differentiation of animals had already taken place in those ancient times. The other phyla not represented are the flat worms, wheel worms, round worms, Bryozoa, and Chordata, the last the one which contains the most specialized of all animals. It is also apparent that the animals living in Cambrian times were not strikingly peculiar, since most of them can be assigned readily to phyla erected on the basis of modern ones.

What, then, are we to conclude? Are we to deny the special creation of the modern fauna, only to find that it is descended from animals specially created some millions of years ago? To answer this question, it is necessary to look further at the fauna under discussion. Diversified as it is, if analyzed further it proves to be simple as compared with that of the present day.

In the first place, the whole phylum Chordata, from fish to bird, is absent.

Although the Arthropoda are numerous, making up more than a third of the fauna, the only class represented is that of the Crustacea, the simplest of the arthropods, and 84 per cent of these are trilobites, the most primitive of the crustaceans. Very late in the Cambrian a few marine arachnids appeared.

The Mollusca are all gastropods, the simplest members of the phylum, and nearly all are of the most primitive type, their shells being simple, bilaterally symmetrical, uncoiled cones (Fig. 10).

Next in importance to the Crustacea are the brachiopods, which make up 27.5

per cent of the fauna. Both the Inarticulata and the Articulata are present, but 80 per cent of the species belong to the former, which is the simpler group. Such of the articulates as are found are of the most primitive type.

The echinoderms are represented only by a few edrioasteroids and cystids, ancestral to all other members of the phylum.

The "Vermes," likewise poorly represented, probably because of lack of preservation, differ from the other animals that have been discussed in that most of the specimens found belong to the highest phylum, the Annulata, and, moreover, to the most highly organized class of the annulates, the Chaetopoda, or bristle worms.

Few specimens of Coelenterata have been recovered from the Lower Cambrian strata, but, curiously, those so far found belong to the Scyphozoa or jellyfish, one of the most specialized groups. These animals are absolutely without hard parts and are composed chiefly of water, yet Dr. C. D. Walcott found a few impressions of them in the Lower Cambrian shales of Vermont.

Spicules of siliceous sponges have long been known from Lower Cambrian rocks, but the oldest complete specimens are of Mid-Cambrian age (Fig. 7). The most specialized of modern Porifera are the glass sponges, so called because their framework is made up of needles of glasslike amorphous silica. The living members of this division are classified according to the shapes of the spicules, whether straight and simple or arranged at various angles in two planes. It is interesting to note that some Lower Cambrian sponges are of the sort with siliceous spicules, and that they can be referred to modern orders on the basis of the forms shown by the elements of their skeletons. Much more common are the calcareous archaeocyathinids, the oldest known sessile organisms (Fig. 8).

The Protozoa, instead of making up the whole of the oldest known fauna, are very poorly represented in it, if they are present at all. A few species of Lower Cambrian Foraminifera have been described from Russia and New Brunswick, but Dr. Joseph A. Cushman, the foremost student of this group at the present time, is doubtful if any of them really belongs to it. A few poorly preserved radiolarians were found in thin sections of rocks collected from the Cambrian of Thuringia, but even their describer was doubtful about them. This lack of specimens, however, cannot be taken as evidence of the absence of Protozoa from the faunas of this age. It must be remembered that the majority of these animals lack skeletons, and that all are small and ill adapted for preservation as fossils. The largest of the radiolarians are less than one millimeter in diameter; most of the Foraminifera are equally small, and many of the latter have exceedingly frail, agglutinated shells which could hardly have been preserved in any abundance in rocks millions of years old.

To complete the survey of primordial life it is necessary to say a word about the plants. Up to the present time no traces whatsoever of terrestrial plants have been found except in the youngest zone of the Upper Cambrian, which means not only

THE FIRST ANIMALS AND PLANTS

that all of the higher types of vegetation but even such lowly things as ferns, mosses, and lichens were absent from the most ancient flora. The Algae (seaweeds) are the only plants known to have been present. A few of the most primitive unicellular forms happen to be preserved because they secreted calcium carbonate. Walcott described other plants, some of which are perhaps red algae, from his famous quarry in the Mid-Cambrian at Burgess Pass. As will be indicated later, there is reason to believe that diatoms and bacteria were already in existence.

No one thinks that the Lower Cambrian fauna contained only 455 species. Probably some already described have been overlooked in the present survey of the literature. Perhaps many are as yet unknown, after seventy-five years of search. The present number may be doubled within the next few years as keener observers split up the species already described or supplement the list from discoveries at new localities. Suppose we add another thousand to accommodate the various soft-bodied animals practically incapable of preservation. Even that brings the list to only two or three thousand species as compared with two or three million supposed to be in existence at the present time. However one looks at the picture, the "primordial" fauna was simple and undifferentiated.

Various writers have been so much impressed by the amount of differentiation shown by the Cambrian fauna that there may be a tendency to overemphasize the amount of time necessary to produce it. This is perhaps because one is apt to think of the great phyla as absolutely distinct from one another, each higher in the scale of organization than its predecessor, as shown in the tables in the textbooks. And one thinks of each step in advance as having been accomplished only after a long period of time. Yet it is a question whether time is a particularly important factor. It is natural to think of the phyla as they are represented today rather than as they were in Cambrian times. For example, the term "chordata" connotes fish, birds, mammals, et cetera, not the lowly backbone-less members of the group, animals so simple that some of them were for years supposed to be invertebrates. To estimate the real amount of differentiation which had taken place by the beginning of Cambrian times it is necessary to compare the most primitive members of the various phyla. If this is done, it appears that there were really only three great steps in the progress which led from the protozoans to the present great diversity of animals. How long it took to organize a unicellular creature we have no idea.

Starting from the unicellular Protozoa (Fig. 13 A, B) there are obviously two possibilities. The new cells produced in reproduction may separate, thus continuing to be Protozoa, or they may remain attached to one another, forming multicellular animals, or Metazoa. This was a fundamental step in the progress, but there is no reason why it should not have occurred as soon as Protozoa began to reproduce. Why it happened may be a mystery, but the introduction of the factor of time does not facilitate the explanation.

Once the metazoan stage was reached, there were again two possibilities. The members of the colony might retain their individuality, although giving up much of their freedom, as in the case of the sponges (Fig. 13 C). That this was a successful plan of coöperation is shown by the abundance of sponges at the present day, but it cannot be called a "great step," for it led nowhere. The other possibility involved the uniting of all the cells into one system, with complete coöperation but loss of in-

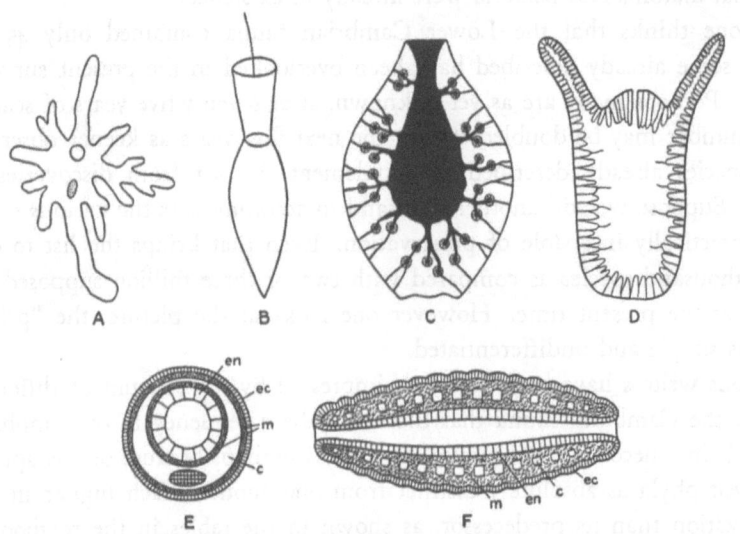

FIG. 13. Schematic drawings to illustrate the fundamental steps in the differentiation of animals. A, amoeboid protozoan, with no definite shape. B, collared protozoan with vibratile cilium. C, a sponge, in which there are nonciliated ectodermal cells, and ciliated endodermal ones within the spherical cavities. D, section of a simple coelenterate, with mouth, digestive cavity, outer ectodermal and inner endodermal cells. The latter differ among themselves in form and function. E, F, transverse and longitudinal sections of a primitive coelomate, showing, *ec*, ectodermal, *en*, endodermal, and *m*, mesodermal cells; and *c*, the body cavity, or coelom. A, B, C, D, redrawn and simplified from various sources; E, F, redrawn after Sedgwick and Wilson.

dividuality. One can imagine various ways in which the cells might have been arranged; what actually happened seems to have been the producton of two layers forming a hollow sphere. The inner layer is the endoderm, the outer one the ectoderm, and the opening to the central digestive cavity is the mouth (Fig. 13 D). Here we have the fundamental characteristics of the Coelenterata. But is this a change which requires time?

Comparing the coelenterates and the sponges, one finds that the primitive members of the former group were, theoretically, freely floating organisms. This belief is borne out by the presence of jellyfish in the Cambrian. Sponges, on the other hand, were sessile, as shown by attached specimens from the same formation. Freedom and progress thus contrast with stability and vegetative growth.

The coelenterates had acquired a digestive cavity. The next great step was the formation of the mesoderm and a body cavity (coelom). This is so profound a change in organization that one must admit that it may have required time, though this admission really expresses only our ignorance of the connecting links between the primitive coelenterates and the primitive coelomates. The transition appears to have been from a pelagic organism to one dwelling on the sea floor, from a swimming or floating to a crawling mode of existence. The physical change was from a more or less spherical form to an elongate one, from spherical to bilateral symmetry (Fig. 13 E, F). The study of modern animals furnishes some clues to the possible history, but since the fossils give no information it will not be helpful to enter into a long discussion.

Once the coelomate stage (body cavity with a digestive tract within it) had been reached, all the "great steps" had been taken. Some groups, such as the bryozoans, brachiopods, echinoderms, and Mollusca, have gone downhill; only two, the arthropods and the chordates, have risen above the status of their ancestors. If one dares to put a summary of these remarks in the form of a diagram, it might be expressed as in the figure below.

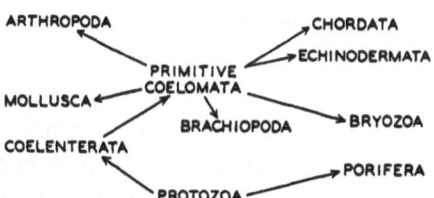

Fig. 14. Diagram to illustrate the relationships of the principal phyla of animals, the "worms" being excluded for the sake of simplicity. Their oldest representatives, the annelids, represent the most primitive coelomates now known.

The inference is that if one should group the animals into super-phyla there would be only four, Protozoa, Porifera, Coelenterata, and Coelomata. There seems to be no reason why the protozoans, sponges, and coelenterates should not have been practically contemporaneous in origin. Although there was a fundamental change, it was a simple one.

The later changes were more complex, but can they be evaluated as to time required? The structure of the brachiopod is much more complex than that of the bryozoan; yet brachiopods were well on their way at the beginning of the Cambrian, whereas bryozoans did not appear till the Ordovician. The primitive echinoderm is more "specialized" than the primitive chordate, but the former left skeletons in the

Lower Cambrian rocks, whereas the latter do not appear in the record till the Upper Ordovician. It should, however, always be borne in mind that the earliest representatives of all phyla were probably without skeletons; hence the true chronology will probably never be known.

Glancing back over what has been said of the Lower Cambrian fauna and flora, it will be seen that the animals and plants of that time were, after all, simple. Backboned animals and other chordates and all the higher plants were as yet unknown. The Arthropoda, Mollusca, Brachiopoda, and Echinodermata, the higher groups of invertebrates, were represented by their simplest types only. On the other hand, the lower invertebrates, Annulata, Coelenterata, and Porifera, had representatives then which were almost as highly organized as any members of the same phyla today. An immense amount of evolution has taken place since the Cambrian. In other words, we are not driven to belief in an ancient special creation, but to further research on the genealogy of organisms.

Where are we to look for the ancestors of the Cambrian organisms? As was said at the beginning of this chapter, the Cambrian contains the oldest assemblage of real fossils as yet known. But the Cambrian strata are not the oldest of the water-laid rocks. They rest upon older sediments formed at a much earlier period in the earth's history. These older rocks are difficult to study, because many — in fact, most of them — have been intruded since their deposition by gigantic masses of hot igneous rocks, and most of them have suffered the stresses of one or more periods of mountain-building. Geologists estimate that there are from 60,000 to 100,000 feet of ancient, pre-Cambrian sedimentary strata, and it is in them that we must look for evidence of animals and plants of ages which antedated the Cambrian. Traces of the earlier faunas and floras have already been found, and it will be our next task to see what has been learned from the numerous searches which have been made for them.

IV

PRE–CAMBRIAN LIFE

> Observe creation mercifully hidden
> either in an imaginary Eden,
> or buried in some absent-minded spasm
> of a self-generated protoplasm.
>
> Humbert Wolfe, *The Uncelestial City*

Ever since Sir William Logan demonstrated that the Cambrian rocks are not the oldest sediments but are underlain by vast thicknesses of water-laid strata, geologists and paleontologists have been searching for evidences of a truly primordial fauna. Some of the pre-Cambrian strata are limestones and shales which appear to be little altered in spite of their incomprehensible age; hence they present a constant challenge to the investigator. So anxious are geologists to obtain fossils from them that anything which remotely resembles an organism is carefully saved and studied in the greatest detail. Although many such objects have been described, few have been unreservedly accepted as real fossils. No recently discovered "Old Master" is subjected to more critical scrutiny or to more acid tests than is a "find" from rocks older than the Cambrian. In this chapter only the more important ones, those which have been accepted as genuine by eminent "authorities," will be discussed.

Paleontologists have always before them odds and ends which, although they seem to be of organic nature, show no definite structure. They used to be called sponges, for lack of a better designation, but are now supposed to be calcareous algae. This is due to their resemblance to the "lake balls" or "water biscuits" which at the present day are formed in lakes by blue-green algae. These so-called algae are minute one-celled plants more nearly related to the bacteria than to the algae. As they reproduce they form long filaments and mats of cells which, during their life processes, cause the deposition of calcium carbonate from its solution in the water in which they live. As each layer of limestone is deposited, it is overgrown by the plants, so that in time a concretionary mass is built up. Such cakes are common in the lakes of Michigan and of western New York, where they were first noticed by paleontologists. When the water biscuits are cut in slices, the sections show a concentric structure, with radially arranged, irregular cavities.

When these objects came to Charles D. Walcott's attention about 1906, he at once recalled having seen many similar structures in the Newland limestone of the pre-Cambrian Beltian series of Montana. He thereupon collected and studied great quantities of the ancient specimens, describing many new genera and species of blue-green algae. To one who has not studied them in detail, Walcott's species appear to range all the way from those which from their resemblance to the Cambrian *Cryptozoön* seem to be organic, through doubtful ones to septarian concretions, with side lines among the ripple marks and shrinkage cracks, ending with what may be a calcareous tufa. None of them shows more structure than a general similarity to a water biscuit, although the describer figured what he took to be chains of algal cells. These are not convincing, since they are replaced by opaline silica, retain their original convexity, and were derived from one of the least organic-appearing of the specimens.

In the case of the Ordovician and more recent deposits attributed to lime-secreting algae, it is possible to identify the plant, at least generically, by the internal structure, since the walls of the skeleton are well preserved. That such is not true of the pre-Cambrian specimens may, of course, be due to the greater vicissitudes which they have suffered. Perhaps the only test which can at present be applied to this class of objects is the apparently simple one of whether or not they are of organic origin. If they are organic, it is more likely that they are calcareous algae than that they are anything else.

Since the publication of Walcott's paper, geologists have found similar masses in the pre-Cambrian strata in the district south of Lake Superior, in the vicinity of Hudson Bay, in the Grand Canyon, and in other parts of the world. Although most geologists and paleontologists have accepted them as algae, a few, including the writer, have maintained a somewhat skeptical attitude toward these "plants." As Professor Olaf Holtedahl of Oslo has pointed out, similar concretions have been found in situations which preclude the possibility of their having been formed by organisms. In any case, there is no reason for applying generic and specific names to such indescribable objects.

Perhaps the most astonishing discovery in pre-Cambrian rocks was that announced by Walcott in 1915. Sections made from a specimen of one of the "calcareous algae" proved, on examination under high powers of the microscope, to have in them minute (0.95–1.3 microns in diameter), somewhat irregular rods which were identified as bacteria. Walcott likened them to the modern *Micrococcus*, but they have been more aptly compared by Henry Fairfield Osborn, so far as superficial resemblance goes, to some of the nitrifying bacteria which exist in soils. Walcott's paper is a brief one, in which he gives credit to Albert Mann of the United States Department of Agriculture for the identification, and leaves it to be accepted on faith that an organism without hard parts, and less than 0.001 millimeter in diameter, would be preserved in identifiable condition from pre-Cambrian times to the present!

The calcareous algae and the bacteria are the only plants yet reported from the Pre-Cambrian. Attention may now be turned to the animals.

Years ago L. Cayeux described and figured many species of radiolarians, foraminifera, and sponges from metamorphosed quartzites in Brittany. They have been widely accepted as pre-Cambrian fossils, although doubts as to their authenticity have been expressed. It now appears that the whole controversy is a tempest in a teapot, for a French geologist has shown that the strata in question are not pre-Cambrian but probably Devonian in age. Whatever the age of the rocks, they are so changed from their original condition that it is doubtful if the so-called fossils in them are actually the remains of organisms. The objects which Cayeux described are almost the only European pre-Cambrian remains which have been supposed to be of animal origin. Most of the discoveries have been made in America. More fortunate than Cayeux, Walcott found what are considered to be real sponge spicules in the Chuar of the Grand Canyon. No figures of them have been published.

Atikokania lawsoni, described by Walcott, was widely heralded at the time of its discovery (1911) as the oldest fossil, since it was obtained from a limestone of the Steep Rock series (Huronian), west of Port Arthur, Ontario. The best specimens are silicified in a calcareous matrix and have a cylindrical or pear-shaped form from one to fifteen inches in diameter. They are described as having inner and outer walls, the space between being filled with radial pillars which are laterally connected to produce a concentric structure. Walcott compared them to Cambrian sponges, particularly to a South Australian genus, which has an inner and an outer wall connected by radially arranged tubes. In later papers he has referred to them as "spongoids."

The Permian dolomites at Sunderland, England, contain specimens which are similar to Beltian calcareous algae and to *Atikokania lawsoni*, but they have been shown by G. Abbott and by Holtedahl to be of inorganic origin, for they are concretionary structures resulting from the replacement of limestone by dolomite. The process began along joints and cracks in the rock. If the "fossil" was produced that way in the Permian deposit, it is likely that the specimens in the Pre-Cambrian were also of inorganic origin.

Walcott described another "fossil" from the Canadian locality under the name of *Atikokania irregularis*. Some excellent specimens of this form have been investigated by the writer. Thin sections show that it is composed of aggregates of quartz crystals embedded in a matrix of limestone. It is, therefore, of purely inorganic origin. Similar crystals occur in pre-Cambrian limestone on one of the shoulders of Mount Edith Cavell in the Canadian Rockies. At that locality they appear to have been formed by silica-bearing solutions which have passed through the limestone.

The pre-Cambrian strata in Montana which have been so productive of "calcareous algae" have long been searched for other evidences of ancient life. Several discoveries were reported from this region by Walcott, who described various trails and burrows, ascribing their origin, probably correctly, to the agency of annelids. More noteworthy, however, are the numerous specimens which the same writer named *Beltina danai*. This species was founded upon numerous fragments which were believed to belong to an arthropod allied to *Pterygotus* or *Eurypterus* (see Chapter

VII). The supposed test is extremely thin and, in most cases, without any definite outline. A few fragments, selected from thousands, do remotely resemble parts of eurypterids. This may be said of four of the specimens figured by Walcott. Not only is the absence of outline an objection to the reference of these specimens to arthropods, but an even more significant circumstance is their total lack of surface marking. This excludes them completely from the Eurypterida, for even small pieces of the tests of these animals show a characteristic series of scales. The fragments are probably of organic nature, however, and they are widespread in Montana and British Columbia. It may be that they are of vegetable origin, perhaps remains of brown algae, though nothing definite can be ascertained from their structure. It seems likely that the "arthropod" recently described by Sir Edgeworth David and R. J. Tillyard from the Pre-Cambrian of Australia is of similar nature, if organic at all.

The alarms rung by the discoveries of pre-Cambrian fossils have been sounded so often without real cause that the paleontologist is perhaps becoming unduly skeptical. From the standpoint of one who has seen all sorts of discoveries accepted without question, it is refreshing to note the inception of a more critical attitude. Paleontology, like many another science, has suffered from the general human tendency to play follow the leader. Paleontologists have been too prone to accept blindly the dicta of "authorities." At one time everybody was describing as fossils the inorganic Eozoöns, just as for the past twenty years everyone has been finding calcareous algae. As a matter of fact, the flora and fauna of the Pre-Cambrian, so far as they are recorded by actual fossils, cannot be said, even by the most credulous, to represent more than four groups: blue-green algae, brown algae, sponges, and annelid worms. If paleontologists were called upon for strictly scientific evidence, they could not prove that their determination of any one of these groups is correct.

It is perhaps possible to get a more just idea of pre-Cambrian organisms if we make a brief survey of the state of evolution of life during Cambrian times. So far as plants are concerned, the evidence is but little more satisfactory than that regarding the similar fossils of the Pre-Cambrian. *Cryptozoön* is common in the Upper Cambrian, but whether the specimens are the secretions of blue-green algae, as is now the popular opinion, or of hydrozoans, as was formerly supposed, no man can say. All agree that they are probably of organic origin. Other Cambrian remains supposed to be of vegetable origin are in unsatisfactory states of preservation, although some of those which Walcott described from the Middle Cambrian may be red and brown algae. Fortunately, the Cambrian fauna is much better known. As indicated in the preceding chapter, trilobites and other crustaceans, gastropods, brachiopods, annelid worms, echinoderms, coelenterates, and sponges have been found in Lower Cambrian strata. From the state of evolution of each of these groups, a reasonable inference as to their antiquity may be made.

The fact that the Protozoa are unknown as fossils in the oldest fauna is no indication that Foraminifera and Radiolaria were not abundant at that time. The animals may have been naked; at best, their minute skeletons are ill adapted for preservation or recovery. More informative are the remains of the sponges. Although they comprise only slightly more than one per cent of the fauna, Walcott has shown that all orders of the siliceous sponges were represented among the fossils of the Mid-Cambrian; hence, we may conclude that their history began far back in the Pre-Cambrian. We are not so sure about the coelenterates. Walcott's Mid-Cambrian jellyfish is probably authentic; hydrozoans of some sort appear in the Mid-Cambrian, and graptolites in the Upper, whereas the corals are represented by a single non-calcareous species in the Mid-Cambrian. Differentiation in this group probably took place during the Cambrian, although there must have been an ancestor in pre-Cambrian times. Since this was written, a jellyfish has been found in the Pre-Cambrian of the Grand Canyon. It has not yet been described.

The presence in Walcott's famous quarry of numerous specimens which seem referable to chaetopod annelids necessitates the postulation of an early pre-Cambrian origin for this group. The Cambrian representatives of the echinoderms are all simple cystids or edrioasteroids, a fact which indicates that this phylum had not existed long. Despite their numerous genera and species, the brachiopods were not in a high state of evolution in Cambrian times. The inarticulates were old, but the articulates probably made their entrance upon the world's stage about the beginning of the Cambrian. There is no reason to suppose that the Mollusca, known chiefly from numerous free-swimming gastropods (hyolithids) and a few simply coiled snails, had any long pre-Cambrian history, for their Cambrian representatives are surely primitive. The trilobites and other Cambrian arthropods suggest a different story. Undoubtedly animals of this sort had been in existence for millions of years before the time of their first actual appearance as fossils in early Cambrian strata.

Summarizing this brief analysis of the Cambrian animals, it may be inferred that the pre-Cambrian fauna consisted of naked Protozoa, siliceous sponges, primitive coelenterates, annelid worms, inarticulate brachiopods, and trilobites, the latter accompanied, perhaps, by creatures resembling their ancestors.

If this inference be sound, why are there so few pre-Cambrian fossils?

Numerous and varied are the answers to this question; six will be considered here.

1. Pre-Cambrian fossils were destroyed during the changes which took place in the metamorphism of the rocks.

2. Daly's theory: pre-Cambrian marine organisms had no calcareous skeletons because of insufficient calcium in the oceanic waters.

3. Lane's theory: the pre-Cambrian oceans were acid; this condition prevented the formation of calcareous skeletons.

4. Walcott's theory: all the pre-Cambrian strata now accessible were deposited on land in fresh water of low calcium content.

5. Brooks's theory: pre-Cambrian organisms lacked hard parts because they lived in the surface waters of the oceans, where skeletons were detrimental because of their weight.

6. The writer's modifications of the Brooks theory: skeletons appeared after the Pre-Cambrian as a result of the adoption of a sessile or sluggish mode of existence.

Taking these up in order, it may be said that the first explanation holds for most pre-Cambrian strata. It is the exception, rather than the rule, to find sediments of that age which have not been so completely altered as to change the original sandstone, shale, clay, or limestone into gneiss, schist, slate, or marble. The recrystallization which accompanied these changes generally destroyed any organic remains. It is practically useless to search for fossils in metamorphosed rocks, although they are sometimes found. There are a few formations, such as the Beltian of Montana, the Keweenawan of Michigan, parts of the Huronian of Ontario, and other strata in Texas, Newfoundland, and China, which appear to have partly escaped the processes of alteration. Their lack of fossils must be explained on other grounds.

The second theory, proposed some thirty years ago by R. A. Daly, has found considerable acceptance. Obviously, soft-bodied animals are ill adapted for preservation as fossils, and no calcareous skeletons can be formed if the water in which the organisms live does not contain sufficient available calcium. The oceans of the present day contain a larger percentage of the bicarbonate of this metal in solution than do rivers and lakes upon the land; hence marine organisms build thicker skeletons. Moreover, there is a direct relationship between the thickness of shells of Mollusca and the concentration of calcium salts in solution in rivers and lakes. For example, the fresh-water shells of New England, where there is little calcium, are thin as compared with those in the Mississippi basin, an area where much limestone is dissolved. Daly noted these facts and found a reason for the lack of calcium in the oceans of the Pre-Cambrian. In brief, he argues that in the absence of an effective scavenging system on the floor of the deep oceans there must have been, over two-thirds of the globe, an accumulation of organic matter, which, on being decomposed by bacteria, produced much ammonia. This, in his opinion, caused the precipitation of nearly all of the calcium in the oceans and the formation of calcareous ooze, an inert compound, unavailable to animals. The only calcium in solution thereafter was the small amount contributed by the rivers, and itself on the way to being precipitated by the same process. It was not until Ordovician times, when scavengers became numerous, that much skeletal material was accessible to animals.

It would be a logical deduction from Professor Daly's theory to suppose that the pre-Cambrian carbonate deposits were of deep-ocean origin. So far as I know, there is nothing about the rocks themselves to indicate this; Daly himself agrees that there is much evidence that they were formed in shallow epeiric seas like those of the Paleozoic. Under such conditions, the dissolved calcium bicarbonate brought in by the rivers must have passed through shallow seas on its way to regions cold enough to cause the water containing it to sink and creep equatorward along the bottom, where alone it could come in contact with decaying organic matter. Such a circulation would necessarily be slow, and inhabitants of the shallow seas would have first chance at such calcium as there was. Furthermore, in James Bay and the eastern Baltic the waters are of low salinity, yet the Mollusca build shells, as do those of the marginal portions of the Black Sea, down to a depth of fifty or more fathoms. Animals form shells even in rivers of extremely low calcium content; if skeleton-secreting cells are present in an organism, they manage to find the needed material in any medium except *aqua pura*.

The third theory, A. C. Lane's, cannot be discussed adequately without wandering far afield into the realms of chemistry. There are good reasons for believing that during the early history of the oceans their waters held so much chlorine and other dissolved and ionized chemicals as to make them acid and thus effectively prevent the formation of calcareous shells. Hence the first skeletons may have been composed of chitin or silica. It is probable that this condition did exist in the early days of the Pre-Cambrian, for such primitive animals as radiolarians and the oldest sponges have siliceous skeletons, and the basis of those of the inarticulate brachiopods and of the trilobites is chitin. On the other hand, immense quantities of limestone were deposited during later pre-Cambrian times, a happening which would have been impossible in an acid sea.

The man who probably expended more time and energy than any other individual in trying to find pre-Cambrian fossils was Walcott. As a result of his discouraging experience, over a period of eighteen years, he came to the conclusion that almost all the accessible strata older than the Cambrian were accumulated on land as fluviatile or lacustrine deposits. The Beltian, he thought, might be partly marine, for he firmly believed in the crustacean nature of *Beltina danai*. He pointed out that in Beltian times the continents were larger than at present and that all the deposits in which fossils have been sought were formed within their margins. The shallow-water origin of the strata is indicated by red beds, ripple marks, and surfaces checked with shrinkage cracks. Such shallow-water characteristics are, however, as often found in sediments of marine as of fresh-water origin. Furthermore, the bodies of water in which the strata were laid down were too large, too permanent, and too similar in pattern to the later marine epeiric seas to present much evidence that they were lakes.

W. K. Brooks was one of the first to suggest a theory to account for the lack of fossils in the Pre-Cambrian. He believed, with many other zoölogists, that the superficial waters of the open ocean were probably the place of the first great expansion of life, if not the region of its inception. He developed the idea that the pre-Cambrian animals lived exclusively in this zone, where they maintained a free-swimming or floating existence in which a heavy calcareous shell would have been an encumbrance, not a help. Such animals eventually came to live on the sea floor, near shore, where the conditions of a new habitat caused them to secrete calcareous shells. As it was originally stated this theory had various defects, one of the most obvious of which is, as Daly pointed out, lack of reason for the sudden discovery of the ocean bottom. Since the time of its promulgation, much has been learned about the secretion of shells, so the writer has somewhat modified his explanation, adhering, however, to his central idea of the importance of activity.

That a definite relationship between activity and skeletal armor exists is obvious. The common symbols of sluggishness are the snail and the tortoise; such animals as the corals and the bryozoans, fixed in one spot, show a maximum of calcareous shell and a minimum of flesh. It is commonly said that armored animals are sluggish because well protected, and that the unarmored are active because they must seek safety in flight. As a matter of fact, cause and effect are reversed in this oft-repeated remark; the truth is that animals are probably armored because of their sluggishness or entire lack of movement. Calcium is present in greater or lesser quantities in all water and in many kinds of food. It enters the alimentary tracts of animals in solution, but within them it is converted into solid calcium carbonate, either by the action of ammonia produced by putrefactive bacteria or by the effect of the nascent methane formed during the digestion of the cellulose of plants. The calcium carbonate so produced is harmless, but small amounts, still in solution, get into the body fluids and reach the protoplasmic cells. Some of these cells apparently pass it on to the excretory system; others cause its precipitation *in situ* and so build a skeleton. All animals and many plants are confronted with the necessity of getting rid of surplus calcium. Those most active best solve the problem. The production of a calcareous skeleton is an involuntary chemical function which will take place in animals in any environment. In one sense it may be thought of as a pathologic condition, brought on by inactivity.

As a check, let us examine the Cambrian fauna for connections between activity and skeleton.

Knowledge of the Cambrian Protozoa is too unsatisfactory to allow any profitable discussion. The sponges were sessile, and those which survived secreted siliceous spicules. The extinct archaeocyathinid sponges were sessile and formed a calcareous skeleton. The jellyfish and most worms were active and had none, but the more sedentary tubicolous worms early began the formation of calcareous tubes. Cystoids and articulate brachiopods were anchored to the bottom, and both secreted calcareous shells, but the wriggling and burrowing inarticulates had only a chitinous covering, slightly impregnated with phosphate of lime. Most of the gastropods of the Cambrian were free-swimming, or at least moderately active, and formed thin calcareous shells. The Crustacea were active and had skeletons which were fundamentally chitinous. In a general way, the rule holds.

An indication that sessile life was a novelty, newly discovered in Cambrian times, is the absence of colonial animals from the Lower Cambrian fauna. Until Mid-Cambrian times reproduction appears to have been by fission, or by a truly sexual process, for all earlier creatures are separate individuals. The first examples of budding reproductions, resulting in the formation of colonies, appear among the sponges and hydrozoans found in strata of Mid-Cambrian age. There are but few of these. Budding was doubtless the result of the late-formed habit of fixation. Colonial life did not become popular until Mid-Ordovician times.

We can only speculate as to the origin of sessile life. If we turn to the list of pre-Cambrian animals, we find that all of them were motile and that predaceous carnivores were entirely absent. This group appears first with the few cephalopods of the late Cambrian, having missed the Pre-Cambrian completely. In the latter era, untroubled by predaceous animals, the swimming and floating organisms must have increased rapidly until there came a time when the upper, sunlit parts of the seas and oceans were overpopulated. This must have forced some individuals to the bottom. It may also be, as E. W. MacBride has suggested, that some of the more sluggish animals would naturally drop to the bottom from time to time, simply because they were too lazy to keep afloat. Those which fell into the dark abyss of the deep oceans mostly perished, for food is scarce outside the zone of sunlight, where alone plants can live. Active animals reaching the bottom in shallow water continued to swim, or learned to crawl about after food. The more passive adhered to the substratum, became relatively inactive, and began the secretion of skeletons because they were no longer able to rid themselves of calcium carbonate.

That the pre-Cambrian animals were all motile seems, therefore, to explain their lack of hard parts, and hence to solve the question why so few pre-Cambrian fossils are found. That more than are now known will eventually be discovered is probable, but it may be predicted that they will prove to have either no skeletons or else thin ones composed of silica or chitin.

It may be well to amplify the discussion of the facts on which what has already become known as the "sessile" theory of the origin of calcareous skeletons is based. Perhaps the sessile part of the preceding modification of the Brooks hypothesis has been overemphasized. The matter is essentially one of degree of activity in relation to the elimination of calcium salts, and the hare and the tortoise are perhaps better examples than the motile jellyfish and the fixed coral. One must also take into account the relative mobility of various parts of the body. It has been suggested that if deposition of calcium carbonate is due to inactivity vertebrates should have no skeletons, since they are active animals. This is a fair inference but not an argument against the idea, for, as will be shown in later chapters, the ancestors of the vertebrates had no calcareous internal skeletons. In fact, they had no internal skeletons at all, although many sluggish benthonic members of the group did develop calcareous external ones. They were succeeded by fish with cartilaginous internal skeletons, a primitive condition which has remained unaltered in some to the present day. The skeletons of others have been partially calcified. But where? Chiefly in the axial region, the least mobile part of a fish. It may be objected that the fish swims by means of rhythmic contractions of the muscles of opposite sides of the body, and that the axial region is not inert and rigid. Perfectly true. But everyone who has read the most elementary books on physics must be aware of the phenomenon of the transmission of vibratory waves along a wire on which there is interference with free movement: the nodes of no motion, and the internodes of movement. It is at the motionless nodes in the cartilaginous axial support that calcifications, vertebrae, are produced, whereas the internodes remain unossified. The calcareous dorsal and ventral spines and ribs likewise follow lines of relatively little motion. This is not the place to enter into a detailed discussion of the skeleton. Suffice it to say that the head, primarily a cartilaginous covering of the brain, is immobile, except for the vertical movements necessitated by the opening and closing of the mouth, and that the originally cartilaginous portions of this region ultimately became ossified in the higher fish. The appendicular system (limbs) of the land-living vertebrates was pre-formed in cartilage in the ancestral fish. Here again the segmented nature of the appendages indicates a nodal and internodal arrangement, regions of no motion alternating with those of mobility.

So much for one of the major objections to the "sessile" theory or, as the writer would have preferred to call it if he had considered it a new theory rather than a modification of the Brooks hypothesis, the "sluggard" theory. It might be well to state in passing that the slug, although one of the most active of the gastropods, is so called because he is, relatively, a sluggard. The term far antedates the naming of the animal.

Perhaps one or two other objections may be forestalled by a few further remarks. For instance, it may be pointed out that the sea anemones and stony corals have the same structure; both are corals and both are sessile, yet the former have no skeletons at all, the latter massive ones. A part, at least, of the explanation of the anomalous condition is to be seen in the fact that most of the sea anemones are inhabitants of shallow or deep cold waters, whereas most corals are tropical or subtropical. The matter of temperature as a factor in the formation of skeletons has been mentioned above and will be again in the discussion of the Ordovician fauna. It may seem that a more acceptable theory could have been built up on the basis of the increasing warmth of the seas after the Pre-Cambrian. Really thick calcareous skeletons do not appear until Ordovician times, the first period in which broad, shallow, warm, epeiric seas were widespread. For example, numerous sorts of arthropods are living today, but only one nonparasitic group, the barnacles, contains sessile creatures. And the barnacles are the only arthropods with really calcareous skeletons. But, as is well known, barnacles are not by any means confined to warm seas. Anyone who has tried to bathe off the rockbound coasts of New England is well aware of the association of barnacles and cold water; I think it was Irvin S. Cobb who once said that he would as soon have his foot bitten off by a shark as stick it in the water north of Cape Cod.

Various lines of evidence indicate that when the oceans were first fit for organic occupancy they were somewhat acid, and that the first plant and animal skeletons were siliceous or chitinous. Somewhat later, when animals had progressed as far as the stage of the primitive brachiopod and trilobite, the waters had become neutral or slightly alkaline but were still "fresh"; that is, soluble salts were present in parts per million rather than in parts per thousand. In the course of time calcium bicarbonate has become more abundant in marine waters, and animals have been confronted by an increasingly difficult physiological situation.

It is possible — in fact, probable — that animals discovered the sea floor from time to time during the Pre-Cambrian. It is unlikely that it would have taken a billion years to overcrowd the upper zones of water in the oceans. Even if the rate of reproduction were slow, it would have taken only a short time to fill the region permeable by sunlight with organisms. But there is evidence that adaptation to benthonic life was a slow process. As has been said, sessile and crawling life started, so far as the record tells us, at the beginning of the Cambrian, but it was not till the Mid-Ordovician, nearly a hundred million years later, that real sluggards became common. This hundred million years was probably a quiet period, as compared with any previous hundred million. The unstable submarine floor may have been repeatedly colonized, but it was probably disturbed by earth movements before the lime-secreting habit had been acquired.

Pre-Cambrian times were the real "dark ages." Many industrious and brilliant geologists have devoted their lives to the study of the rocks then formed, but, though they have learned a great deal about them, they have been unable to establish a satisfactory chronology. Authentic history begins with the oldest rally fossiliferous strata.

V

THE ORDOVICIAN FAUNA

Rich with the spoils of Nature.
Sir Thomas Browne, *Religio Medici*

The fauna of the Ordovician differs from that of the Cambrian in its greater diversity, many more classes of animals being present. The new groups appearing at this time are largely of kinds which secrete thick calcareous skeletons, shells more apt to be preserved than the thin chitinous tests of the Cambrian animals. Fossils are rare in many Cambrian rocks, but some Ordovician limestones are literally made up of them. If one happens to be in the neighborhood of Cincinnati, Ohio; Richmond, Indiana; Lexington, Kentucky; Nashville, Tennessee; Minneapolis, Minnesota; or other places that could be mentioned, it is worth while to take a trip into some of the old quarries on the hillsides or walk along the natural exposures beside the streams. Even the experienced collector is amazed by the countless millions of organic remains which make up the greater part of the Ordovician strata. At most of the localities mentioned, the rocks are soft and break down readily on exposure to the atmosphere. The fossils, being harder than the enclosing matrix, weather out, and one can find specimens as perfect as those on a modern beach. To a person who has a natural bent in that direction there is probably no pleasure more keen than that of wandering among these debris of ancient ocean floors, picking up the shells of animals which lived countless centuries ago, never knowing what treasure the next step will reveal.

The most important of the new arrivals in Ordovician faunas may be enumerated. Among the coelenterates are the colonial organisms, graptolites and corals. A few species of graptolites have been described from the Upper Cambrian, and a little evidence for the existence of corals during the Cambrian has been found, but both were virtually newcomers. The echinoderms increased greatly in variety, starfish, crinoids, blastoids, and sea urchins being represented. Bryozoans came upon the scene in profusion, and their numerous coral-like skeletons are highly prized by geologists for their value in correlation. Many kinds of gastropods augmented the molluscan population, and pelecypods, cephalopods, and chitons made their first appearance. New groups among the arthropods were the bivalved water fleas, or ostracods, the eurypterids, and the sessile barnacles. Remains of fishlike animals have been found in the Upper Ordovician strata at Canyon City, Colorado, in the Black Hills, and in Wyoming, but the specimens are all badly preserved. Nevertheless, thin sections show that these fragmentary fossils really represent early Chordata. By the end of the Ordovician nearly all the classes of invertebrate animals except the insects and some other specialized arthropods had appeared, but the flora had as yet made little advance upon the lowly position which it occupied in Cambrian times. The present data, however, are incomplete, for no fresh-water or terrestrial deposits older than the late Silurian have as yet been discovered.

The Ordovician was the time of the first abundant secretion of carbonate of lime by animals. This may have been due to one of two causes. It may be that the water of the original oceans of the globe was comparatively fresh — that is, that it contained little dissolved mineral matter. As time went on, the breaking down of the rocks on land under the influence of weathering would cause great quantities of soluble material to be carried to the sea. It is known that the amount of sodium chloride (common table salt) in the sea has constantly increased throughout the ages; it may be that the amount of dissolved calcium has likewise tended to accumulate. The other possible cause for the increased precipitation of calcium carbonate by animals in Ordovician times may have been a change in climate from cold to warm. Water containing carbon dioxide can retain in solution much more calcium than pure water can, and cold water absorbs and retains much more carbon dioxide than warm water does. Animals living in cold water, consequently, are much less bothered by calcium than those living in warm water and as a result secrete thinner shells. It is possible that the general temperature of the oceans and seas was higher in Ordovician than in Cambrian times. Certain it is that, taking the world as a whole, there is much more limestone in the Ordovician than in the Cambrian series of rocks. Moreover, it may be noted in accordance with the above principle that the chief places for the formation of limestone at the present time are in tropical and subtropical regions. During Ordovician times limestone was formed both under the equator and in northern Greenland within a few degrees of the pole, a fact that indicates uniformly warm conditions throughout the world. (The reason for the equable climate of the whole earth during the greater part of the Ordovician may be seen on consulting a palaeogeographic map on which are delineated the lands and seas of the time. It was a period when the lands were of low relief and when there was a widespread flooding of the continents; the whole extra-polar world enjoyed a warm oceanic climate.) Hence it is not surprising that sluggish animals surrounded themselves with layers of the easily precipitated mineral. This point is stressed, despite some repetition, because it is too often stated that animals build hard parts to protect themselves. As was said in the previous chapter, the formation of external skeletons began in Cambrian times before there were any predaceous carnivores, and consequently before there was any need for protection. In the Ordovician, shells did serve as armor, and such animals as had them were endowed with a natural advantage over others. Hence one may note the operation of both the Lamarckian and Darwinian evolutionary processes. The environment (water with dissolved calcium) forced shells on the animals. These served, when the time came, as a means of protection, and the animals best protected survived. A good example of this is seen in the great multiplication of the calcareous-shelled brachiopods, which completely overshadowed the chitinous inarticulates in the Ordovician.

Among the most important animals of the Ordovician seas were the graptolites. To understand this group of fossils it is necessary to dwell for a moment upon the lowly modern animals known as hydroids. These are numerous both in salt and fresh water at the present day but attract little attention, since, if seen, they are usually supposed by the casual observer to be small aquatic plants. Their small tubular bodies are radially symmetrical, having at the upper end a mouth surrounded by slender, fingerlike tentacles. It opens into a simple undivided digestive cavity. They are, therefore, coelenterates, closely allied to the corals. They differ from the latter, however, in lacking radial partitions within the body cavity and in having a life history in which a sessile, plantlike, asexual generation alternates with a free-swimming or floating one. In the fixed generation a slender, shrublike form is produced by successive budding. This asexual colony eventually produces buds which, instead of forming ordinary feeding polyps, become little jellyfish that break off from the parent colony and float away. The jellyfish produce the sexual elements. When an egg is fertilized, it develops into a larva which settles on some foreign object, where by repeated budding it grows into an asexual colony. This sequence is not followed by all members of the class, however, for some kinds, such as the *Hydra* of fresh water and the *Sertularia* so common along the Atlantic coast, produce special buds which are sexual. From them come young larvae without the intervention of a jellyfish stage. The graptolites are similar in appearance to *Sertularia*, and probably had a similar structure and much the same habits.

Many layers of the black shales which form long narrow belts in the Great Valley of the Appalachians from eastern New York to Alabama show figures which suggest pencil markings. On closer examination it can be seen that these streaks have on one or both margins serrations which suggest the teeth of a saw. Linnaeus, a century and three-quarters ago, saw such markings on the black shale of southern Sweden and called them *Graptolithus*, a word which has furnished a name for this group of fossils. Their leaf- and stemlike nature led the early paleontologists to class them as plants, but comparison with modern hydroids has revealed their true nature. Most of them are so flattened that it is practically impossible to learn what their real structure was. But, although they are largely confined to shale, it fortunately happens that some specimens preserved in an uncrushed condition have been found in limestone and chert, from which they have been freed by the use of chemicals. Gümbel, Holm, Wiman, Sollas, Bulman, and other European paleontologists have made elaborate studies which have furnished information that has explained the

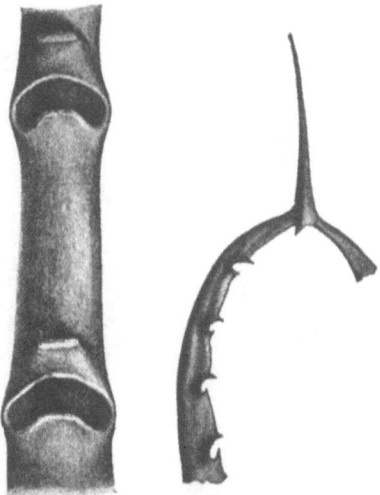

Fig. 15. A well-preserved graptolite (*Isograptus*), twelve times natural size, with a portion enlarged to sixty-two times natural size. Note the details of the bilateral thecae, features which cannot be seen on flattened specimens. From O. M. B. Bulman.

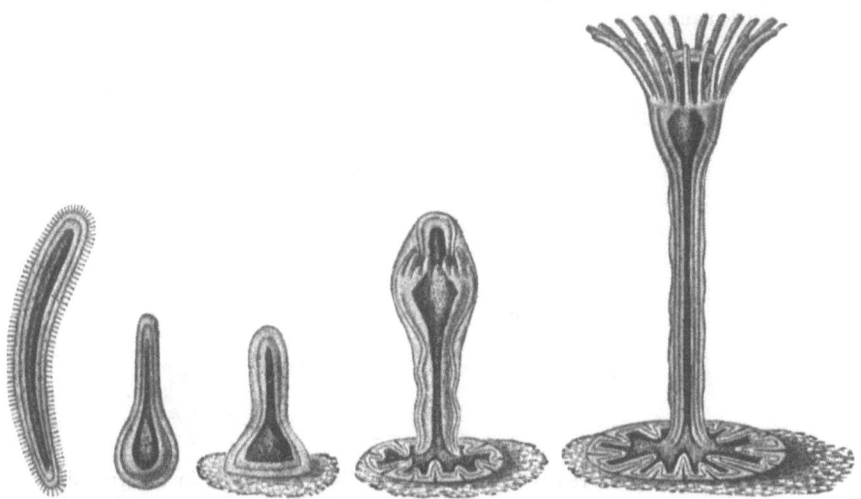

Fig. 16. A series showing the early growth stages of the modern hydroid, *Eudendrium*. At left, the free-swimming ciliated planula; next, the cilia lost, the larva settles to the sea floor; there the basal disk is formed; later, enlargement takes place, tentacles are formed, and (at right), the thin chitinous envelope bursts open. All from G. J. Allman.

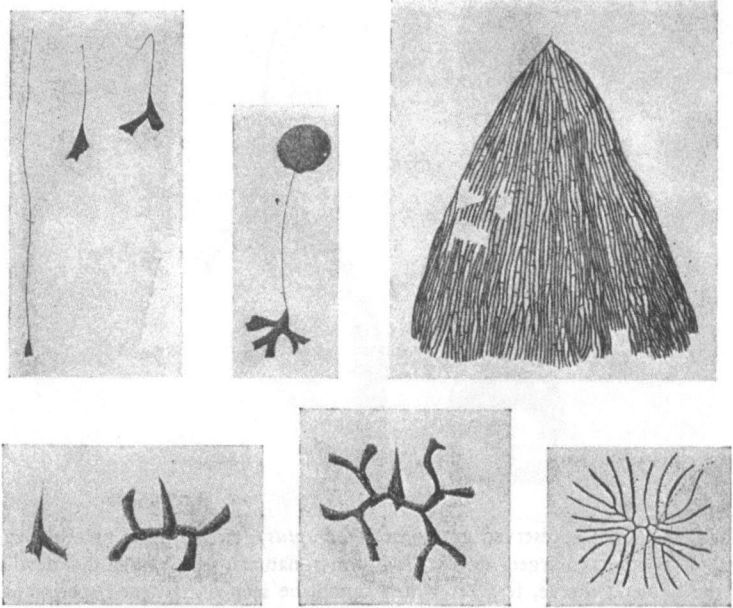

Fig. 17. Above, a pendent dendroid graptolite, *Dictyonema*, showing stages of growth, and an adult rhabdosome. Below, a many-branched axonolipan graptolite, with young specimens in the early stages of growth. From R. Ruedemann.

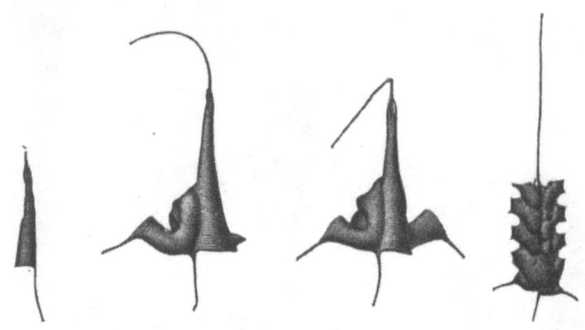

Fig. 18. Well-preserved specimens showing the early stages in the growth of an axonophoran graptolite. At left, a sicula with virgella; next, a sicula and first daughter theca; third, the stage in which a second theca has budded from the first; and at right, the stage in which the later thecae climb up the nema. From O. M. B. Bulman.

common flattened specimens (Fig. 15). A considerable knowledge of the morphology and developmental history of the group has resulted.

All graptolites form colonies which have the general appearance of recent hydroids (Fig. 16). All seem to have been attached and to have had a basal disk or rootlike expansion for that purpose. From it springs a thin filament, or nema, which enlarges at the top to form the sicula, a thin-walled, chitinous cup containing the first zoöid. From the first cup, or theca, arise one or more young individuals from which other buds grow out along certain definite patterns to form colonies of various shapes. The daughter zoöids, like the parent, inhabited small bilaterally symmetrical receptacles which have a thickened ring about the aperture, one side of which in most cases is produced into a spine. The thecae, when flattened, form the sawtoothlike markings of the margins of the specimens in the shale. The bushy graptolites, the Dendroidea, have three kinds of thecae. There are large open ones, called hydrothecae, which house the nourishing individuals; budding thecae, which do not open outward, and which give rise to the other kinds; and the so-called bithecae, of unknown function. Most of the graptolites, however, appear to have only one kind, the hydrothecae, which contained the feeding animals.

As immature graptolites are common, the development of colonies, technically known as the astogeny, has been carefully studied in each of the three subdivisions, or orders, into which the group has been divided. The shrublike Dendroidea show a simple plantlike growth, the sicula first budding into a composite theca which consists of a nourishing and a budding individual (Fig. 17, above). The latter gives rise to other buds which produce elongate, straight, or irregular stipes, connected at intervals by crossbars which strengthen the colony. Most members of this order had an upright form of growth, and, to support the colony, secondary skeletal matter with rootlike expansions was secreted about the nema. Some, however, had a slender nema, and must have grown downward instead of upward.

The simply branched forms included in a second order, the Axonolipa, grow from the sicula by the production of a bud which turns to one side, at right angles to the parent, whereas its daughter theca crosses to a position directly opposite to it (Fig. 17, below). From these thecae new individuals arise which may develop into a single pair of stipes, or may by further dichotomous budding produce four, eight, sixteen, or even as many as sixty-four branches, not connected by lateral supports. All species in this order have a long, slender nema entirely incapable of supporting the colony in an upright position. It is therefore inferred that such graptolites were pendent.

Entirely different is the colonial development of the third order, the Axonophora. The sicula gives rise to a single theca, which turns abruptly aside and upward so that its aperture faces in the direction opposite to the original one (Fig. 18). Later buds keep the same orientation, producing a simple stipe with thecae on one or both sides of the nema. During growth the nema continuously elongates, becoming enclosed in the skeleton to form the axial support or virgula. Some representatives of this order are known to have had a bladderlike float; beneath this were clustered a number of reproductive sacs, in which larvae developed to the sicula stage. Some of the siculae remained attached to these gonangia, and the result was the production of a compound colony (synrhabdosome) with many simple colonies (rhabdosomes) pendent from one float.

The habits and habitat of the graptolites have been much discussed, but what is now known of their structure, development, and distribution leaves little opportunity for difference of opinion. Such of the Dendroidea as have thick, rootlike supports appear to have lived in an upright position, growing like seaweeds on rocks, stones, shells, or other objects on the sea floor. All other kinds hung downward, many of them probably attached to seaweeds, whereas those with floats drifted about in the surface waters at the mercy of currents. Even those attached to seaweeds were in many cases a part of the migrant fauna, for great numbers of their hosts must have broken loose and drifted about. This interpretation of their mode of existence is borne out by their geographical distribution, for some of the species are almost cosmopolitan, being found in such widely separated regions as Scandinavia, eastern America from Newfoundland to Alabama, Utah, Idaho, British Columbia, the Yukon, and Australia. It is to their rapid dispersal by oceanic currents that their great value in correlation is due.

Why, if most of the graptolites led a floating existence, are their remains so generally confined to black or other shales of the finest grain? If they drifted about in the surface waters of the seas, they might be expected to die or be killed everywhere, and fall to the bottom where all sorts of sediments were being accumulated. It is true that specimens are occasionally found in limestone, silt-stone, and sandstone, but they are not abundant except in shale. Various possible explanations have been advanced by students of paleoecology, the science descriptive of the habits, habitats, and associations of extinct animals. The writer accepts the opinion championed by the first great student of this group, Lapworth, that the black shales of the Ordovician and Silurian represent muds accumulated at a considerable distance from shore, and that the presence or absence of graptolites depends upon the nature of the bottom on which they fell. It is probable that these animals were of more or less universal distribution in the surficial waters of the oceans and epeiric (intracontinental) seas to which they had access, and that on death or detachment from their supports they sank to the bottom everywhere. Since the nema by which they were supported was very slender, every storm must have taken toll of great numbers of victims. Those which fell on a relatively firm bottom where starfish, cephalopods, snails, and other carnivores were present were quickly devoured, for their small size and thin tests made them an easy prey even to small animals. Such enemies were, however, few or nonexistent in the soft mud of the offshore region. This is attested not only by the theoretical conclusion that animals would flounder and smother on a bottom composed of slimy ooze but by the actual fact that the shales containing graptolites seldom contain any other fossils than those which, like the graptolites, were planktonic (floating) in habitat.

Enough appears to be known of the group to allow a brief summary of its history to be made. Bushy, upright, bottom-living forms were the first to appear, a few

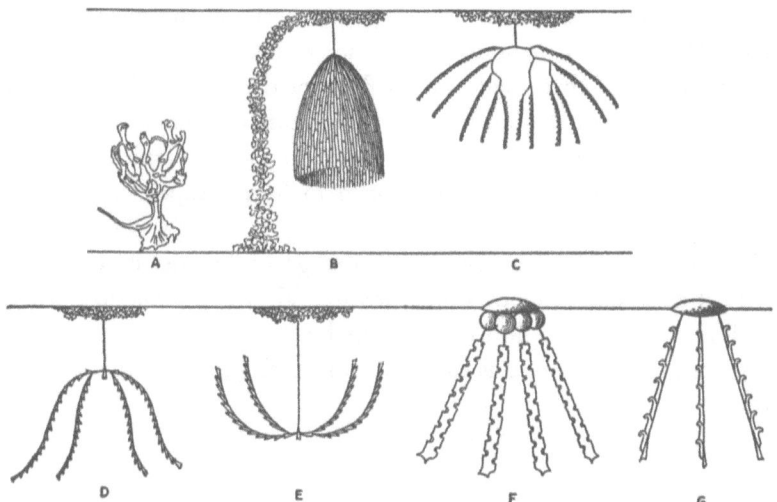

Fig. 19. Diagram to illustrate some of the stages in the evolution of the graptolites. A, basal portion of an upward-growing dendroidean. B, a pendent dendroidean. C, a many-branched axonolipan. D, an axonolipan with few stipes and with thecae turned downward. E, same, with thecae, and therefore stipes, turned upward. F, an axonophoran, with thecae turned upward. G, last and most specialized stage, an axonophoran with simple rhabdosomes, but thecae secondarily turned downward. An elaboration of a series by H. F. Cleland.

specimens having been found in the Upper Cambrian (Fig. 19 A). Larvae of these seem to have settled in various places, some of them becoming accidentally attached to seaweeds. Those which became fixed on the under surfaces of fronds and those whose hosts were fragments of seaweeds which became detached and floated away found themselves inverted, and so grew downward instead of upward, thus originating the pendent types (Fig. 19 B). Throughout the early history of the group there was a continuous reduction in the number of stipes. This may have been the first response to inversion and a somewhat unfavorable environment (Fig. 19 C–F). The zoöids appear to have tried to turn toward the light; that is, they were, in the language of the behaviorist, positively heliotropic. Among the early and mid-Ordovician Axonolipa there is a general tendency for the stipes to bend upward as each new bud turned more and more toward the light (Fig. 19 D, E). The upright position was regained by the scandent Axonophora, in which all the thecae except the sicula open upward (Fig. 19 F, G). This was the most successful group, for, appearing in the upper part of the Lower Ordovician, it persisted till the end of the Silurian, whereas most of the Axonolipa disappeared at the end of the Ordovician. Had it not been for the accidental attachment to floating objects, it is likely that the graptolites would never have become abundant, for those which lived on the bottom, although existing until Devonian times, apparently were kept in check by the animals that preyed upon them. Those which went to the surface escaped their enemies and throve, especially after many of them had evolved floats of their own. But their doom was sealed toward the close of the Silurian, by the increase in numbers of actively swimming animals. These could eat graptolites faster than new colonies could grow, and the unfortunate unprotected hydrozoans seem to have escaped one sort of enemy only to fall prey to another.

Considerable space has been devoted to this group because it is one of the few entirely extinct assemblages about whose rise and fall it is possible to draw relatively reasonable inferences. It also illustrates the methods of the paleontologists in dealing with an extinct group and, in this case, with material which on the whole is rather unpromising.

VI

PETRIFIED BUTTERFLIES

> Methinks I see thee gazing from the stone
> With those great eyes, and smiling as in scorn
> Of notions and of systems which have grown
> From relics of the time when thou wert born.
>
> From a recently discovered poem, "To a Trilobite,"
> by Timothy A. Conrad, first professional American
> paleontologist

Ever since people began to take an interest in "natural curiosities," the trilobites have excited the interest of those who have seen them in the rocks or in collections. Before the time of scientific study they were known as "petrified butterflies" or "flat fish," and their symmetrical forms, their elegantly ornamented surfaces, and the comparative rarity of really complete specimens have made them the favorites among all invertebrate fossils to the present day. Moreover, these are truly ancient animals, which have been extinct for millions of years, although they were the dominant group in the period of the oldest really fossiliferous rocks.

The name trilobite, or three-lobed stone, refers to the fact that longitudinal furrows down the back divide the surface of the shell into three lobes (Fig. 20). A more important tripartite division, however, is one in a transverse direction, since there are two conspicuous shields, situated at the anterior and posterior ends, connected by a segmented median portion. The cephalon or head shield of most bears a pair of more or less elevated compound eyes. The segments of the median portion, known as the thorax, are movable upon one another, like those in the body of a lobster or crayfish, and the hinder shield, or pygidium, has transverse creases or furrows that indicate an incipient segmentation.

Most trilobites are small, the average size being perhaps about two or three inches. Some are only five or six millimeters in length when fully grown, but there are several giants more than twenty inches long. There was no steady increase in size among them, as there was among the Mesozoic reptiles and Tertiary mammals. Some of the largest are found in Mid-Cambrian strata, still more in the Mid-Ordovician, including the largest of all, length twenty-seven inches, an ancient inhabitant of France and Portugal. The Silurian has produced no real giants, but a considerable number of them appeared during Lower and early Middle Devonian times. Later trilobites are all small.

The trilobites show all the sorts of visual organs to be found among the Arthropoda. Some were entirely sightless. Many paleontologists think that in some groups this condition was primitive, whereas in others it can be shown that the blinding is secondary, for all gradations of degeneration, from forms with good eyes to those without any, can be traced. Some have a pair of simple eyes, with one lens apiece, and possibly a median simple eye, or ocellus, like that of many modern crustaceans. Still others have a group of three minute simple lenses on each cheek. Throughout the great superfamily Phacopidacea the large conspicuous eyes are of a type which is commonly spoken of as compound, but which might better be called composite or

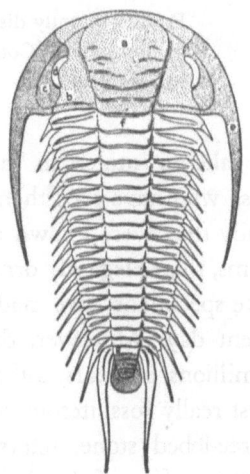

FIG. 20. A Mid-Cambrian trilobite, *Paradoxides,* to illustrate the parts of the dorsal shield. Stippled portion at anterior end, cephalon; at posterior end, pygidium. *a* and *b*, cranidium (coarsely stippled); *c*, free cheeks (finely stippled); *a*, glabella with glabellar furrows; *b*, fixed cheek; *d*, visual surface of eye; *e*, genal spine; *f*, first thoracic segment; *g*, last thoracic segment (thorax not stippled). The lines at the boundaries between the finely and coarsely stippled areas on the cephalon are the facial sutures. Original drawing by Eugene Fischer.

aggregate. The eye is really made up of several simple ones, each with its own cornea. This is the sort of eye possessed by most arachnids. Each little lens is perfectly distinct, separated from its fellows by narrow ridges. The composite eye differs in this respect from the true compound one, in which there is a common cornea and each minute element is so small that one has to put a specimen under the microscope to see the lenses. The latter is the type possessed by most trilobites. Some of these eyes are small; others are relatively enormous, the whole of the cheeks being given up to them. A climax is found in *Symphysops,* in which the eyes of opposite sides of the head are continued around the front.

All trilobites are longer than wide, most of them with a cephalon both longer and wider than the pygidium, although there are some which are isopygous, that is, have subequal terminal shields. The number of segments in the thorax varies in the group as a whole from two to more than forty-four, although the number of species with more than twenty is small. There is a distinct relationship between the length of the pygidium and the number of thoracic segments. If the pygidium is long, there are but few free segments; if it is short, there are many. This is due to the fact that the thorax is formed by the freeing of segments from the anterior end of the pygidium. The growing point, as in all arthropods, was immediately in front of the anal opening, which was situated near the posterior end of the pygidium. As in all arthropods, increase in size was accomplished at the time of the periodic shedding of the shell (molting), and it was at this time that new segments were introduced into the posterior shield in front of the anal opening and others freed at its anterior margin.

The process of molting of modern arthropods is a critical one, for the animals must slough not only the whole external shell but a part of the lining of the anterior and posterior portions of the alimentary canal. Even the surficial covering of the eyes comes off. Growth proceeds to the point at which the animal becomes too big for its coat; then the shell bursts open down the back or along a margin, and the inhabitant crawls out of its old case. It is then in what we call the soft-shelled condition — we watch the season for soft-shelled crabs — but the thin chitinous skin soon hardens, becomes more or less impregnated with salts of calcium, and the hard-shelled state is quickly resumed.

Practically all trilobites with compound or aggregate eyes had a special provision for shedding the shell. Around the top of each eye where the edge of the visual surface joins the so-called palpebral lobe, there is a narrow groove which can be traced forward until it unites with its fellow groove on the other side of the cephalon. Traced backward from each eye, the grooves extend outward to the lateral margins, or backward to the posterior one. These grooves are the facial sutures. At the time of molting, the middle piece of the head (the cranidium) parted from the lateral portions (free cheeks), and the animal crawled out through the ample opening thus presented (Fig. 22). Most of the specimens commonly found are cranidia, free cheeks, individual thoracic segments, and pygidia, the various members into which the shell disintegrated during ecdysis. Trilobites which are secondarily blind have obvious facial sutures. Most of those which seem to be primitively blind have them also, but they do not show on the upper surface, being marginal or submarginal in position. The free cheeks of some have become firmly ankylosed to the cranidium. Such trilobites seem to have molted as the horseshoe crab does, by splitting the shell along the margin of the head.

Only the dorsal surface was protected by a firm shell. On the under side there was but a single plate, the hypostoma, which was situated beneath the central part of the head. The remainder was covered by a thin membrane. To protect their delicate lower surfaces, most trilobites, in the days after the Cambrian, acquired the habit of enrolling themselves. Many died in that position, and these furnish some of the best-preserved specimens.

As has already been stated, the group was well represented even in Lower Cambrian times and perhaps reached its culmination in differentiation of species and abundance of individuals in the Upper Cambrian. Trilobites remained abundant until the initial days of the Mid-Devonian and then declined rapidly both in variety and abundance throughout the remainder of the Paleozoic. They survived until the Permian, where they are represented by perhaps a half dozen species. There are probably many reasons for their extinction. Other scavengers, particularly fishes, made competition for food increasingly severe. In the soft-shelled condition they themselves were excellent food, practically unprotected, a ready prey to the carnivorous animals which became more and more numerous after the end of the Cambrian. They appear to have withstood the attacks of the cephalopods, but failed before the keener-witted fish, which became gradually more clever and active while their prey was growing more stupid and logy.

Since they are extinct, there has been a great deal of speculation about their affinities to modern animals. They were early recognized as arthropods because of their segmented bodies and compound eyes, and as they are always found associated with the sorts of animals which nowadays live in the sea, it was deduced that they were marine. It is now possible to prove their true relationships and to gain some plausible ideas as to their habits, for within the last forty-five years a series of fortunate discoveries has revealed the limbs on the under side of the shell. Although trilobites grew in a multiplicity of shapes, all appear to have possessed the same simple type of limbs (Fig. 21). Each animal had numerous pairs arranged all along the body from end to end, with little differentiation among them, those under the head shield being but slightly different from those under the middle or posterior part of the body. Each leg was bifurcated near the body, the upper branch (exopodite) forming a thin, fimbriated organ, perhaps an external gill, whereas the lower one (endopodite) was a six-jointed slender walking leg, much like that of a thousand-legged "worm" or a wood louse. In some forms the segments of the lower branch were more or less flattened in a way that suggests that they may have been used in swimming. In all cases the limbs were attached to a series of processes which extended downward from the under side of the shell, beneath the two longitudinal furrows which give name to the group. Besides these divided appendages, all trilobites appear to have had a pair of many-jointed feelers (antennules) attached to the anterior part of the lower surface of the head; some of them possessed another pair at the posterior end.

The first thing which strikes one in looking over a collection of specimens is that the shells are all relatively broad and depressed. Nowhere in the group do we find narrow, compressed animals like the sandfleas or shrimps. The upper surface is gently convex, the lower one concave. Moreover, all the organs were confined to the central one of the three lobes, the part outside the longitudinal furrows serving largely as a protective covering for the appendages. The broad form, suggestive of an overturned boat, was obviously well adapted to give the animal buoyancy, so it may readily be supposed that the creatures found it easy to keep afloat and easy to swim. On the other hand, it is equally obvious that they would not have made rapid progress in swimming, for they must be compared to a flat-bottomed scow

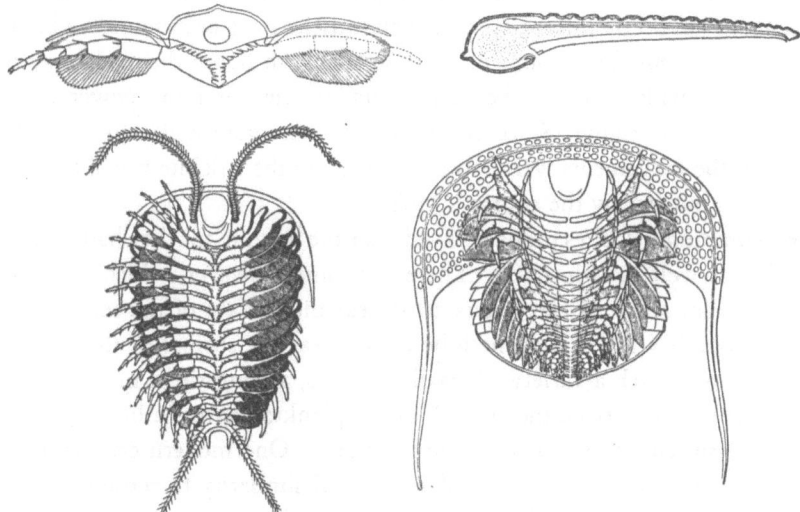

FIG. 21. The anatomy of trilobites. At left, *Neolenus*, a generalized Mid-Cambrian form; schematized transverse section of a thoracic segment above, restoration of the entire ventral surface below. The endopodites are purposely left off the right side, to show the exopodites more clearly. At right, above, diagrammatic longitudinal section through the axial lobe of a trilobite, showing (stippled) the stomach and alimentary canal, and the segmented heart above it; beneath, a restoration of the lower side of the specialized *Cryptolithus*. From P. E. Raymond.

rather than to a keeled yacht. They lacked entirely the ideal lines of the swiftly moving fishes. The appendages, likewise, were not adapted to producing rapid and powerful strokes; their structure indicates that the trilobite proceeded rather slowly and sedately, with a kind of dog paddle. If the depressed form suggests buoyancy, it is also indicative, as is known from the study of modern animals, of bottom-living habits. Probably all could crawl by the use of the segmented portions of their legs.

There is, however, a suggestion that the broad pygidium may have been used as a sort of flipper, which by spasmodic strokes gave a quick backward impetus to the body that sent it off in darting flight when occasion made escape necessary. The lobster executes such movements with its broad telson, and the construction of the trilobite is comparable. Action of this kind requires ample musculature, of the sort every epicure who has dissected the lobster knows that animal to possess. It might seem that nothing would be known of the muscles of the trilobite, since these are perishable tissues, but although the muscles themselves are lacking it is possible to reconstruct them from the scars of their attachments on the inside of the tests. Many of the fossils are found in an enrolled condition, and if they could roll themselves up, it is obvious that they must have had the power of straightening out again. Hence they must have had two sets of longitudinal muscles, one dorsal and one ventral, and, as has been said, the scars of such muscles have been found. The size can only be inferred from the amount of room available to house them. It may be that trilobites with a wide median lobe had strong muscles, and the power of darting backward (Fig. 24, at right). Since the animal had no organs with which to steer, it is probable that the movements were rather erratic, and the trilobite was likely as much surprised as its pursuer by the course it took.

The swimming trilobites probably lived near the bottom, for they had legs adapted for crawling as well as swimming. There are some, however, which seem to have floated and swum about in the waters at or near the surface of the sea, as members of the plankton. If a net of fine mesh be towed near the surface of the ocean at the present day it will catch a variety of small creatures, among them many crustaceans. In spite of their great variety, the animals of the plankton agree in having thin, transparent, or translucent shells, many of them spinose. One modern crustacean in particular has received attention because of its general similarity to certain trilobites. It is a broad, flattened, spinose animal with large eyes. During the day it swims about in relatively deep water, out of reach of strong sunlight, coming to the surface only at night. Some of the spinose trilobites probably had these same habits; a particularly good case might be made out for one known as *Robergia*, which has a thin shell and large eyes, and is always found in fine-grained black shales, associated with graptolites, themselves a part of the plankton.

Although most trilobites could probably both swim and crawl, some seem to have lost the power of swimming and to have lived entirely on the bottom of the sea. These animals are readily recognized by their elongate, wormlike form, the small pygidium, and the great number of segments between the two shields (Fig. 23). Some of the most ancient species show this form best, indicating that the habits were differentiated early. Others were not merely crawlers, but went further and burrowed in the mud or sand. Burrowing seems to have led to blindness, as might be expected from a complete adaptation to such a mode of life. *Cryptolithus* is the best example, because the most fully known. It is a small animal, from half an inch to two inches long, with a proportionately large and rather prettily ornamented head, the three mounds on which suggested its more commonly known name of *Trinucleus* (Fig. 21, at right). Around the margin of the cephalon is a broad brim which long

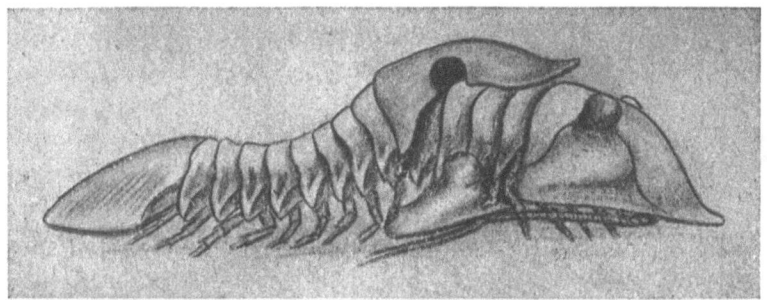

Fig. 22. Restoration, to show the probable appearance of a trilobite in the process of molting. Free cheeks were cast off along the facial sutures, and the animal crawled out. Original drawing by Charles J. Fish.

Fig. 23. An elongate olenellid trilobite crawling on an irregular sea floor. Original drawing by Charles J. Fish.

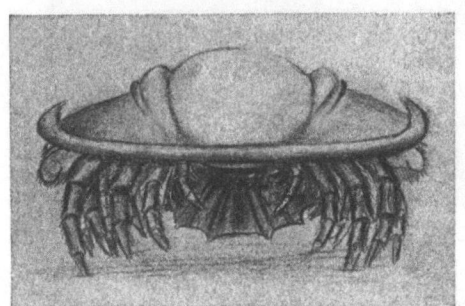 

Fig. 24. At left, a *Neolenus* walking toward the observer, on the "toes" of its bowed endopodites. At right, above, an *Isotelus* jumping backward, using its pygidium as the lobster does its telson; below, swimming forward, by up and down movements of its posterior shield. Original drawings by Charles J. Fish.

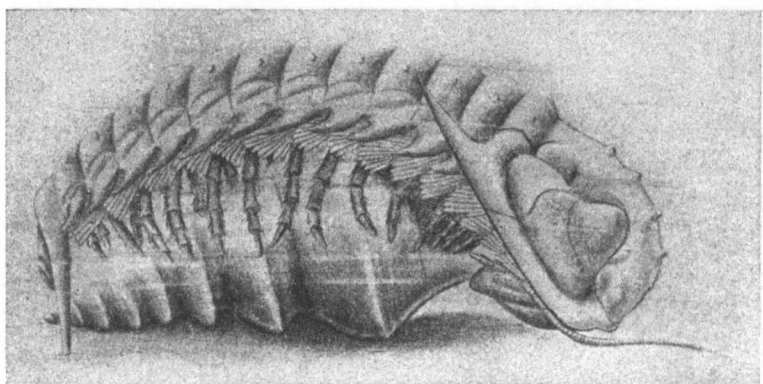

Fig. 25. A trilobite, *Ceraurus*, shown in feeding position above a gastropod. Original drawing by Charles J. Fish.

ago suggested that the creature may have buried itself, like the modern horseshoe crab, by pushing down the brim and digging out the sand with its legs. The appendages, when discovered, confirmed this suggestion, for they are stout, bowed at midlength, and armed at the ends with numerous spines. Most striking of all, the feeling organs, instead of reaching forward, as in all other genera, were turned backward beneath the shell, so that they were out of harm's way in burrowing. But not all burrowers were blind, for in several groups there are species which have their eyes at the top of long, immovable stalks. These forms probably buried themselves up to the eyes and lay hidden with only their periscopes visible, perhaps awaiting their prey or observing the movements of their enemies.

It has occasionally happened that specimens of fossil fishes and some other animals have been found which had within them undigested but petrified remains of their last meal, so that it has been possible to get direct information as to the kind of food taken. The feeding habits of the trilobites, however, must be inferred from the structure of the jaws and the alimentary canal. As for the former, the animal was well supplied, for the inner segments of each leg were adapted to serve as jaws (Fig. 21, at left). We are used to thinking of these useful organs as being within the mouth, but they are external in many of the lower animals, and in this case some are far from the mouth. The basal segments of opposite pairs of legs projected inward, so that they met on the median line, and each was armed with numerous strong, sharp spines. The jaws probably waved backward and forward in unison, so that food seized by any pair of them would be swept forward to the mouth.

Just what these animals ate is unknown, but since their jaws were neither large nor powerful, although numerous, it is probable that they lived upon small animals or plants or upon such tender matter as decaying organisms. Modern crustacea are great scavengers, and it seems that their distant ancestors had the same habits. One can imagine a boat-shaped figure, coasting along the sea bottom, feelers outstretched fore and aft, eyes watching to port and starboard. Something is encountered and the animal settles over it, thus, by one action, reserving that particular morsel for its own refection and bringing its numerous jaws into position for work. The great underlip at the front is lowered, spreading the mouth open and making a trap past which food is not likely to pass. The distal end of the pygidium or its spines sink down into the mud, and dinner begins. Back and forth rasp the spiny jaws, and a continuous stream of food moves forward to fill the hungry stomach (Fig. 25).

Not all trilobites fed in this way. *Cryptolithus*, the burrower, was a mud feeder. Every schoolboy is familiar with the earthworm, which eats its way through the ground, taking out from the soil such organic matter as it can digest and voiding the remainder. Such were the habits of *Cryptolithus*, and probably many another of its kin. Many years ago, Count Sternberg, Barrande, and others who delved among the wonderful stores of fossils in Bohemia described the casts made by the mud which filled the digestive tube at death. By these and other fortunate circumstances we have come to know something of the internal anatomy (Fig. 21, at right, above). Such specimens show that the alimentary canal was exceedingly simple. A short throat passed upward to the saclike stomach, which occupied the whole of the middle of the head. The stomach tapered into a narrow intestine extending back along the median line to the anal opening, on the lower side of the posterior end of the pygidium. Interesting deductions can be made from the circumstance that the median lobe of the head afforded lodgment for the stomach. That region, known technically as the glabella, is short and narrow in the more ancient trilobites; hence it may be inferred that they lived on concentrated animal food. In later ones this lobe increased in size till in some of the Devonian forms it made up most of the head. Since bulky vegetable food requires more feeding and larger digestive chambers, it is supposed that the trilobites with large glabellae and, therefore, large stomachs were vegetarians. The middle of the head is, however, a most unfortunate location for the stomach, for the larger that organ, the less room for a brain, so we must ascribe to these unfortunates an increasing stupidity. In fact, one is tempted to impute a part of their downfall to the fact that there was no chance at all for cerebral development. For those who wish to point a moral, the case is obvious.

Little more is known of the internal anatomy of trilobites than what has been set forth. On either side of the head there are glands which connect by lateral ducts with the stomach, a fact which suggests that their function was the secretion of digestive juices. A unique Russian specimen preserves the heart. It is a long, segmented, tubular body, like the heart of a worm or of one of the modern crustacea. Its most remarkable feature is that it should have been preserved at all. An early attack of arteriosclerosis is almost the only explanation.

Trilobites, then, seem to have been active, if stupid and inoffensive animals. They floated, swam, crawled, and burrowed, ate mud, carrion, seaweeds, and microscopic animals and plants. They played their part in the economy of the world in which they lived and performed their humble service as scavengers countless years ago. Perhaps their greatest contribution is the aesthetic pleasure the contemplation of their elegant shells has given to countless collectors and students of fossils.

VII

THE AQUATIC ARACHNIDS

*And along came a spider,
And sat down beside her ...*

You know the rest of it. This is merely a subtle way of reminding you that a spider is an arachnid. But the spiders are not the only arachnids. "Daddy longlegs" is another, and so is the miserable "chigger," which the paleontologist collects involuntarily as he crawls over the deeply weathered Paleozoic outcrops of the South, hunting for fossils. All these are short-bodied animals, unlike the ancient trilobites. Entirely different in general appearance is the elongate scorpion, splendidly equipped posteriorly with a poisonous stinger. Although a terrestrial animal, it retains most of the features of its ancestors, which were entirely aquatic.

The only primitively aquatic arachnids living today are the horseshoe crabs of the shores of eastern North America and southern and eastern Asia. They are commonly known by the generic name of *Limulus*, correctly *Xiphosura*, although there are two other genera of xiphosurans. The horseshoe crab resembles a crustacean, and was long supposed to be one. It has a chitinous shell, compound eyes, and is aquatic, whereas the other modern arachnids are terrestrial and have aggregate eyes. The first appendages are, however, claws, not tactile organs. This is the simple way of distinguishing crustaceans from arachnids: first pair of appendages, tactile, Crustacea; first pair, claws, Arachnida. There are, of course, other obvious differences, such as that nearly all crustaceans have some biramous appendages, whereas those of arachnids are uniramous.

Limulus is conspicuously trilobate longitudinally, especially in its larval stages. It has several of the characteristics of trilobites, so many, in fact, that most zoölogists think that it descended from them. The chief difference in addition to those already cited, is that all the thoracic segments are ankylosed and form a rigid median shield. The appendages on the head are segmented, ending in claws or bladelike plates adapted for digging; these could have been derived from the inner of the two branches of the trilobitan limb (Fig. 26). Beneath the median shield, the legs are broad and flat. Dr. Leif Stoermer has recently shown that they have many of the structures of the trilobitan exopodite. In *Limulus* they serve as gills, which may have been their function in the trilobites also.

The oldest definitely identified arachnids are found in Upper Cambrian strata, chiefly in Wisconsin. The best-known genus is *Aglaspis*, a form which much resembles *Limulus*, except that the segments between the head and the telson are free from one another. Dr. G. O. Raasch has found specimens showing appendages and has proved that the first pair are claws. The Middle Cambrian at the Walcott quarry contains animals of the same general shape and segmentation as *Aglaspis*, but these have tactile antennules and some of the appendages are biramous. They must, therefore, be classified as crustaceans, but it is possible that they were ancestral to the Upper Cambrian arachnids, which, in turn, were distant ancestors of the horseshoe crab. Unfortunately, no members of this line are known from Ordovician strata. The Upper Silurian of Scotland has furnished one well-preserved specimen much like a small horseshoe crab with free thoracic segments, and others are known from the Devonian. The group suddenly became rather abundant during Upper Carboniferous times, particularly in Great Britain and Illinois. Scattered specimens have been found in various other places, some of them in Permian beds (Fig. 26, at right). In fact, during the late Paleozoic the xiphosurans reached their culmination in variety, though not in size. All are small, an inch or two in length. They show an advance upon the condition of the Silurian specimen in that the body segments of most are coalesced to form a median shield, although some show obvious segmentation. It is curious that all of the late Paleozoic specimens have been found in freshwater deposits. The Upper Cambrian *Aglaspis* was marine; the Upper Silurian *Neolimulus* was found in a formation about which there is question. It contains some marine fossils, but also many others whose habitat is in dispute, as we shall see when we come to discuss the eurypterids.

The subsequent story of the horseshoe crabs is quickly told. A few have been found in the marine Triassic; there is nothing more till the Upper Jurassic, when the modern *Limulus* appeared. Numerous beautifully preserved specimens have been found at Solenhofen. Some of them are larger than the modern Asiatic species but only half as large as the form so common along our New England shores.

This is a strange history, with its transition from marine to fresh-water life, followed by a return to the sea. It seems probable, however, that some groups of fish may have had a similar experience.

The xiphosurans may have been the first arachnids, but they made little impression on early Paleozoic life. Their first cousins, the eurypterids, had the honor of being the only important arachnids during Ordovician, Silurian, and Devonian times. They were not shy or unobtrusive organisms, for some of the largest were from five to nine feet in length. On the average, however, they were much smaller, ten to twelve inches being the length of a good-sized specimen.

Like the horseshoe crab, the eurypterid had an anterior shield with a pair of compound eyes and, near the median line, a pair of simple eyes or ocelli (Fig. 29). The appendages beneath this shield are also like those of *Limulus*, there being six pairs, the first of which are pincers by which food could be passed to the mouth

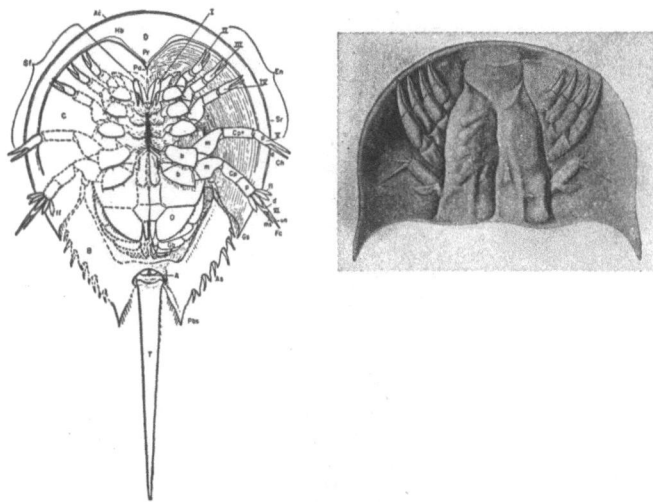

Fig. 26. Ventral surface of a modern *Limulus*, showing the six pairs of appendages on the anterior shield and the flattened ones on the mid-shield. I, anterior pincers (chelicerae); VI, the "pusher" legs, with their rosettes of flattened blades; O, operculum. From K. E. Caster. At right, ventral surface of the Permian *Paleolimulus*, with appendages. The chelicerae are not preserved, but note the blades on the sixth legs. Three and a half times natural size. From C. O. Dunbar.

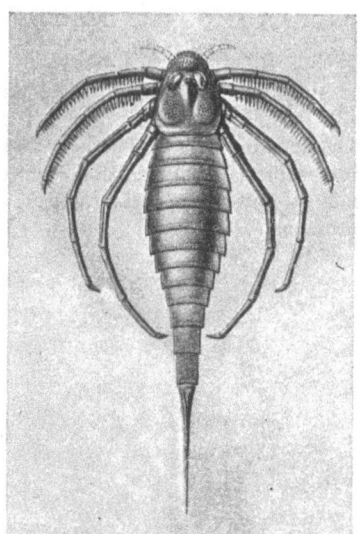

Fig. 27. *Stylonurus*, showing the simple unspecialized appendages. From Clarke and Ruedemann.

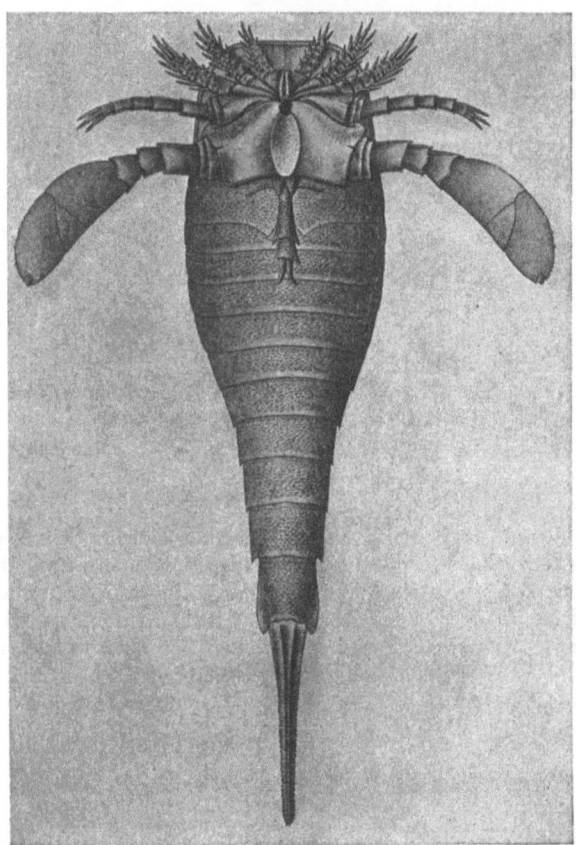

Fig. 28. The ventral surface of *Eurypterus*. Note the chitinous ventral plates, as contrasted with the thin membrane of the trilobite. The anterior pincers are turned backward to the mouth. The next three pairs are diggers, and the last pair oarlike paddles. The large basal segments of this pair are the jaws. Between them is the operculum; back of it is the supposed ovipositor. From Clarke and Ruedemann.

(Fig. 28). As in the trilobites and *Limulus*, the mesially directed projections of the appendages served as jaws. The limbs are variously modified in the various genera, four pairs behind the pincers being adapted for crawling or digging. Except in a few genera, the posterior pair, the sixth in the series, were the longest and largest, broad, flat, oarlike; useful in swimming. The trunk contains twelve free segments, without visible appendages. At the posterior end is a *Limulus*-like telson, a long spike in most, but broad and flattened in a few. The anterior part of the body consists of segments which are distinctly wider than those on the posterior portion. The first two on the upper side are covered by a single plate below, a fact which shows that they are not continuous chitinous rings about the body. The skeleton actually consists of a series of dorsal plates (tergites) and ventral ones (sternites), the anterior pairs of which are not united at the lateral margins. When the ventral plates of some unusually well-preserved specimens were removed, remains of gills were found above them.

Eurypterids do not show so many superficial resemblances to trilobites as do the horseshoe crabs, but they have an obscure trilobation, compound eyes, and free body segments. Unlike the trilobites, they have a chitinous ventral covering, most of which has already been described. On the head, behind the last and principal jaws, is an oval plate known as the operculum; behind it, on the median line, is an elongate organ whose function is really unknown, but which is supposed to be an ovipositor. This organ is pointed at the posterior end on some specimens, but on others that seem to be the same species, it is bifurcated. Since the latter individuals appear in general to be somewhat wider than the former, it is supposed that they were females. It is not uncommon, among modern arthropods, that the females are larger and broader than the males, but this is by no means so universally true that it can be accepted as a ready way of determining sex. Whether or not proportions have any significance, it seems probable that the organs on the median line do indicate differences in sex. Which is which is immaterial. The important point is that as early as Mid-Ordovician times, sex differentiation had achieved recognizable physical characteristics. These are, however, primary, not secondary features. The latter appear much later. Just when, nobody knows.

The structure of the test (shell) of the eurypterid affords considerable evidence as to its habits. The flattened body indicates that it was a bottom-living (benthonic) animal. The pincerlike claws at the front show that it was a predaceous carnivore. All eurypterids were able to crawl, as is shown by the structure of the second to fifth pairs of appendages. Most could swim, as is indicated by the flattened shape of the sixth pair. *Eurypterus* itself, and some of its relatives, may have had the habits of *Limulus*. It could crawl and swim and, on occasion, burrow beneath the surface by the aid of its appendages and spikelike telson. But *Pterygotus*, the "Seraphim" of the Scottish quarrymen, was purely a swimmer and crawler, as is indicated by the marginal eyes and broad telson (Fig. 29). It could hardly have burrowed without injury to the cornea. Other forms, such as *Eusarcus* (Fig. 30), also had marginal eyes. Probably they were principally crawlers, for the short curved telson would have been no assistance in swimming. The interpretation of the habits of most of the various sorts of eurypterids is easy, but there is a line which culminated in the five-foot-long Devonian *Stylonurus* of the Catskills with appendages which seem useless (Fig. 27). All except the first pair are long and slender, each segment being unusually long. There are no oarlike paddles, and the telson is a spike. These facts make it evident the animals were not swimmers; on the other hand, the legs were too long and clumsy to have been useful in crawling. Nevertheless the group had a long and successful history.

There is considerable difference of opinion as to whether the Ordovician and Silurian eurypterids lived in salt or fresh water. The conventional view, based on geological occurrence, is that they were originally marine but that in the late Silurian they became adapted to life in fresh water, where they persisted until their extinction in Permian times. We do not know their pre-Ordovician history, but infer that they sprang from some Cambrian arachnidian stock closely allied to *Aglaspis*. One of the most conspicuous of the Mid-Cambrian fossils from Burgess Pass is an eurypterid-like creature which Walcott named *Sidneyia* after his young son, who found the first specimens. Although the shell is eurypterid-like, the animal had tactile antennules, and the appendages are divided like those of a trilobite. It is chiefly remarkable for a pair of extraordinary claws. This creature, however, was a crustacean, and so specialized that it could not have been ancestral to the eurypterids. Its presence in Mid-Cambrian rocks proves that there was at that time a line of crustaceans in which evolution was tending toward the elongate body form of the eurypterids. There is even a differentiation in the trunk between anterior broad and posterior narrow segments. This sort of animal suggests, although it by no means proves, a marine origin for the eurypterids. Did the ancestors migrate into fresh waters as early as Cambrian times?

Dr. Marjorie O'Connell, while studying under Professor A. W. Grabau, wrote a long paper in which all the then known occurrences of eurypterid remains were discussed. The evidence cannot be reviewed in detail here, but her conclusion was that all were inhabitants of fresh water. She pointed out that no complete specimen had ever been found in the Ordovician, which might indicate that the animals lived in rivers and that only their broken and macerated fragments were carried to the sea, to be recovered by paleontologists from marine deposits.

The climax of eurypterid differentiation was in the Upper Silurian. Rocks of this age have produced the best and largest specimens, and in them are the four famous localities for such animals. At all of them the conditions that existed while the strata were being deposited were probably somewhat unusual, for, although

Fig. 29. *Pterygotus,* the largest and most aggressive of the eurypterids, and probably the largest arthropod which ever existed. Note the huge size of the pincers, the feeble development of the diggers, the small size of the paddles, and the large flat telson. From Clarke and Ruedemann.

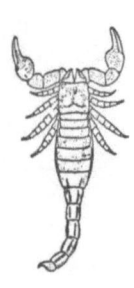

FIG. 30. At left, a Silurian *Eusarcus*. At right, sketch of a modern scorpion to show similarity of trunk and curved telson. *Eusarcus* from Clarke and Ruedemann, scorpion after R. I. Pocock.

mollusks and a few other marine animals are found in the same layers, the faunas are admittedly peculiar. The writer has been fortunate enough to have visited all the sites, and finds that the eurypterid-bearing rocks were not all formed under the same conditions. The first site is a tiny, now disused quarry a half mile southwest of the center of the hamlet of Kichelkonna on the western margin of the island of Oesel (now Saaremaa) in Estonia. Here the specimens are wonderfully preserved in a light gray dolomite. So remarkable is the preservation that some individuals retain their color markings; practically all have the original chitinous shell. It is thinner than the thinnest tissue, and is the despair of the collector, for as the newly excavated specimens dry, it breaks, curls up, and blows away. Gerhard Holm, an eminent Swedish paleontologist, discovered how to remove the shells from specimens found at this locality, and how to preserve them. To him is due a great deal of our knowledge of the morphology of the eurypterids. Curiously enough, he never disclosed the secret of his technical skill. Typical marine beds are found both below and above this eurypterid bed. The lithology is the same in all. Another famous locality is south of Lesmahagow, not far southeast of Glasgow, in the district where the banks of the Clyde are really "bonny." Here the rock is a dark gray shale, which contains so many marine mollusca that there can be no question but that it was deposited in salt water. It does grade, higher up, into ostracoderm-bearing beds about which there may be some question. The third and fourth localities are in New York state, one in Herkimer County, south of Utica, and the other in the southern part of the city of Buffalo. Most of the good American specimens to be seen in museums came from one or the other, although neither is now an active producer. The rock at both is an argillaceous dolomite. Few other fossils accompany the eurypterids, although Dr. Rudolf Ruedemann has pointed out that such as have been found are marine.

In Herkimer County and at Buffalo the *Eurypterus*-bearing beds are almost the youngest Silurian strata, but in the region between, centering around Syracuse, are the great Upper Silurian salt deposits and gypsiferous shales. These beds, perhaps, suggested the idea that all Upper Silurian deposits are abnormal. To account for the salt and gypsum it has been necessary to consider this part of New York to have been, at the time, a region of embayments, nearly cut off from the sea, in which evaporation produced concentrated brines from which salt was deposited. Into such embayments may have been swept such eurypterids as could not maintain their position in the tributary rivers. Death in the salty water would account for the abundance of well-preserved, complete specimens.

It is impossible to refute certain arguments. If the remains are fragmentary, it is because the animals lived in rivers, and only remnants of molted specimens reached the sea. If the individuals are complete, it is because they were killed when they were washed out of their normal habitat into saturated brines.

The discussion of these localities may serve to show that conditions were not the same in Oesel, Scotland, and New York. At all of them well-preserved specimens are found. If only the one at Lesmahagow were known, it might have been inferred that it was the increasing freshening of the waters which caused the destruction of the eurypterids.

Whatever their trials and troubles, eurypterids survived them and went on to new triumphs in fresh waters during Devonian times. Perhaps they grew stronger through adversity, for their shells were no longer carried into the sea, even after death. We have no idea what caused their final extinction. Post-Devonian specimens are rare, and relatively small, even though the group survived till the Permian. Perhaps the early amphibians and reptiles ate them.

Although the eurypterids have vanished, they have left us a souvenir, the scorpions. If one examines a scorpion, preferably a dried or pickled one, he sees that it is practically a eurypterid (Fig. 30, at right). The anterior shield bears the same number of appendages, the first a pair of pincers, all the others walking legs. But the eyes are aggregate, not compound. The trunk has the same number of segments, divided, as Sir Ray Lankester, the most distinguished naturalist England has produced in recent years, liked to say, into tagmata. That is, the anterior ones are enough broader than the posterior to define separate regions. At the posterior end is a telson, a short segment with a curved spine, through which the amiable animal delivers its poison. Beneath the ventral plates of the anterior segments of the trunk are the breathing organs. They are similar to those of the eurypterids, but modified for the absorption of oxygen from the air, not from the water. They are, therefore, not gills but lungs, denominated, because of their leaflike structure, book lungs. Air is admitted to them through paired circular openings known as stigmata.

The oldest known scorpions are three individuals found associated with eurypterids, two in Europe and one in New York. They are known to be scorpions and not eurypterids because they have pincers on the second pair of limbs and "combs" on the first trunk segment. The latter appendages are peculiar to scorpions. These specimens were for long celebrated as the oldest air breathers, but Dr. R. I. Pocock of London has shown that they lacked stigmata and hence were probably as fully aquatic as their associates. If one compares the skeleton of one of these Silurian scorpions with that of the contemporary eurypterid, *Eusarcus*, he is readily convinced of the close relationship between the two groups (Fig. 30). It must be admitted that the connecting links have not as yet been discovered, but there can be no doubt where the ancestry of the scorpions lies. No Devonian members of the group have yet been found, but Carboniferous scorpions have stigmata. They are unquestionably air breathers. The successful branch of the family was the one which learned to do something new.

Once the arachnids got out into the air, evolution was rapid. Spiderlike animals have been found in the Lower Carboniferous, and true spiders, as well as various related forms, are known from the Pennsylvanian. Nothing is yet known of the ancestry of these groups. Their secrets are still locked in undiscovered layers of the Devonian deposits. Specimens found in the Rhynie chert of Scotland show that mite-like creatures were in existence as early as Mid-Devonian times, but their aquatic ancestors have not yet been discovered. Almost all modern arachnids are air breathers. It is somewhat exasperating that so little is known of their origin. But it is not really surprising, for comparatively few terrestrial animals die in situations where they have a chance of being preserved as fossils. The more completely terrestrial the animal, the less chance the paleontologist has of finding its remains. Judged on this basis, one would suppose that the remains of the flying insects would be much rarer than those of the wingless arachnids. But such is not the case. Hundreds of specimens of fossil insects may be found in strata which yield only a single spider or spiderlike creature. The inference is that the early air-breathing arachnids lived in moist tropical and subtropical jungles where the decay of all organic material was rapid. Many of them probably wandered far from the streams and swamps which seem to have been the haunts of many ancient insects. Little is known of their ancestral food habits. The fondness of spiders for the juices of insects may have been initiated by feeding upon helpless larvae, found in the decaying logs of the jungles. Perhaps the question:

> "Will you walk into my parlor?"
> Said the spider to the fly,

was not asked until comparatively recent times. Flies wrapped in webs and individuals partially devoured have been found associated with spiders in the Oligocene amber of the Baltic but in no older deposit.

VIII

THE RADIATES

Their strength is to sit still.
Isaiah, xxx, 7

Although we no longer class together all animals with radial symmetry, it is convenient, for the purposes of this book, to treat of them in one chapter. Coelenterates and echinoderms have it in common that in general they thrive best in warm water, and that both have contributed greatly to the formation of limestone. In speaking in these general terms I am, of course, referring chiefly to the attached coelenterates, the corals, and to the sessile echinoderms, the cystids, the crinoids, and the blastoids.

The animal or polyp of a coral is simple, but not quite so simple as the hydroid mentioned in the chapter on graptolites. It is, in fact, nothing but a living tube, with a mouth surrounded by tentacles, but there is a short oesophagus leading into the body cavity. The latter is divided by incomplete radial partitions (mesenteries) into a series of alcoves which afford a certain amount of privacy to the digestive processes (Fig. 12 D). These infoldings of the walls greatly increase the amount of endoderm, whose cells attend not only to the processes of digestion and assimilation but also to reproduction. As in the sponges, many of the cells have cilia, for each is practically a protozoan. The animal is not, however, a mere colony of protozoans, for there is a division of labor. The components are not all alike. Some are sensory, some digestive, some reproductive. Although the polyps have nothing remotely resembling a brain, or even a central nervous system, they get along very well. All are carnivorous. Perhaps you have seen pictures of a sea anemone swallowing a fish, an animal a thousand degrees above it in the social scale. It is not necessary even for the anemone to pursue its prey; its strength is to sit still.

But sitting still in warm sea water saturated with calcium bicarbonate has its disadvantages. The animal has to get rid of the surplus calcium. The coral does this by secreting, or excreting, a skeleton beneath, rather than around its body. The larva, after a short free-living existence during which it passes through a part of its development, settles on some hard object upon which the remainder of its life is to be spent. The lower surface then puckers up into a series of radial folds, alternating in position with the mesenteries. In the grooves so formed, minute irregular bodies (spicules) of calcium carbonate are secreted. These eventually coalesce, producing a set of radially arranged plates (septa). With further growth, new folds appear at the base, and new septa are formed between the primary ones. As the process continues, more spicules are added at the outer ends of the septa, along the outer surface of the tubular polyp. Thus the ends of adjacent septa become connected, and a circular outer wall (theca) is produced. Later in life most corals continue to form new septa and new thecal tissue, so that a conical skeleton is formed. It should be noted, however, that this is not a house into which the animal can withdraw. The coral never gets beyond the stage of building the foundation. To make sure that there shall be no retreat, the animal builds more or less horizontally arranged plates extending from wall to wall (tabulae) or from septum to septum (dissepiments). The living coral is always on the uppermost story of his pedestal; one can scarcely call it more.

Some corals reproduce sexually only, producing simple cup corals, which were much more numerous during the Paleozoic than they are today. At the present time these are found most commonly in cold water, either deep-sea or, if shallow, in subarctic or cold-temperate regions. As has already been pointed out, reproduction by budding was unusual in Cambrian times. It is shown there by only a few animals, sponges in the Mid-Cambrian, graptolites in the Upper. But by the Mid-Ordovician it became a commonplace, probably the result of many generations of fixation. Perhaps the warmer waters and more abundant food had something to do with it.

Most modern corals and a large proportion of the Paleozoic ones are colonial in habit. As new buds form, they build their skeletons upward and outward, producing more or less hemispherical, moundlike masses. Much more rare in the Paleozoic are colonies of the "staghorn" type so common at the present day. The individual cups (corallites) in the colony (corallum) may be circular if the buds are sufficiently divergent. Most, because of rapid budding and consequent crowding, are polygonal in section, three, four, five, six, or seven sided. Some achieve the optimum for closely arranged polygons, the hexagonal section. The square and the hexagon are the only patterns shown by colonies in which all the corallites have the same shape, but these are less common than corallites which have a variable number of sides.

The stony corals are divided into two great groups, the septate, with conspicuous radial septa, and the aseptate, without them. To the first belong the common reef-building corals; to the second, the sea fans, dead-men's-fingers, and the precious coral. Both were represented in the Paleozoic, although the aseptate forms then living differed from those of today. Although the septate kinds of the Paleozoic and Mesozoic seem more alike, they are really structurally different. The Paleozoic ones have four primary septa, to which others are added in multiples of four; hence they are called Tetracoralla. In modern corals there are twelve (specialists please read this number as two, four, six, or twelve) primary septa, and the increase is in multiples of six. Naturally they have been named the Hexacoralla. This difference in numbers may not seem to be a matter of fundamental importance, but it really is, as may be seen if the mode of introduction of new septa be studied. In the tetracorals they originate as outgrowths from two sides of one of the primary septa (the cardinal) and from one side of each of two lateral ones (the alar). It may be seen by consulting the accompanying figures that, although the symmetry appears to be radial, as the corallite is viewed from above, it is fundamentally bilateral, as is shown by the structure of the cardinal side (Fig. 31). Young hexacorals likewise have bilateral symmetry, but it is lost so soon after the introduction of the first septa that the radial arrangement seems to be primary. It has already been pointed out that the graptolites differ from modern hydroids in being bilaterally symmetrical. The same is true of the early septate corals: in fact, they remained bilateral till the end of the Paleozoic. The evidence indicates that radial symmetry is secondary, the result of fixation. Spherical symmetry may be more primitive than bilateral, but the days of spherical symmetry in adults appear to have been over before the beginning of the Cambrian.

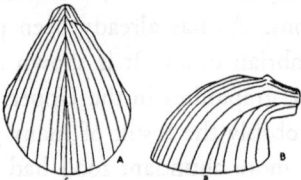

Fig. 31. Diagrams of a tetracoral showing, A, the divergence of septa from the primary one (*c*) on the cardinal side, and, B, from the alar (*a*).

The aseptate Paleozoic corals differ from most modern members of the group in that their skeletons were external, not internal. No sea fan, gorgonian, or bit of precious coral has been found in Paleozoic rocks. In their place are Tabulata, such as the honeycomb-like *Favosites* (Fig. 38), the chain coral, *Halysites*, and others familiar to everyone who has ever seen a textbook of geology. Attempts to explain their relationships to modern corals have been and still are being made. We need not go into the matter. Perhaps they were the "lower classes" in the corallian social scale. Each individual produced a multitude of offspring who remained at home, forming large colonies. With so many mouths to feed, no individual got enough nourishment to grow large, for the number of buds was enormous. Some coralla of *Favosites* are five or six feet in diameter. They compare well with the "brain corals" of the present day, forms in which the incomplete separation of the daughter polyps from the parent produces convolutions superficially similar to those of the human brain.

It is interesting to note that the corals, which had a modest beginning in the Ordovician, became sufficiently abundant in Silurian and Devonian times to build structures analogous to modern coral reefs, though no cases are known in which such reefs corresponded exactly to modern barriers or atolls. It may be that this adaptation caused the extinction of tetracorals and Tabulata. It is known that the reef-building corals of the present day thrive only in waters whose mean annual temperature does not depart greatly from 68° F. A drop of a few degrees will kill them, if maintained for any considerable period. The reef corals also succeed best in relatively shallow water, that is, from just below low-tide level down to a depth of about twenty to twenty-five fathoms. Not all hexacorals are of this type, and not all of them require such a relatively high temperature. Some live in the deep sea, and some in high latitudes. Corals are found near Woods Hole, Massachusetts, and off the Newfoundland banks, and at least one species grows abundantly off the coast of Norway in rather deep water. These, however, are not reef-builders.

Professor R. A. Daly has set forth considerable evidence to show that during the Pleistocene glacial period, most of the reef-building corals of the Pacific islands were killed and that the luxuriant growth on the present reefs is due to a later re-colonization. The great masses of ice which covered the polar regions, extending southward to a latitude of 40° on the continents, must have had a chilling effect upon the oceans. That it was a time of general refrigeration is indicated by the presence of local glaciers even in the Hawaiian Islands. Since a small change in temperature affects the corals so seriously, life on many of the reefs may have been extinguished during the Pleistocene.

In the light of this it seems probable that the almost total change in coral fauna at the end of the Paleozoic was due to the Permian glaciation. At that time unusually cold conditions existed both north and south of the equator, as is indicated by glacial deposits in South Africa, South America, India, and in the small area near Boston made known by Robert W. Sayles. The accumulation of ice and the general lowering of temperature must have had their effect upon the oceanic waters. The tetracorals and Paleozoic aseptate corals may have been exterminated. These animals doubtless throve best in warm localities, for their skeletons are most abundant in limestone, and it is well known that at the present time marine limestone is accumulating chiefly in warm water.

It may be supposed that skeletonless Hexacoralla were in existence during the Paleozoic but that they occupied the colder waters, holding the same position in the community that the subarctic corals do today. Since they were accustomed to a low temperature, they would easily survive the chilling effects of the Permian glaciation. In fact, this may have been the moment of their great opportunity. The destruction of the earlier warm-water corals left vast regions suitable for colonization. Current-carried larvae no longer perished because of transportation into warm seas. Their ecological position was vacant. Food was still abundant. The time which comes once in the lifetime of every race, as well as in that of every man, had arrived. The naked actinians spread into every sea.

Is there any indication that this happened? Possibly a clue, to which the writer called attention some years ago in an article which excited much interest but which has never been favorably reviewed. The oldest known coral is an actinian, that is, a skeletonless sea anemone, of the Mid-Cambrian. Walcott described it by the name of *Mackenzia* as a sea cucumber (holothurian), but Drs. A. H. and H. L. Clark soon showed that it was not an echinoderm (Fig. 32). Like other marvelously preserved specimens from Burgess Pass, its photographic imprint shows that it had internal mesenteries and the remains of sixteen retracted tentacles about the mouth. To all intents and purposes it is identical with *Edwardsia*, a sea anemone still common in the cool waters about the English coast. Despite its sixteen tentacles and mesenteries, *Edwardsia* is considered by students of corals to be a member of the Hexacoralla, because the more advanced members of that group pass through an *Edwardsia* stage in their individual development. Since the Mid-Cambrian *Mackenzia* and the modern *Edwardsia* have the same important characteristics, it seems probable that similar corals must have existed ever since Mid-Cambrian times. Their lack of skeleton seems to be due to the fact that they were inhabitants of cold water. But suppose that during the cold-water period of the Permian, the *Mackenzia-Edwardsia* type of animal spread all over the earth. As the waters warmed up again during the Mid- and Late Triassic, they would have been forced to secrete calcium carbonate. Might not the skeletons have been of the hexacorallan type, since even now the Hexacoralla pass through an *Edwardsia-Mackenzia* stage?

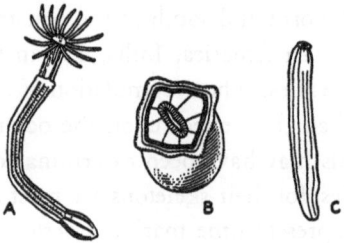

Fig. 32. A, diagram of the modern actinian, *Edwardsia*. B, larva of a modern actinian in the eight-mesentery stage. C, sketch of *Mackenzia*, with its sixteen contracted tentacles. From P. E. Raymond.

The history of the aseptate Paleozoic corals is even more obscure. The most successful type, the favositid, culminated and perished in Mid-Devonian times. Relatives struggled on till the late Carboniferous. Perhaps the cold waters of the Permian finally "did them in." One can see that the muddy waters of some of the shallow late Devonian seas may have exterminated *Favosites*, but it is remarkable that the larvae of such a virile stock should not have managed to get into the clearer waters. There may have been "barriers" of which we as yet know nothing.

Except for their radial symmetry, echinoderms have no resemblance at all to corals. Although many are sessile, they do not reproduce by budding. There is an alimentary tract hung within the body cavity, as in all coelomates, and there is a central nervous system. The skeleton, which consists of calcareous plates, is internal rather than external, though close to the surface. It is deposited by special cells which build up cribriform plates composed of spicules of calcium carbonate. Although the plates of modern echinoderms are porous when deprived of their organic matter, each plate or spine has the structure, though not the form, of a single crystal; when broken, a typical calcite cleavage is shown. This applies to the fossils as well. So far as is known, no other sorts of organisms have this property. Often it is possible to identify the fragments of a badly preserved specimen as an echinoderm by its rhombohedral cleavage.

The starfish is the most familiar member of the group. Although free throughout adult life, it is sessile for a short interval during the larval stages. A brief discussion of its structure will assist the interpretation of the ancient forms. It shows conspicuously one of the most characteristic of echinodermal structures, the water-vascular system. On the upper surface, between two of the rays, is a perforated plate, the sieve plate or madreporite, through which water is drawn into a tube which leads to a circum-oral vessel (the stone canal) on the lower side. Extending along a median groove on each ray, externally, is a branch from the stone canal carrying water to the "tube feet," which are the prehensile and locomotory organs. Each tube foot is connected through a pore between adjacent pairs of rafterlike plates (ossicles) with a small spherical reservoir within the arm. The ossicles, like the tube feet, are paired on either side of the median groove, forming a long, narrow, outwardly tapering region known as an ambulacral area. At the tip of each ray, at the end of the median groove, is a small plate bearing a pigmented organ, probably sensitive to light and hence called an eye. From it the plate through which it projects has been named the ocular.

Naturally, all directions from the center of a radially symmetrical animal are radial, but technically the ambulacral areas of the echinoderm are considered radial, whereas the regions between them are interradial. Thus the madreporite of the starfish is interradial, the oculars are radial in position. The terms "dorsal" and "ventral" have little or no meaning when applied to the echinoderms. The mouth is beneath (ventral) and the anal opening on the upper side (dorsal) in the starfish, but in many others these organs are on the same side, dorsal in some, ventral in others. It is, therefore, customary to speak of the side having the mouth as actinal; whatever is opposite to it is abactinal.

The starfish living at the present day are of two general kinds: some, such as the common ones of the New England waters, have a flexible skeleton consisting of small plates embedded in a tough integument; others, inhabitants of warmer regions, have larger plates which greatly reduce the mobility of the arms. Both kinds have existed since the Mid-Ordovician, the time when the group made a sudden appearance.

An early narrow-rayed line of starfish culminated in Mississippian times in the true brittle stars, ophiurans, in which each pair of plates from opposite sides of the median groove has been combined into one so-called "vertebral" ossicle.

Starfish do not have branching arms, but some modern ophiurans, the basket stars, do. A basket star, turned actinal side up, bears considerable resemblance to one of the feather stars, free-swimming crinoids (Fig. 33, at left). The latter animals are unknown on American shores. Probably more Americans are familiar with fossil than with recent crinoids, for quarries at Ottawa, Ontario; Warsaw, Indiana; Burlington and Keokuk, Iowa; and localities in New York, Ohio, Tennessee, Missouri, and elsewhere have filled the collections of museums and private individuals with beautiful specimens. In fact, fossil crinoids were known to naturalists before recent ones

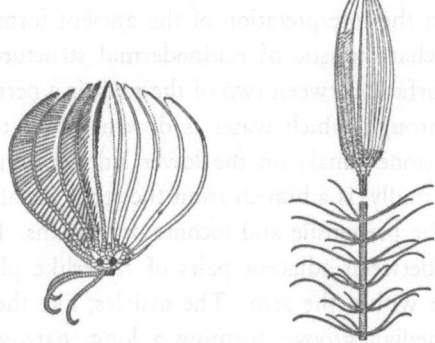

Fig. 33. Diagrammatic representations of, at left, a modern free-swimming crinoid; at right, a modern stalked crinoid. From A. H. Clark.

were discovered. All the localities mentioned above furnish only the old-fashioned stalked crinoids. Feather stars are of relatively modern invention, dating, so far as is now known, from the Jurassic. Such free-swimming forms are now most common in shallow tropical waters, though one hardy form is found in the Irish Sea. Their stalked cousins, so abundant in Paleozoic shallow seas, are now few in number and confined to restricted areas, although not to deep water, as is generally supposed.

Modern crinoids, whether free or fixed, have the same structure. The mouth is on the upper side, in the midst of the appendages. From it radiate five primary grooves, with branches extending onto the arms; they are lined with cilia which capture food and carry it to the mouth. Branches of the arms may be few or numerous, but all are made up of calcareous plates, connected at the ends, deeply grooved above. Along their sides are upright outgrowths, the pinnules, which are not food-getters but which bear the sexual products (Fig. 33). The main portion of the body, chiefly a spirally twisted digestive system, is enclosed in a calyx consisting of two or three horizontal rows of calcareous plates. Those from which the five primary arms originate are known as the radials (Fig. 34, at right). Below, and alternating in position with them, hence interradial, are the basals. Below the basals of some crinoids there is a third row of five plates, radial in position, the infrabasals. The infrabasals, if present, or basals otherwise, rest upon the dorsocentral abactinal plate of free crinoids or upon the upper columnal of the stalked ones. The column itself is made up of a series of coin-shaped plates with a central opening or lumen. They are attached to one another by ligamental tissue situated in radial grooves, a method of union which gives flexibility to the stalk. Some crinoids, chiefly the modern ones, have lateral appendages, cirri, on the stalk or on the dorsocentral plate. They serve the free crinoids as anchors during temporary "tie-ups."

Modern crinoids of the stalked variety have been in existence since Triassic times. All the Paleozoic ones (there was greater variety in those days) differ from them in that the mouth does not open at the surface but is completely covered by a series of plates forming a vault known as the tegmen. Like the later crinoids, the Paleozoic ones gathered food on the arms, but it passed along food grooves which led through lateral openings at the bases of the arms to the concealed mouth (Fig. 34). Some of the Paleozoic crinoids have a simple calyx, composed of two or three circles of plates. But more of them have the primary calyx (patina) enlarged by the inclusion of proximal portions of the arms. Primarily, there are but five arms; if more issue from the sides of the calyx, then its upper portion has been secondarily enlarged by the incorporation of arm plates (brachials) and interradial plates (interbrachials) (Fig. 34, at left). The most abundant Paleozoic crinoids are the Camerata, forms with large secondary calices including many brachial and interbrachial plates below the places of attachment for the free portions of the arms. They also have large, highly vaulted tegmens. Perhaps the most conspicuous feature is the elongated anal pyramid or tube. The anal opening of many is within the circle of arms, a strange arrangement for which their ancestors, the cystids, were responsible. Naturally this was no place for an excretory organ, at the very center of the food-collecting grooves. There wasn't, however, much that the unfortunate animals could do about it. But accidental variation, or inheritance, or an orthogenetic urge, or something, attempted to improve conditions. The anal pyramid, primarily a circle of five or ten plates, rose higher and higher, till at last it towered far above the arms. From its lofty and imposing height it could now shower out the waste where all branches of the food grooves, rather than merely the inner portions of them, could collect it. Evolution is not in all cases intelligently directed. Perhaps this practice of using the same food repeatedly may have been a factor in the sudden downfall of the crinoids.

The crinoids, like the starfish, appeared first in the Middle Ordovician but reached their climax during the early part of the Mississippian, when they were not only abundant but highly varied. This, however, was not the only time when they were abundant, for strata of various ages consist more or less completely of disjointed columnals and plates. Since the skeleton is held together largely by ligamental tissues, the parts naturally became separated after death. One of the reasons why the early Mississippian strata contain so many well-preserved specimens seems to be that among the camerate crinoids there was a constantly increasing tendency for the adjoining plates of the calyx and tegmen to become cemented along their edges and thus more resistant to disintegration.

Why so many kind of crinoids disappeared in Mid-Mississippian times is not known. Before the Mid-Pennsylvanian they had become rare, and they have remained in a subordinate position since, although after achieving freedom from the stalk in Jurassic times they "picked up" a bit. Even the free crinoids may have passed their maximum, for their Mid-Cretaceous representative, *Uintacrinus*, is by far the largest known. Individual calices are as much as three inches in diameter, and arms have been traced to lengths of forty inches. Kansan slabs completely covered with beautiful specimens of *Uintacrinus* with intricately intertwined arms tell of Mid-Cretaceous disasters; not, surely, due to panic in fire or flood; nevertheless, the result of some unreasoning crowding.

The nut-shaped blastoids are the most stereotyped of echinoderms. All are alike in that they have a slender, but rarely preserved, column, a calyx composed of three or five basals, five deeply cleft radials (forked plates), and five small interradial "deltoids." Radiating from the mouth are five broad, ambulacral areas, which during the life of the animal bore many simple pinnules (Fig. 37). Surrounding the plate-covered mouth are paired openings connected with the water-vascular and genital system. An unpaired opening, or in some a specially modified member of the paired ones, is the anus. The water-vascular system, which left but slight marks upon the skeleton of the crinoid, was highly developed in the blastoids.

Much has been, and much of general interest could be, written about the group, but there are limits, even to books. True blastoids appear first in the Silurian. The various blastoid-like animals of the Ordovician are so complicated in structure that they must for the present be left to the specialist. There are several Middle and Upper Devonian species, but it was not until after the decline of the crinoids in the early Mississippian that this group became common. They apparently took over the vacated ecological position in the warm epeiric seas during the later part of the Lower Carboniferous. But their opportunity came too late for them to make much of it. Although individuals became so extraordinarily numerous that they cause some late Mississippian strata to resemble conglomerates, and although the major genus, *Pentremites*, has hundreds of species, yet there is really no great variety among the blastoids. After their brief day of glory in late Mississippian times, they returned to their former insignificance. Indeed, they became extinct in all but a few regions.

Oldest and most important, and, to some, the most fascinating of the echinoderms, are the cystids. Among them the scientifically imaginative can see the ancestors of all

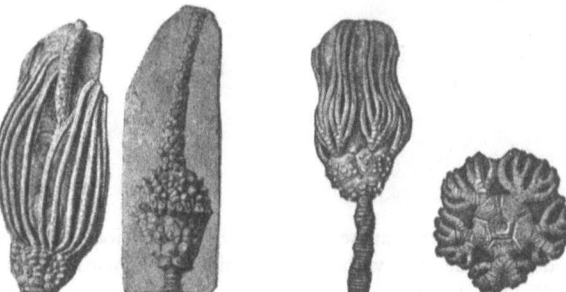

FIG. 34. Camerate crinoids from the Mississippian of Iowa. At left, a calyx with arms and anal tube; next, a similar crinoid lacking arms, showing tegmen and anal tube, and the openings at bases of the arms through which food passed to the concealed mouth; third, and at right, a somewhat simpler type, drawn in lateral and lower (dorsal) views, showing the form of the calyx and method of branching of the arms. From Wachsmuth and Springer.

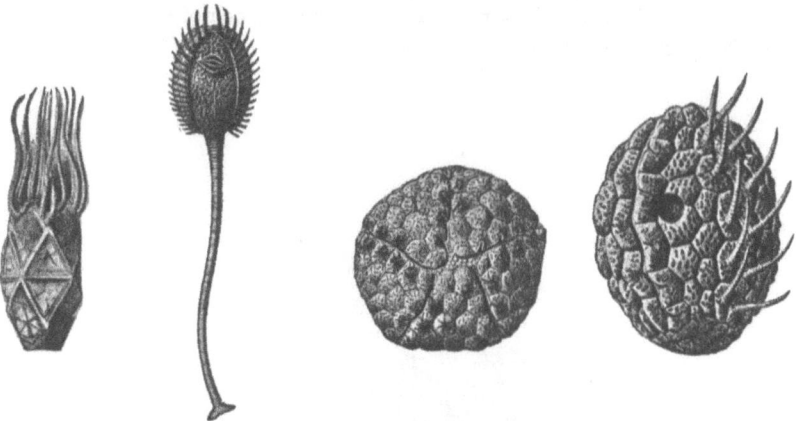

FIG. 35. Three kinds of cystids. At left, *Cheirocrinus*, one of the Rhombifera, with erect arms on the summit of the calyx; next, another of the Rhombifera, with sessile food grooves; at right, actinal and lateral views of one of the Diploporita, with numerous pores in pairs, and sessile food grooves. All from Otto Jaekel.

 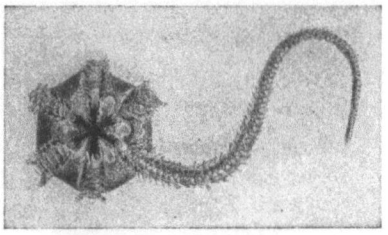

Fig. 36. At left, a bit of Ordovician sea floor, with four edrioasteroids. About half natural size. At right, a modern brittle star, with only one of the five arms shown. Original drawing by the late Professor William Patten.

Fig. 37. Three views of a Mississippian pentremite, to show the large plates of the calyx, the five ambulacral areas, the central position of the mouth, and the openings around it. From J. F. Pictet, *Traité de paléontologie*.

Fig. 38. A branching kind of *Favosites*, one of the most common genera of tabulate corals.

other members of the phylum; but it must be admitted that, if absolute proof of this theory is demanded, the record now known is not fully convincing.

The oldest cystids (Lower Cambrian) are sac-shaped. But a sac has no particular shape; that depends upon what is in it, and whether or not it is full. Such a description fits the early cystids to a nicety. They had no particular shape, no symmetry. They can be called echinoderms only because their skeletons were made up of plates and because each had a water-vascular system. This last is not expressed in ambulacral areas radiating from the mouth, for true cystids had no ambulacral grooves, and the position of the mouth of many is difficult to locate. The presence of a water-vascular system is shown among the oldest and most primitive of them by numerous perforations in every plate (Fig. 35, at right). In the earliest it was not confined to particular areas but penetrated the whole skeleton.

Evolution amongst the cystids is best expressed by the statement of certain tendencies. First, there was a tendency toward the reduction of the number of plates. Second, there was a tendency toward constriction of the sac at the attached end, to form a column. That the stalk of the crinoid is a restricted area of the primitive body cavity is indicated partly by the fact that it contains a "heart" and other organs, and partly by the fact that the distal columnals of some of the Paleozoic forms are made up of five, ten, or more, radially arranged pieces. Third, there was a tendency toward an orderly arrangement of the plates, beginning at the region of attachment and spreading toward the free (actinal) end. Fourth, the food-gathering area, originally a simple mouth, tended to spread. It did so in two ways: either by the production of food grooves along arms which are outgrowths or constrictions of the upper end of the original sac, or by the spreading of sessile food grooves over the surfaces of the plates (Fig. 35). Fifth, the number of openings for the water-vascular system tended toward reduction. In the later cystids (Rhombifera) they are confined to particular areas, where they are systematically arranged (Fig. 35, two specimens at left). There was originally no connection between the food grooves and the water-vascular areas. One small group of cystids lost the latter system entirely. Only a few of the older crinoids show external traces of it. In the starfish and blastoids, on the other hand, it is conspicuously developed, and combined with the food grooves in the ambulacral areas. Sea urchins have a highly developed water-vascular system, but no food grooves, although the tube feet occupy the ambulacral areas. Holothurians, the most specialized echinoderms, have an echinoid type of water-vascular apparatus.

Various cystids show various combinations of the results of the tendencies listed above. One line, characterized by loss of the water-vascular system, great reductions in the number of plates, and the development of food grooves on the arms, led to the crinoids. Another, in which there was less reduction of plates, but a restriction of pore-bearing ones to five particular areas, seems to have started the sea urchins in Ordovician times. Still another group, the edrioasteroids (Fig. 36, at left), in which food grooves and water-vascular systems were combined, may have been ancestral to the starfish, and they in turn to the brittle stars (Fig. 36, at right). The sea cucumbers, or holothurians, with much reduced skeleton, may have come from the same stock, although many believe that they are more closely related to the primitive sea urchins. Although the early blastoids remained attached and actinal side up, they must have had the same ancestors as the starfish.

All of which is more or less speculation. How fully justified, further research alone can tell. As for the cystids themselves, they appeared first in the Lower Cambrian, where they are rare. Rather specialized, flattened, possibly motile forms are found in Mid-Cambrian deposits. Unfortunately for Dr. F. A. Bather's ideas, outlined above, of the origin of the various other echinoderms from the cystids, the really primitive members of the group have so far been found only in Ordovician rocks. That, however, is no real objection to the acceptance of the theories, for there are numerous cases where simple, primitive animals have continued to live long after their more specialized derivatives have disappeared. The cystids seem to have reached their culmination in abundance and variety during the Middle Ordovician. Except at a few localities they are extremely rare in Silurian and Lower Devonian strata, and the last of them are found in rocks of Mid-Devonian age.

A great deal is known about the history of the sea urchins and something about that of the sea cucumbers, but there is no space to record it here. The echinoderms are so numerous, so beautiful, and so scientifically important that a whole book would do them less than justice.

IX

THE BEGINNINGS OF THE CHORDATES

> You strange, astonish'd-looking, angle-faced,
> Dreary-mouthed, gaping wretches of the sea,
> Gulping salt water everlastingly,
> Cold-blooded, though with red your blood be graced,
> And mute, though dwellers in the roaring waste . . .
> Leigh Hunt, "To a Fish"

The animals of the earlier Paleozoic had no backbones. It is believed that vertebrates were derived from some sort of invertebrate, but the problem of their origin still baffles zoölogists and paleontologists. Zoölogists have attacked it from the standpoint of comparative morphology; that is, they have studied all sorts of vertebrates, with the intention of learning which attributes are primitive and which specialized. Paleontologists, in their search for missing links, have studied fossils rather than living animals. In Devonian strata they have found many puzzling objects which, although they show some of the features of vertebrates, do not fit readily into the modern categories of such animals. In order to understand these curious creatures one must first investigate modern fish and learn something of their fundamental characteristics.

Most fish have a cartilaginous or bony skeleton consisting of three parts. One of these, including the skull and backbone, is the axial skeleton. The backbone is segmented, the parts represented by numerous biconcave vertebrae, each separated from its neighbor by cartilage. The study of fish embryos reveals that no backbone is present in young stages, its place being filled by a continuous unsegmented rod known as the notochord. This organ is not formed of strong tissue but is made up of watery cells enclosed in a tubular sheath; it may be likened to a rubber container surcharged with water. Although not in itself strong, it gives the embryo a longitudinal support around which the vertebrae develop, growing inward from the outside, until in the adult they subdivide it, enclosing its remnants between their concave ends. The anterior end of the notochord is beneath the brain, a part of the base of the skull developing around it. Since, in the ontogeny of the individual, the notochord precedes the backbone, it is probably more primitive than the vertebral column, whence the name Chordata, rather than Vertebrata. The axial support is, therefore, the distinguishing feature of the phylum. Two sections of the skeleton, the visceral and appendicular portions, are pendent from it.

The visceral skeleton is composed of from seven to nine inverted arches, although some sharks have vestiges of more. The first arch is double, forming the upper and lower jaws. Next is the hyoid arch, a support located in the throat. Beneath the posterior part of the skull are from five to seven similar arches which sustain the gills.

The appendicular portion consists of the limbs and their supports. In modern fish this system has two parts, the anterior pectoral, and the posterior pelvic paired fins, with their girdles. As will be seen, however, some of the ancient ones possessed more than two pairs. The pectoral girdle of the bony fish is made up of several bones and is attached to the back or sides of the skull. The pelvic girdle, on the other hand, is poorly developed; absent, indeed, from modern ones. It has no connection with the backbone, even in sharks, where it serves to support the relatively small posterior pair of fins. These general facts about the nature of the skeleton suffice for current purposes, so we may now turn to a more general consideration of the fish world.

At the present day there appear to be three principal groups of fish, most easily distinguished by the nature of their scales. The shark (Fig. 39) or skate has no continuous external skeleton, but numerous individual, more or less thornlike scales embedded in a tough skin. In section each scale shows an internal pulp cavity surrounded by a hard compact substance like the dentine of our teeth, traversed by numerous branching but minute canals radiating from the pulp cavity. The external layer consists of extremely hard, dense enamel, the hardest of all organic tissues. Under the microscope this substance appears uniform in texture, although it may be traversed by extensions of the canals of the dentine, which do not, however, branch in the enamel. Such scales, called placoid, may or may not be mounted on a bony base. Since they are found only in sharks, skates, and their allies, they characterize a subdivision of the fishes, known as the elasmobranchs. Another type of scale is found in the ganoids (Fig. 39). These fish, represented at the present by the gar pikes and sturgeons, possess a covering of scales, diamond-shaped or, more rarely, circular, overlapping one another like tiles and in most cases interlocking by toothlike processes. When examined in section each scale shows a bony substratum coated externally with a substance known as ganoine, enamel-like in structure. Hence the ganoids are often spoken of as the enameled-scaled fish. It should be noted, however, that there is no pulp cavity or true enamel. The bony fish, teleosts, have thin flexible scales or none at all. Their scales are referred to as cycloid or ctenoid, according to whether their margins are smooth or possess comblike projections.

The skeleton of the elasmobranch consists almost entirely of cartilage, or of cartilage impregnated with calcium carbonate; that of the ganoid is also largely cartilaginous, especially among the more ancient ones. As the evolution of the ganoids is traced, however, it is found that with the passage of time the skeleton became more and more ossified, whereas the scales became thinner and thinner, the change marking a gradual transition from the ganoids into the teleosts, whose skeleton is completely bony. Most modern writers consider the ganoids as a subdivision of the teleosts.

Fig. 39. Above, a modern shark, showing position of mouth, gill slits, and heterocercal tail fin. Water color by J. Burkhardt, 1864. In middle, a modern ganoid, showing scales and heterocercal tail fin. Wash drawing by Jos. Dinkel, 1834. Below, skeleton of a modern bony fish (teleost), with homocercal tail fin. All from drawings, of the time of Louis Agassiz, in the Museum of Comparative Zoölogy, Harvard University.

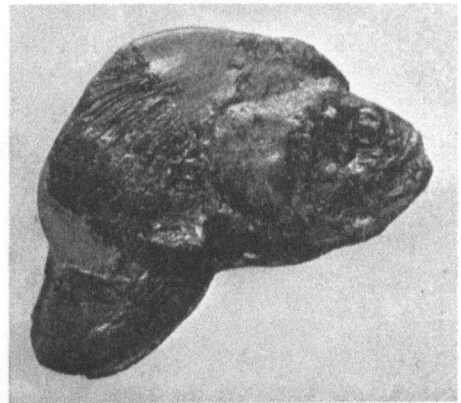

FIG. 40. Skull and pectoral fins of a *Portheus*, a large Cretaceous teleost. This skull, eleven inches high, is in the Museum of Comparative Zoölogy, Harvard University. It was found in northern Queensland.

FIG. 41. *Phareodus*, a large Eocene teleost from Wyoming. Length about twenty-two inches. Related fish are found in tropical fresh waters today. From a specimen in the Museum of Comparative Zoölogy, Harvard University.

Another variation in fish is in the nature of the tail fin. Three principal types exist. In the first the fin extends completely around the posterior end of the fish, equal parts being above and below the backbone. This type, the diphyceral, was formerly believed to be the most primitive, but recent studies do not confirm this hypothesis. It is found, however, in some of the most ancient fishes, as well as in modern ones which have a specialized, elongate, eellike form. It is symmetrical, but another common sort, the heterocercal, is unsymmetrical, the backbone extending into its upper lobe, whereas the lower one, which may be the larger, receives support only from the fin rays (Fig. 39, at top and middle). Practically all sharks and skates, and most of the ganoids, display the heterocercal tail fin. Finally, there is the homocercal type, exhibited by the teleosts, in which the fin lies mostly behind the posterior end of the vertebral column. These fins are symmetrical, many with a reëntrant angle at the posterior end, as seen in profile (Fig. 39, at bottom).

From these descriptions it seems obvious that specimens of the three principal sorts of fish should be easily recognized as fossils. Since the only really hard parts of the elasmobranchs are the teeth, scales, and fin spines, few of them are preserved in their entirety, and they are known chiefly from fragments. The ganoids, on the other hand, having a relatively strong external covering, may be completely preserved, but reveal little of the internal skeleton. Although the thin scales of the teleosts may be lost, their bony skeletons are readily recognizable (Figs. 40, 41).

The vertebrates are the most familiar of the chordates, but certain living as well as extinct animals have no backbones and yet show vertebrate rather than invertebrate characteristics. We must, therefore, list those characteristics in which the Vertebrata appear to differ from the Invertebrata. These are: a backbone developed about or replacing a notochord; a tubular central nervous system which lies above the vertebral column, which is in turn above the alimentary canal; an anterior expansion of the nerve cord consisting of a tripartite brain; a variable number of visceral arches with gill slits between them; internal supports for the paired appendages; and bone, dentine, and enamel.

It may seem from the nature of their work that paleontologists would be little concerned with brains. However, although they are themselves soft tissues, brains are surrounded by cartilaginous or bony structures with cavities which, on the destruction of their contents, may be filled with mud or other substances. Such natural casts of the cranial cavities occur among fossils in rocks as ancient as the Silurian, and their study has in recent years enabled us to understand some previously obscure fossils. Fundamentally, the brain is a hollow object, an anterior enlargement of the tubular nervous system, divided into three parts. The anterior or forebrain has at the front the olfactory lobes, behind which are located the cerebral lobes. The midbrain is connected with the power of vision and bears the optic lobes. At the anterior end of the hindbrain is the cerebellum. Behind it is the *medulla oblongata*, which tapers backward into the main nervous cord. The tripartite brain really, then, has five principal areas, although at its first formation only three parts are discernible.

The paleontologist must be able to recognize bony tissue, which has a structure readily identifiable when thin slices are studied under the microscope. A section cut transversely to the longer axis of a bone shows circular openings which are known as Haversian canals. These are surrounded by concentrically arranged lamellae, among which are situated numerous narrow openings or lacunae; from the latter radiate minute tubular canals crossing the lamellae transversely, producing a more or less checkered appearance. Such structures, although not entirely unknown among the invertebrates, are characteristic of true bones, and enable the paleontologist to decide whether a doubtful specimen belongs to the chordate phylum.

The oldest fossils which have been identified as allied to the vertebrates are enigmatic creatures with strong external and no internal skeletons. Although in this respect they resemble invertebrates, the posterior part of the body is fishlike, covered with scales. Because of the massive plates with which the anterior part of many is protected, these organisms have been grouped together under the name of ostracoderms, or crusted-skinned animals. Although they have been known and studied by paleontologists and zoölogists for many years, only recently have their relationships begun to be understood. There seem to be five types, which will be described briefly, with an attempt to show their relationships.

The most ancient well-preserved ostracoderms are found in mid-Upper Silurian strata in northern Europe and in Pennsylvania. They are small forms known as *Pteraspis* and *Paleaspis*; for our purposes they may be referred to as the pteraspids. *Pteraspis* is the better known genus (Fig. 42). It appears to have been a slender animal, from three to seven or eight inches in length, the anterior half encased in an oval buckler composed of a few thick plates, whereas the posterior part bears thin scales, rarely preserved. The eyes are lateral, and the mouth, only recently discovered, is a transverse slit just beneath the anterior margin, bordered in front by a plate bearing a large median and two lateral processes which may be interpreted as having had the function of teeth. Behind the mouth are sixteen elongate scales which may have been capable of movement against the anterior plate, thus producing a sort of chewing apparatus. These are, however, not teeth or jaws in the true sense. No internal skeleton and no appendages have been found.

Many specimens of a much larger ostracoderm, *Drepanaspis* (Fig. 42, at right), have been found in roofing slates of Lower Devonian age near Gemunden, Germany. *Drepanaspis* has a broad, flattened shield with large median and lateral plates, connected by a mosaic of smaller ones. The posterior part, which is mostly tail fin, is short, covered with overlapping scales. The transverse slit-like mouth is bordered by numerous plates covered with small denticles. The ocular orbits are small, widely separated, situated near the anterior corners of the shield on the dorsal side.

Drepanaspis is allied with the pteraspids not only by the similarity in the position of the principal plates of the buckler and the location of the eyes and mouth but also by the microscopic structure of the shell. In thin section, both show a bony layer overlaid by compact dentine with pulp cavities and ramifying canals. W. L. Bryant has recently demonstrated that the oldest known ostracoderms, described many years ago from material collected from an Upper Ordovician sandstone at Canyon City, Colorado, have plates with the same histological structure. Although ostracoderms occur in vast numbers at Canyon City, all the material so far collected is extremely frag-

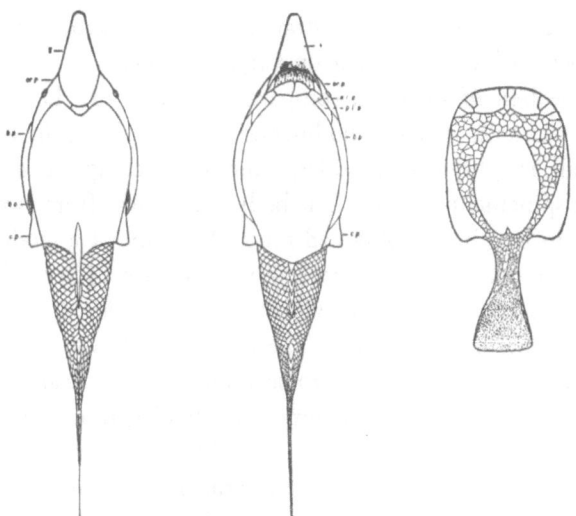

FIG. 42. At left, restorations of the dorsal and ventral surfaces of *Pteraspis*, to show the few large plates of the anterior shield. One-third natural size. From E. I. White. At right, the ventral surface of *Drepanaspis*, to show the numerous small plates between the few large ones. One-ninth natural size. From H. C. Stetson.

mentary, not one complete plate having been recovered. When better specimens are found, they will probably prove to have a form much like that of the pteraspids. It should be noted that in the latter the plates of the buckler are fewer in number and more solid than those of the Devonian *Drepanaspis*, a fact which suggests that in the evolution of this group there was a tendency to produce a more flexible external skeleton by the increase in number and decrease in size of the plates.

A second group, called ostracoderms merely because its relationships have been obscure, is exemplified by the genera *Thelodus* and *Lanarkia*. High in the Silurian in western England is a thin layer which has long been known as the Ludlow bonebed, because of its typical development in the vicinity of that picturesque town. The bone-bed contains few if any true bones but is full of toothlike scales. Similar scales are found in other localities, particularly in Upper Silurian strata on the island of Oesel. Most of these isolated scales are considerably water-worn; hence the various species of *Thelodus* which have been described with them as a basis are really unidentifiable. Mr. David Tait, the veteran collector and geologist of the Geological Survey of Scotland, succeeded in finding complete specimens of this genus and of the related *Lanarkia* in the Upper Silurian of the Lowlands of Scotland. They were described first by Dr. R. H. Traquair, but recently all the old and much new material has been scrutinized by Mr. H. C. Stetson in the paleontological laboratory at Harvard. Mr. Stetson has also been fortunate enough to discover large and fairly complete specimens of *Thelodus* in Silurian strata north of St. John, New Brunswick, and scales in the Upper Ordovician at Canyon City, Colorado.

Even with the best of material, comparatively little can be learned about the anatomy of these animals. There is no buckler of large plates, the entire body being covered with small, roughly cubical scales, some of which appear to be connected to those adjacent by spinelike processes. The body, which was from two or three inches to a foot or more in length, probably had much the shape of a modern catfish, being somewhat blunt-headed. Although there is no evidence of true fins, unsupported lateral expansions back of the middle, which may have served as balancing organs, suggest skatelike habits of locomotion. The eyes are anterior and lateral, far apart as in *Drepanaspis*, and the mouth seems to be a transverse slit near the anterior ventral margin of the head. It is bordered by somewhat smaller, more thornlike scales than those covering the remainder of the body. Whether they were toothlike in function remains to be determined. At any rate, the creatures were agnathous, jawless.

The most important fact about these animals has been learned from thin sections of the scales. The histological structure is exactly that of a tooth. Each has a large pulp cavity surrounded by dentine, the whole capped by a layer of enamel. The scales are therefore of the true placoid type, which would indicate that *Thelodus* and *Lanarkia* belong to the elasmobranchs. The pteraspids and *Drepanaspis* are so closely related by their general features and the microscopic structure of their plates that they must be placed in the same group. Since *Thelodus* is now known to have lived in Ordovician times, it is probable that the forms now known represent at least two genetic lines, descending from an as yet unknown Cambrian or Ordovician forebear. There is a suggestion that *Thelodus* may have been a member of the line which gave rise to the true sharks; the other lineage (pteraspids) probably died out during the Devonian.

The third group to which attention should be directed is that of the cephalaspids, long known but only recently understood. The oldest occur in mid-Upper Silurian in Shropshire, in Scotland, and on the island of Oesel. Scotland and Spitzbergen have supplied most of the specimens of Devonian age, although some have been found in Canada. Until the last decade only a few good specimens were known, but Norwegian expeditions to Spitzbergen brought back great numbers of them, which have been cleverly prepared and studied by Dr. E. A. Stensiö of Stockholm, with results of far-reaching importance.

The head shield of a cephalaspid is shaped like the curved blade of a halberd, with backward directed lateral projections (Fig. 43). It protected only the head and anterior part of the body, the remainder being fishlike and covered with long, narrow scales. The head shield seems at first sight to be composed of numerous polygonal plates joined together at the edges, this appearance being emphasized in some by the presence of small knobs, one in the center of each polygon (Fig. 45, at left). Investigated in thin section, this condition is traceable to the presence of canals which radiate from centers within the test. This suggests the structure seen in plates of echinoderms, particularly cystoids and crinoids. There is no evidence, however, of suture lines bounding the plates of the greater part of the shield. There are three areas covered by small, thin plates. Two of these are lateral, whereas the third, on

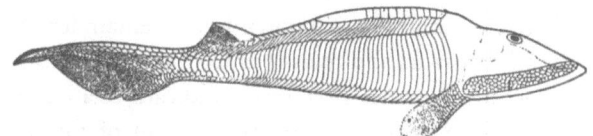

Fig. 43. A restoration of *Cephalaspis*, by E. A. Stensiö. Reproduced by permission of the Trustees of the British Museum (Natural History).

the median line, is called the "postorbital valley" because it appears in the fossils as a depressed area behind the eyes. The orbits are large, close together, near the middle of the top of the shield. On the median line, just in front of the eyes, is a small opening; behind it, and between the eyes, is another smaller one. These openings were explained some years ago when Professor Carl Wiman of Upsala obtained from Oesel a weathered specimen which exhibited a cast of the brain. It is a typical primitive tripartite brain, the anterior end of which lay beneath the anterior of the two openings just mentioned. Since this is the location of the olfactory lobes, the opening is to be interpreted as that of the nose. The posterior opening is above the hind part of the forebrain, and corresponds to the position of the epiphysis leading to the median or pineal eye. Dr. Stensiö, by making numerous skillful dissections of specimens from Spitzbergen, has succeeded in working out the central nervous system in great detail (Fig. 45, at right). He has not only developed the endocranial casts but has identified the ten pairs of cranial nerves so characteristic of the vertebrate. There is, then, no possible doubt that the cephalaspids are allied to the vertebrates, in spite of the absence of appendages, jaws, and internal skeleton.

It will be noted that there are few characteristics common to the cephalaspids and the two other groups described above. Although bone is present in the test, there is nothing so far known which suggests the presence of such tissues as true dentine or enamel in their shells.

The fourth group is the one which Traquair named the anaspids, because, like *Thelodus*, they lack a buckler. In all other respects, however, they are totally unlike that primitive elasmobranch. First found about thirty years ago in the Upper Silurian (Downtonian) of Scotland, these fossils were at the time entirely misunderstood. Our real knowledge of them dates from Professor Johan Kiaer's monographic description (1925) of the specimens which he collected in strata of the same age in the Ringrike district of Norway. Stetson has increased our information by his studies on material from the original localities in Scotland.

The anaspids are unique among the ostracoderms in that they have a really fishlike shape (Fig. 49). All are small, from one to eight inches in length, with a body covered by numerous elongate scales. The tail fin is unsymmetrical, the main portion of the body extending, curiously enough, into the lower, rather than the upper, lobe. This is the reverse of the usual heterocercal tail, and Kiaer suggested that it be called hypocercal. Such a condition is known in the remainder of the animal kingdom only among the ichthyosaurs and marine crocodiles, a good example of the truth that similarity of structure does not necessarily indicate relationship. The head is remarkably short, there is no neck, and the arrangement of the scales suggests that of the bands of muscle in modern fish (Fig. 50). They are directed forward on the back, backward on the sides, and forward again on the belly. A single pair of lateral spines indicates the presence of pectoral fins; above these on either side is a row of circular openings probably connected with the organs of respiration. The mouth is terminal, surrounded by small plates variously arranged in the different species, but without true jaws or teeth. The eyes are far forward, close together, high on the head, and there are two small openings on the median line which may, from analogy with those of the cephalaspids, be identified as the narial and pineal apertures. Behind the latter is an area of small plates corresponding to the "postorbital valley" of the cephalaspids. It is, therefore, a fair inference that the anaspids had a vertebrate type of brain, and that they are more closely allied to the cephalaspids than to any other group. They are discussed at greater length in connection with the theories of the origin of the vertebrates.

The last and probably the most specialized group of ostracoderms is that known as the Antiarcha. The best known of these are *Pterichthys*, found chiefly in the Middle and Upper Devonian red sandstone of Scotland, and *Bothriolepis*, a rather widely distributed fossil in strata of the same age in Quebec, eastern New York, and Pennsylvania (Fig. 44). These are the original and real "ostracoderms," for the anterior part of the body is encased in a shield of thick, solid plates. It is rather amusing to find that the latest workers exclude them from that group. They differ

Fig. 44. Dorsal and ventral views of *Bothriolepis*, as reconstructed by Professor William Patten. From Patten's *The Evolution of the Vertebrates and Their Kin*, by permission of P. Blakistons Son and Company, Inc.

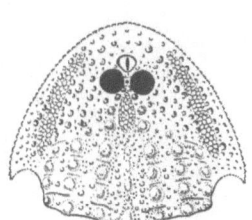

Fig. 45. At left, the head-shield of the Silurian cephalaspid, *Auchenaspis*. Note the narial opening in front of orbits, the pineal opening between them, and the postorbital series of small plates. From J. Kiaer. In middle, cranial roof of an anaspid, showing the same features. From J. Kiaer. At right, a dissection of a Devonian cephalaspid, showing casts of the cavities occupied by the brain, principal nerves, and other organs of the head. From E. A. Stensiö.

Fig. 45A. A diorama, now in the Rochester Museum, showing common animals in the Late Devonian Sea of western New York. There are many kinds of glass-sponges, several starfishes, a nautiloid cephalopod, and the primitive shark, *Cladoselache*. Reconstruction by George Marchand Studio. Courtesy of the Rochester Museum of Arts and Sciences, and Irving Reimann.

from all others in having a head shield which is movably connected with a boxlike case covering the anterior part of the body, and in possessing a pair of jointed appendages. The posterior part of the body is fishlike, with a dorsal fin and a heterocercal tail. This part of the animal is protected only by the thinnest scales, and would hardly be known except for the discoveries made by Professor William Patten at Scaumenac on Chaleur Bay in Quebec. The eyes occupy opposite ends of a transversely placed opening in the top of the head, separated by a plate not ankylosed to the shield and hence rarely preserved. The mouth is ventral, near the margin, bounded laterally by three pairs of plates which may be variously interpreted. Patten, who discovered and described them, believed that they represented jaws which moved inward toward the median line, but it has more recently been suggested that they may have moved up and down. The external covering has the histological structure of bone.

The specimens from Scaumenac are of particular interest because they are so well preserved that a few carbonized remains of the internal organs are found, revealing something of the anatomical structure. Sections show what seem to have been gills and a stomach, but no remains whatever of vertebrae, notochordal sheath, gill arches, or any appendicular support. Although this is negative evidence, it strongly supports the view that there were not even rudiments of an internal skeleton in the Antiarcha.

To summarize what has been said above, the various fossils which have been referred to the ostracoderms represent five distinct groups. Two of these, typified by *Thelodus* and *Pteraspis*, are shown by the microscopic structure of the scales and plates to be allied to the shark-skate group, and should be transferred to the elasmobranchs. The cephalaspids and anaspids are probably related to each other. The form of the brain shows that they are true chordates. In the opinion of the writer, the anaspids are the more primitive forms, possibly ancestral to the ganoids, whereas the cephalaspids were somewhat specialized bottom-dwellers which left no descendants. Dr. Stensiö's argument that they were the ancestors of the modern cyclostomes is of interest but cannot be discussed in this book without entering into too many technicalities. The last group, that of *Bothriolepis* and its allies, shows no relationship to the others, but occupies an isolated position. There is considerable evidence to show that it is related to the arthrodires, a group of chordates not closely allied to any other known animals.

Although they are not classified with the ostracoderms, some of the arthrodires are so ostracoderm-like that they may be described in connection with them. One of their principal characteristics is that their armor consists of two portions: a buckler over the head, and another over the neck and anterior part of the trunk. These articulate with one another by a hingelike joint, just as the two shields of the Antiarcha do. The best known arthrodires are the small *Coccosteus* (Fig. 46) of the Mid-Devonian of Scotland and the huge *Dinichthys* and *Titanichthys* of the Upper Devonian of

Ohio. The Scottish animal is most like the ostracoderms, for the anterior part is covered by a shield whose sculptured plates show it to have been largely external. The posterior part of the body, however, not only is fishlike in shape, with a heterocercal tail, but has a series of neural and hemal arches throughout its length which enclose a notochord and form the rudiments of a backbone.

Remains of *Dinichthys* are rather common in a black shale in the vicinity of Cleveland, Ohio (Fig. 47). Although its large platelike bones have been known for a long time, recent collections and studies have added much to our knowledge of

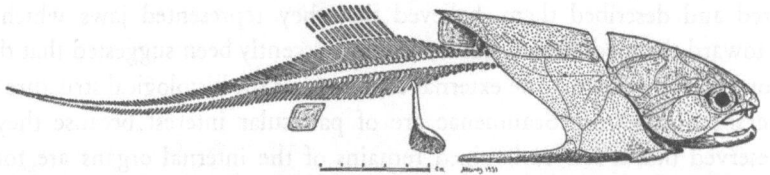

Fig. 46. A reconstruction of the Mid-Devonian arthrodire, *Coccosteus*. Note the space for a notochord between the neural and haemal spines. From A. Heintz.

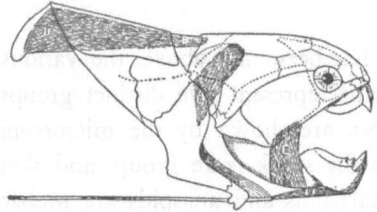

Fig. 47. The skull of the huge arthrodire, *Dinichthys*, to show the "teeth" and, diagrammatically, some of the principal muscles. From A. Heintz.

this extraordinary carnivore. Some of the shales in which the bones occur are so near to the center of Cleveland that they were in danger of being completely lost as collecting places through the expansion of the city. At the crucial moment, however, public-spirited citizens raised sufficient money to allow their exploration by means of a steam shovel. As a result of this modern and thorough instrument of search, new material has been obtained which allows an accurate restoration of these extinct animals. A great fish, with a skull nearly three feet across, and a body ten or a dozen feet long, *Dinichthys* was in marked contrast to the little *Coccosteus*. Unlike those of the latter, the shields were internal, entirely covered by flesh and scaleless skin. The jaws were fearful, armed with trenchant teeth more efficient than those of most sharks. Yet in reality *Dinichthys* had no teeth. It bit by working the whole top of the head instead of snapping shut its lower jaws. The latter may have been held rigidly while one set of muscles lifted the head shield, and another pulled it down. Sharp-bladed projections of the bones of both upper and lower jaws served as teeth. Although they are toothlike in form, thin sections show that these projections are nothing more than dense bone, without dentine or enamel.

The arthrodires are the only creatures in the whole animal kingdom which actually use their jawbones for biting. They were not senile and toothless grandsires, mumbling their food, but were capable of quickly slicing in twain any of their contemporaries. *Dinichthys* and *Titanichthys*, the latter possibly fifteen to twenty feet long, were truly the rulers of the late Devonian or early Mississippian seas. Yet this group suddenly disappeared, at the zenith of its powers. The too rich and powerful are always on the verge of extinction.

X
THE ORIGIN OF THE VERTEBRATES

Causa latet: vis est notissima.
"The cause is hidden, but the result is known."
Ovid, *Metamorphoses*, IV, 287

It was pointed out in the earlier pages of the preceding chapter that there are many fundamental differences between the two great divisions of the animal kingdom, the Chordata and the Invertebrata. Those who try to trace the path along which evolution has proceeded find in this region only the faintest trail. As the discussion of the ostracoderms has shown, these fishlike animals were not full-fledged vertebrates. Although in some respects they stand between the vertebrates and the invertebrates, nevertheless they give few clues to the particular phylum from which the higher group sprang. The problem of discovering the connecting links belongs both to the zoölogist and paleontologist. It is still far from solution, but some progress is being made. In this place all that can be done is to present the current theories, with some of the evidence for and against them.

The one which at present has widest acceptance is derived entirely from the study of certain living animals, all, unfortunately, entirely without hard parts, and so not likely to be preserved as fossils. The line proceeds downward from primitive fish through *Amphioxus*, thence to the larval tunicate, then to *Balanoglossus*, and finally to some unidentified echinoderm. Since the presence of a notochord is the most important factor, this theory is commonly called the "chordate" or, because of the prominent place that animal occupies in it, the "*Amphioxus*" theory.

Amphioxus, the simplest living true chordate, is a small, slender, cigar-shaped animal, from one to two inches in length. It has a worldwide distribution in warm shallow waters, where it lives partially buried in the sand, the mouth protruding. It is nevertheless capable of swimming, even of active movements. It has no distinct head; the mouth is ventral, a little behind the anterior end, with paired gill slits back of it. Although there are no real fins, a median dorsal fold and a pair of ventro-lateral ones suggest rudiments of such organs. The notochord extends from the anterior to the posterior end and constitutes the entire axial support, for the creature has no skeleton. Above the notochord is a tubular nerve cord, as in the vertebrates; below it, a straight alimentary canal. A conspicuous feature of the animal is the regular segmentation of the muscles, which may be plainly seen because it is translucent. They have the form of V-shaped segments, with the angle pointing forward. The segmentation, having an outward expression, is much more obvious than that of the higher vertebrates, in which it is shown only by internal characteristics, such as the repetition of vertebrae in the spinal column, nervous ganglia, and the like. The muscles of a fish (Fig. 50), however, have the same arrangement as those of *Amphioxus*, as one may observe on any Friday.

Although it possesses only a few characteristics of the higher vertebrates, those it has are so important that no one now denies *Amphioxus* a place in the same phylum with them. It is generally believed that the fish arose from an animal much like it, even though the connection cannot be traced. Furthermore, the search for its ancestors has not yet met with success. The one contact between it and the invertebrates is found in the fact that the larva is ciliated, swimming by means of hair-shaped outgrowths. The possession of a ciliated larva is common among the invertebrates but entirely unknown among the craniate vertebrates; however, this is only a vague connecting link with the former subkingdom. It is possible that *Amphioxus* may have been derived from some member of the Hemichordata, a group in some respects more primitive. These animals, the tunicates or sea squirts, are commonly included as links in the chain of this theory because during their larval stages segmented muscles, gill slits, and a notochord are present. So far as they are now understood, they really throw no light on the problem.

There is one group of living animals which may represent a simpler type of chordate than the larva of the tunicates. The best-known member is *Balanoglossus*, a wormlike marine creature which lives in fragile tubes in the region between high and low tides (Fig. 48). At the anterior end of the "worm" is a long, sensitive proboscis, capable of great extension and contraction. Behind the collar is the trunk or body proper, on the anterior part of which are numerous pairs — there may be as many as fifty — of openings or gill slits, much like those of *Amphioxus*. In the interior of the proboscis is an axial rodlike structure which arises as an outgrowth from the alimentary canal above the mouth. Because of its position, its origin, and the nature of its cells, this rod has been considered a notochord. In another genus, *Harrimania,* there is a similar but much larger structure, strikingly like a true notochord. According to S. F. Harmer, *Balanoglossus* is "provided with a dorsal, tubular, central nervous system, and although it does not extend beyond the limits of the collar," it shows noteworthy resemblances to that of vertebrates. This protochordate is therefore considered the simplest existing representative of the group from which the vertebrates were derived.

Even *Balanoglossus*, simple as it is, does not serve to show positively from what phylum the chordates were derived. A suggestion, however, is furnished by the form of its larva. Students of zoölogy have learned that the very young, free-swimming larvae of all members of the same group are much alike, and have therefore come to depend to some extent upon the young as indicators of relationship. In this case, if one were to judge from the form of the adult, *Balanoglossus* would seem to be closely allied to the worms. The free-swimming larva, however, is unlike that of the worms but rather resembles that of the echinoderms. In fact, it is so much like the latter that when first discovered it was supposed to be the young of some member of the latter group. Although it is not possible to go into detail about these larvae, the diagrams (Fig. 48) will illustrate the resemblances between those of the echinoderms and *Balanoglossus*, and the difference from those of the worms.

In addition to the similarity of the larvae, there are other ways in which *Balanoglossus* appears to be allied to the echinoderms. It has a system for the circulation of water comparable to their water-vascular system, and H. H. Wilder has found that its five body cavities are probably represented in the echinoderm larvae. No proof of connection between the echinoderms and the chordates is yet forthcoming, but we

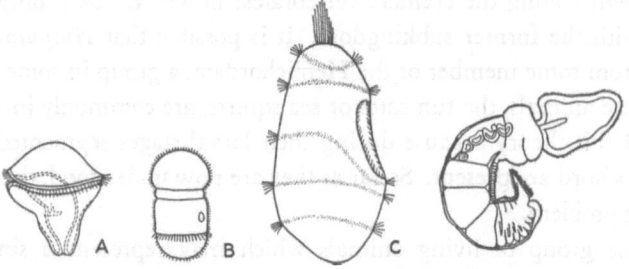

Fig. 48. A, trochophore larva of an annelid, compared with (B) the free-swimming larva of *Balanoglossus*, and (C) that of a crinoid. A, after B. Hatschek; B, after W. H. Bateson; C, after Goette and Thompson. At right, a diagrammatic representation of *Balanoglossus*, simplified after W. H. Bateson.

have the interesting suggestion that a bilaterally symmetrical ancestor gave rise to the radially symmetrical, lowly echinoderm and the progressive and finally dominant vertebrate. Here is a place where paleontology should come forward with definite evidence, but it has as yet failed to do so.

Another idea which has been more or less popular among zoölogists is that embodied in the "annelid theory." The fundamentally segmented nature of the vertebrates naturally suggested the idea that they arose from the annelids, a view not yet entirely abandoned. The chief ground for the hypothesis is the similarity which exists between embryonic vertebrates and adult annelids. For example, the vertebrate embryo shows distinct repetition of parts, and even in the adult the vascular system, the nervous chain with its paired ganglia, and the general relationship of the principal organs to one another show considerable similarities to those of the annelids. Some authors have maintained that the bundle of fibers which in some annelids supports the nervous chain is comparable to a notochord, although histologically it is of entirely different structure. Moreover, both Chordata and Annelida have their internal organs in a body cavity, the coelom, and there is agreement in the structure of the kidney tubes, or nephridia.

The most obvious difficulty encountered in trying to derive a vertebrate from the segmented worm is the general reversal of all parts of the body that is necessary to make the change. The anterior ganglion of the nervous system of the annelid is above the mouth, but the two cords which are the principal links in the chain pass around the anterior end of the alimentary canal, continuing beneath it along the ventral part of the body. The main blood vessel lies above the gut, with another beneath it, the blood flowing forward in the dorsal and backward in the lower one. To transform the annelid into a vertebrate it is necessary to reverse the dorsal and ventral sides, abandon the old mouth, produce a new one on the opposite side, close the old terminal anus and form another further forward, change the solid to a tubular nervous system, and develop a notochord between the central nervous system and the alimentary canal. Taken all at one time, this seems difficult of accomplishment, but we are assured that such things are physiologically possible.

The theories which have most interested paleontologists are naturally those in which fossils play a part. Two eminent zoölogists and investigators, Professors W. H. Gaskell and William Patten, have attempted to connect the vertebrates with the arthropods. Such suggestions may be called, for convenience, "arthropod" theories. Both men brought to the support of their causes a wealth of learning, labor, and ingenuity which entitles them to greater consideration than is commonly accorded. Since the theories are similar, although by no means identical, the discussion of one will suffice. The writer has chosen Dr. Patten's, partly because its exposition is somewhat more recent, and partly because its genial author was good enough to discuss it freely with him.

The arthropod theory is, in brief, that the ostracoderms stand midway between invertebrates and vertebrates. It does not attempt to connect any particular ostracoderm with any particular invertebrate, but takes up general rather than particular features of the two groups. Professor Patten was doubtless influenced, however, by his studies of *Bothriolepis* and the general resemblance of its skeleton to that of the arachnids of the group Eurypterida (Fig. 44). *Bothriolepis* resembles the latter in that both have an external segmented armor, jointed lateral appendages with external but no internal supports, and, according to Patten's interpretation, similar jaws, since he deduced that they moved toward each other in a horizontal plane. Neither has any internal skeleton. *Bothriolepis* has only one pair of appendages, but in the arthropods there is a tendency toward reduction in the number of limbs, those which are retained being on the anterior part of the body. Hence it is logical to infer that the one pair of the Antiarcha may be the sole survivors of the six pairs of the eurypterids.

The arthropod theory resembles the annelid theory in requiring an introduction *de novo* of a notochord and the reversal of dorsal and ventral sides, with its accompaniment of the production of a new mouth and the closing of the old one. There is some indication of a notochord in representatives of the arthropods; for example, a scorpion Patten found in Ceylon has what appears to be a real notochord, even the histological structure being apparently the same as that of the similar rod in the vertebrates. That the mouth of the vertebrate is a relatively new structure is indicated by its rather late appearance in the embryo; moreover, the fact that it is lined with ectodermal tissue proves that it is an inpushing from the outside. The cloacal chamber is likewise lined with ectoderm, indicating a similar origin for the anus. Patten pointed out that the tube of the alimentary canal, where it passes through the nervous system in the arthropod, is subject to constriction in the higher groups, in which there is an increase in the size of the brain because of the bunching together of ganglia in this region. Thus in many insects this tube is so narrow that solid food cannot be taken, a circumstance that causes the animals to develop suctorial habits. The process goes still further in the adults of many Lepidoptera (butterflies and moths), in which the tube is so completely shut off that they do not feed at all after the larval stage is passed. There is, therefore, reason to expect the closure of the old mouth in arthropods. Many anatomists have pointed out that the hypophysis at the base of the vertebrate brain is a blind sac which may be situated where the alimentary canal formerly passed through the central nervous system. Patten saw in the so-called "dorsal organ" of some crustaceans the beginnings of the new mouth. The "dorsal organ" is a sac on the median line which leads downward nearly to the alimentary canal. A slight change, involving only an opening into the digestive tube, would enable it to function as a mouth.

It is not possible to enter fully into all the evidence which Patten gathered to support his thesis, nor is it possible to refute his arguments. The chief reason for the rather general lack of acceptance of the arthropod theory is that all the animals used as connecting links are specialized, and most zoölogists believe that simple members of a higher group never arise from specialized ones of a lower grade. *Bothriolepis* appears to be one of the most specialized of all the ostracoderms, and the oldest of its relatives, found in Lower Devonian strata, have the same structure. The eurypterids themselves are complexly organized, and there is no evidence that they gave rise to anything but the scorpions. As has been pointed out in the previous chapter, the only possible descendants of the Antiarcha are the arthrodires, which are much too highly specialized to be considered as the ancestors of fish. It is generally believed that all the armored ostracoderms (pteraspids, cephalaspids, and Antiarcha) were slow-moving, bottom-living forms, adapted to a mud-grubbing existence. There is little likelihood that they could have thrown off their armor and their sluggish habits, to be transformed into swiftly moving fish.

The most recent of the theories proposed by paleontologists, the anaspid theory, although it cannot at the present time be called complete, is founded upon observations which do not fit in with the schemes outlined above. It therefore appears necessary to state them separately, as the writer has done for several years in his lectures, although they now appear *in extenso* for the first time. The tentative outline is that the earlier scaled anaspids may possibly have been derived from bilaterally symmetrical Cambrian ancestors of the echinoderms. From the scaled anaspids originated the nearly naked *Lasanius*. In it an internal skeleton appears as a series of visceral

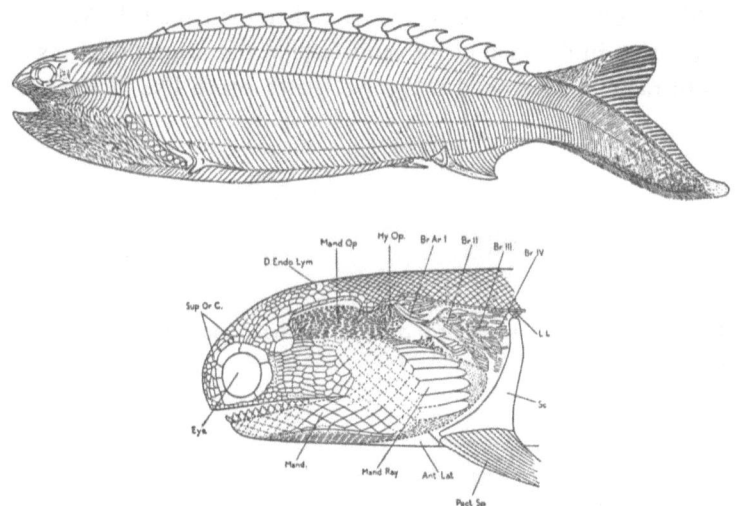

FIG. 49. Above, an anaspid, showing the fishlike form, the hypocercal tail, and the arrangement of the slender scales of the body. From Kiaer. Below, the head and anterior fin-support of an acanthodian, for comparison. Note in both the large, anteriorly placed eye with ring of sclerotic plates, the irregular arrangement of scales on the head, the position of the external fin support (Pect. Sp.), and the fact that the latter is connected with an internal element (Sc.). Mand. Op., Hy. Op., mandibular and hyoid regions; Br., Ar. I, Br. II, Br. III, Br. IV, branchial region; mand., mandible. From D. M. S. Watson.

arches, and there is probably a cartilaginous brain case and a notochord with its sheath. A new exoskeleton, consisting of scales with ganoine, was gradually evolved.

The vertebrate-like characteristics of the scaled anaspids have been pointed out in the previous chapter. They are: the possession of a tripartite brain belonging to a nervous system above the gut; a fishlike form, including a tail which was an effective propeller, indicating the power of rapid, fishlike movement; and a pair of spines in a location which suggests the presence of pectoral fins. Furthermore, the arrangement of the scales indicates a set of muscles like those of fishes (Fig. 50). On the other hand, the scaled anaspids are similar to invertebrates in the lack of an internal skeleton, the presence of an apparently permanently open terminal mouth, the amorphous structure of the scales, which contain no bone, dentine, or enamel (this may be due to poor preservation); and the presence of plates around the mouth and the anus. All these features suggest an echinodermal ancestry. The same relationship is indicated by the branchial openings. They are circular apertures perforating a plate, hence not like the gill slits of vertebrates but suggestive rather of the openings of the water-vascular system so characteristic of the cystoids. Some of the Cambrian and Ordovician cystoids have a linear series of such openings. It must be confessed that the relationship to echinoderms is suggested rather than proved. We have already seen that there exists in the shell of the cephalaspids, the nearest relatives of the anaspids, a series of radially penetrated polygonal plates, much like those of the cystoids.

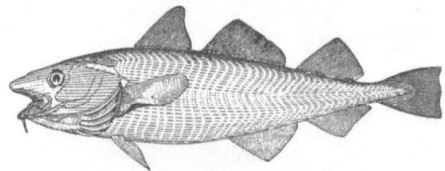

Fig. 50. Sketch of a skinned codfish, to show the arrangement of the muscles.

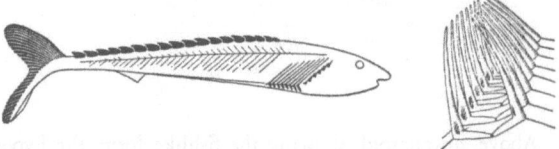

Fig. 51. At left, outline reconstruction of *Lasanius*, showing dorsal scales, hypocercal tail, and internal "branchial" skeleton. The longitudinal line and its branches merely suggest the arrangement of the external scales. From O. M. B. Bulman. At right, diagrammatic reconstruction of the internal skeleton. From H. C. Stetson.

From the scaled anaspids the scaleless form, the Scottish *Lasanius*, was derived. It differs from the other members of the group in having so thin an external covering that the scales are retained only by the largest and best-preserved specimens. Enough of these have been found, however, to show that the outward form and structure were similar to those of the better-known scaled genera. Its most remarkable feature is the possession of eight pairs of slender rods, the explanation of which has only recently been attempted (Fig. 51). Mr. H. C. Stetson obtained material pertaining to this genus from the Downtonian strata at the original localities in Scotland, and the interpretation here presented is that worked out by him. The rods appear to have formed an internal structure, comparable in general form to the gill arches of a shark. They were arranged in pairs, on opposite sides of the body. The long, spine-like dorsolateral portion of each broadens at the lower end, whence a straplike branch projects inward and backwards toward a similar one from its mate on the other side. During the animal's life these ventral processes were probably joined by median cartilages. The broad base supports a slender spine which projects outward and somewhat backward. The first two elements, the dorsal rod and the ventral strap, were doubtless beneath the skin, for they show no trace of surficial ornamentation. A small portion of the broadened base and the slender outer spine, on the other hand, show delicate striations, which indicate that they were outside the body.

It appears, therefore, that *Lasanius* had an internal skeleton of basketlike form, the lower elements connected across the body, the upper ones separated. In seeking its origin, it will be noted at once that the internal elements lie in the same position as the external scales of the other anaspids. They may have originated as external scales later drawn within the skin by the pull of the muscles, but it is more likely that they were formed *de novo* as cartilages between adjacent segments of the muscles. The primary function of this structure is reasonably clear. The outer spines doubtless supported a lateral fin below the median line on each side of the body. Whether the internal parts had another use, connected with gills, is more problematical. In front of the first of the series of rods just described there is on each side a series of six short acicular plates arranged in a diagonally ascending row. The base of each has a semicircular outline, as though it fitted above a circular opening. The plates lie in the position of the branchial plate of the scaled anaspids; hence it is probable that they indicate the location of the gill openings of *Lasanius*. If that be the case, as is the opinion of Kiaer and others who have studied the question, then a pumping movement produced by the up-and-down motion of the fins would have aided in augmenting the flow of water in at the mouth and out through the circular apertures. If gills or gill-like sacs were present, they would probably develop first directly opposite the openings, but as they enlarged they would tend, because of the forward motion of the animal in swimming, to push backward, and so come to be supported by the basket. Hence a connection between the rod structure and the initiation of gill arches may be inferred.

Lasanius was probably the most specialized and certainly the most active of the anaspids. Its external covering of scales had been reduced to a state in which they were no longer an adequate support for the muscles. It is therefore probable that the animal had some sort of internal stiffening rod. Some specimens show traces of what may be interpreted as the remains of a notochordal sheath, so that the transition from such a creature to one which could actually be called a fish is not a difficult one.

In fact, transitional forms may possibly be recognized in certain fossils, the oldest complete specimens of which are found in the Lower Devonian of Scotland. They are the acanthodians (Fig. 49). Many specimens have been collected and studied in considerable detail by paleontologists, who have found that the oldest members of the group are much like anaspids in the structure of plates about the mouth, the anterior position of the large eyes, and the posterior location of the narial opening. They differ, however, in that they have toothed jaws, which seem to correspond in position to the cartilages of the anterior visceral arch. The scales have the histological structure characteristic of the ganoids, and the tail is heterocercal. Some think that the Acanthodii are ganoids; others have supposed them to be sharks. Dr. D. M. S. Watson's recent studies confirm the supposition that they are the oldest gnathostomes (jaw-bearing animals), although there is still difference of opinion as to their relationship to other fish. The most striking difference from the anaspids is that they have a fishlike type of scale. The amorphous nature of the exoskeleton of the scaled anaspids suggests that it was an inheritance from an invertebrate ancestor and was merely the hardened outer layer of the skin. *Lasanius* appears to show that this old skeleton was in the process of being lost. It is possible that a new type of scale might be formed later from the inner layer of the skin after the ganoid fashion.

All the Acanthodii are characterized by a series of small paired fins, each supported by a single anterior spine, which is homologous with the external lateral fin spine of the anaspids (Fig. 52, at right). Behind the spine the membrane was unsupported in the early acanthodians, but somewhat stiffened by cartilages and rays in the later ones. As may be seen by inspection of the accompanying figure (Fig. 52, at left), the arrangement of the elements in the pectoral fin of *Acanthodes* suggests that of a crossopterygian (Fig. 54). In passing, it is interesting to note that these skeletal structures arose between the membranes of a fin supported at its anterior margin and hence within the area of least motion. The fact that the pelvic fins of nearly all the species are in front of the mid-length suggests the gradual posterior shift of fin spines arranged as those of *Lasanius* were. Of particular interest in this connection are the acanthodians of the genera *Euthacanthus* and *Climatius*, from the Lower Old Red Sandstone of Scotland. They have three or, in some cases, four pairs of spines, without internal supports, between the pectoral and pelvic fins. These genera, therefore, retain six of the eight pairs of external spines of *Lasanius*. The shoulder girdle consists of a dorsal rod with a ventral straplike piece on each side and carries the fin spine externally, exactly as in *Lasanius*.

Some of the acanthodians were definitely notochordal, for specimens show dorsal and haemal arches. Watson states that some of the later forms have rudimentary centra, but they are too poorly preserved to be understood. In fact, bony centra capable of preservation appeared in fish much later than in amphibians and reptiles, a natural result of their aquatic habitat. Few seem to realize that the trend in the evolution of fish was not toward a "higher" animal, a terrestrial creature, but toward a better and better adaptation to their own environment.

The acanthodians, therefore, show many characteristics which would indicate that they were derived from the anaspids. Most of them are too specialized to be

THE ORIGIN OF THE VERTEBRATES

placed in an ancestral position to the ganoids, but it may be that there was some animal with acanthodian characteristics which formed a connecting link. At any rate, we get from them and *Lasanius* a hint as to the origin of the internal visceral arches and the mode of reduction of longitudinal fins with numerous supports through a multiple finned stage to the two pairs which survive in later fish. Even the acanthodians had lost the internal supports of the spines between the pectoral and pelvic fins. Watson does not believe that early acanthodians were ancestral to early crossopterygian ganoids. The writer, being no specialist and therefore interested only in general conclusions, does. Yet Watson himself says, in discussing gill arches in a paper published late in 1937: "It is evident that ultimately a Teleostome-like arrangement of V-shaped gill arches, separated by extremely long gill slits and bearing long gill

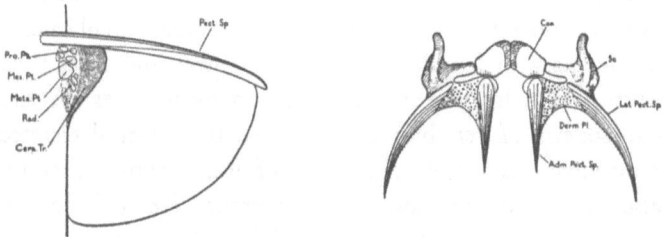

Fig. 52. At left, restoration of an acanthodian fin. At right, reconstruction of the pectoral girdle of an acanthodian. Cor., coracoid; Sc., scapula; Derm. Pl., dermal plate connecting pectoral spines. Both from D. M. S. Watson.

filaments projecting outward into a gill chamber covered by an operculum, is achieved." In other words, the acanthodians reached a primitive ganoidal condition.

We have now surveyed briefly the evidence for and against four of the suggestions which have been made to account for the origin of the vertebrates. We have seen that at least two of these theories point toward the echinoderms as the stem forms. It may be worth while, therefore, to summarize the characteristics which the echinoderms share with the chordates.

The radiate symmetry and the plated skeleton of the echinoderms give these animals an outward appearance so little like that of a typical vertebrate that it may seem absurd to suggest that the two groups had a common ancestry. Although many of the former are freely motile organisms, the evidence gained from the study of their anatomy, embryology, and history indicates that they owe most of their peculiarities, such as the radial symmetry, the water-vascular system, and the coil of the gut, to the fact that their ancestors were sedentary. All of the motile forms had "pelmatozoic" (sessile) forebears, and show this by retaining a fixed stage in their life histories. The earliest echinoderm was probably an animal with mouth at one end, anus at the other, a straight alimentary canal, and no skeleton. From it may have been derived two groups, one of which adopted an active, the other a sessile, mode of existence.

One progressive tendency inherited by both groups was that toward activity, for from the primitive sessile cystoids, crinoids, and blastoids free-moving descendants have sprung.

The great similarity of the larva of *Balanoglossus*, the simplest living chordate, to the larva of the echinoderms has already been mentioned. Another resemblance to the vertebrates lies in the fact that the skeletons in both groups are internal. This can be seen most plainly perhaps if a young crinoid be examined. The fact that the skeleton is internal is there so obvious and the larval plates of the column with their axial canal are so like vertebrae that the similarity is striking. There is also considerable resemblance between the method of formation of the plates of the echinoderms and the bones of the vertebrates. The skeleton of the echinoderm is deposited in the mesenchyme, where there are amoeboid cells whose pseudopodia fuse into a latticelike tissue. Where the pseudopodia meet, the protoplasm secretes small calcareous spicules which gradually increase in length; adjacent ones meet and fuse, thus building a reticulate structure which in time is augmented to form plates. A similar process goes on in the mesoderm of vertebrates, except that the material secreted, instead of being calcium carbonate, is largely phosphate of lime. Some other organs of the echinoderms also connote the vertebrates. Both groups are coelomate; that is, their internal organs are contained in a body cavity. The primary nervous system of both is derived from the ectoderm. It remains superficial in the crinoids and starfish, but in other echinoderms it sinks below the surface, as it does in vertebrates. Even the ligamental tissue which connects adjacent plates is an intercellular substance, secreted in the mesenchyme and like the cartilage of the vertebrates.

It is, of course, not wise to push these similarities too far as proofs of consanguinity between the two groups. Nevertheless, they seem to indicate a sufficiently close relationship to remove any feeling that it would be impossible to derive the vertebrates from the same ancestors as the echinoderms.

In the light of what is now known about the ostracoderms it is possible for the paleontologist to make a few comments on the theories outlined above. One point which impresses him is that perhaps the notochord is not so important a structure as has been supposed. The condition found in Silurian and early Devonian cephalaspids and anaspids suggests that the tripartite brain and the position of the central nervous system above the gut were achieved before the notochord was evolved. This rests, it is true, upon negative evidence, the lack of any trace of notochords in the early ostracoderms. Since the notochord is composed of soft tissue, one would not expect it to be preserved. On the other hand, there does not appear to be any mechanical necessity for an internal stiffening rod in animals with strong external skeletons to which muscles can be attached.

So far as is now known, the three groups into which the ostracoderms are divided show no connecting links; yet all three of them led to animals with notochords. Since it is not likely that a notochord would be evolved independently by each group, this indicates that all came from a single stock and that fundamentally their potentialities were the same. The stem of the three lines must be sought in ancient rocks, not younger than Mid-Ordovician, more likely Cambrian, and possibly even pre-Cambrian in age.

The anaspids are of interest in connection with the *Amphioxus* theory. The scaled anaspid is somewhat *Amphioxus*-like, particularly in its prominent expression of the segmentation of the muscles. In most respects, however, it is a more highly organized animal. Its pectoral and tail fins are more fishlike than those of *Amphioxus*, and the brains of the two are entirely different, *Amphioxus* having no vertebrate-like brain — in fact, practically no head or brain at all. It seems improbable that the anaspids should have descended from any animal with the combination of a poor brain and a highly developed notochord. More likely *Amphioxus* is a retrograde descendant of an anaspid-like creature. With our present knowledge of the fossils it seems unlikely that any scaleless animal like *Amphioxus* will be found in the ancestral line. Difficult as it is, it is easier to connect the anaspids with true fish than to trace a relationship of *Amphioxus* to them.

The fossils throw little or no light on the annelid theory. Annelid worms are known to have been fully evolved as early as Mid-Cambrian times, but no fossils have been found which indicate any connection between them and the ostracoderms. It may be that the postorbital valley of the cephalaspids and anaspids represents the position of a recently closed-off mouth, and it is true that both anaspids and annelids are richly segmented, but neither of these characteristics has much weight in showing relationships.

XI

THE RISE OF THE AIR-BREATHING VERTEBRATES

Anaximander says that men were first produced in fishes, and when they were grown up and able to help themselves were thrown up, and so lived upon the land.
Plutarch, *Symposiacs*, Book VIII, Question IX

More than half of recorded time had passed before animals emerged from the water and tried life on land. During millions of years, invertebrates and fish were the sole inhabitants of the globe. Generation succeeded generation, the newborn animals in each resembling their parents so closely that a contemporary observer, had there been one, would have noted no change, no improvement. Yet there were changes, even though they became apparent only after the lapse of thousands of years. We see the same process in our children. To us they seem the same from day to day, but Grandmother exclaims: "How Ruth has grown since last Thanksgiving!" If we stop to think of it, we see two reasons for Ruth's change. One is inherent, produced by development with increased age; the other is the effect of environment, partly physical and partly mental. Climate, food, and associations all have their effects.

Our attempts to read the records of which fossils are the symbols cannot be completely successful, for we cannot even guess the effects of the mental environment. The hopes and fears of the Devonian fish are not measurable. Yet they were not inanimate clods, not rocks or minerals, but living animals. "An oyster may be crossed in love." We know that they had brains, that they had desires and needs and fears and associations with other animals, and that these must have had their effects upon the creatures themselves. Since he cannot evaluate this factor, the paleontologist is apt to ascribe changes to the physical environment alone. We have an example of this in the theories which have been proposed to account for the emergence of the fish from the water, and we shall see it repeatedly in further discussions of terrestrial animals, for change — progress, if it may be so called — was much more rapid in the years which followed the Silurian than in those which preceded it. The fundamental cause of this change was the transition from life in the water to that on land. The exodus from the aquatic environment and acquisition of organs permitting life on land were epochal events in the history of the vertebrates. From the standpoint of the theories of evolution, we are interested to know whether this step was taken in response to an "inward propelling force," an urge to essay land life, or whether it was motivated by external circumstances.

THE RISE OF THE AIR-BREATHING VERTEBRATES

In the passage from life in water to life on land two systems of organs must have been particularly affected, namely, those of locomotion and respiration. Fish breathe by means of gills, paired vascular pouches on the sides of the head, in which the blood is separated by a very thin membrane from the oxygen dissolved in the water. Adult amphibians, on the other hand, oxygenate their blood by means of air taken into the lungs.

Fish swim partly by use of fins on the tail, partly by sinuous movements of the body, but in addition to the tail fin other similar outgrowths are present which serve to assist in balancing, guiding, and, to a lesser degree, in paddling. Of especial importance in the present discussion are the paired fins, for it is believed that the limbs of the terrestrial vertebrates were derived from them. There are several kinds of modern fish in which the paired fins are known to be used more or less as legs. Thus some are known to creep about on the bottom in shallow water, and even to jump above the surface by the use of the fingerlike rays of the pectoral fins, and the "climbing perch" of Ceylon holds its place on the roots of trees by use of the same organs. No one of these animals, however, shows any structures which even suggest the typical limb of the land vertebrate.

A comparison of the fin of any modern fish with the internal supports of the leg of a terrestrial animal will show that it is impossible to derive one from the other. The upper arm of a tetrapod (four-legged animal) has a single bone, the humerus; the lower arm contains two, the radius and ulna; the wrist has two or three rows of short bones or cartilages, the carpals; the palm a row of long ones known as the metacarpals; and in the fingers are several phalanges. The leg shows the same arrangement, although the bones are given different names. The upper is the femur; the lower ones are the tibia and fibula; the ankle contains the tarsals, beyond which are the elongate metatarsals and the phalanges of the toes. Since even the most ancient four-legged creatures, the amphibians, have this typical arrangement of the bones, they are of no help in furnishing links with the aquatic ancestor. It is necessary, therefore, to approach the subject from the other side, that of the fish.

In Devonian times there were two kinds of fish whose fins had a segmented internal axial support: the "lobe fins" or crossopterygian ganoids, and the Dipnoi or lungfish. With the possible exception of the pleuracanthid sharks, these are the only fish ever in existence which had a fin that by any modification could be transformed into a leg of the tetrapod. At one time it was thought that the lungfish might have been the ancestors of the amphibians, but they have been ruled out by their specialized, fan-shaped dental plates. There is no possible way in which such teeth could have been transformed into the sharp conical ones of the amphibians. Even the oldest known Devonian dipnoans had already lost the teeth of the marginal bones of the jaws. Once lost, they could not be regained.

This leaves only the lobe fins as a possibility, for the sharks mentioned in the preceding paragraph appeared on the scene too late to be considered. For the past twenty years attention has therefore been focused on the crossopterygians. Two genera are especially well represented by specimens, the *Osteolepis* of the Mid-Devonian of Scotland and *Eusthenopteron* (Fig. 53), a common form in the Upper Devonian sandstones at Scaumenac in Quebec. A related fish from the Upper Devonian at Blossburgh, Pennsylvania, named *Sauripterus*, has also proved important in this con-

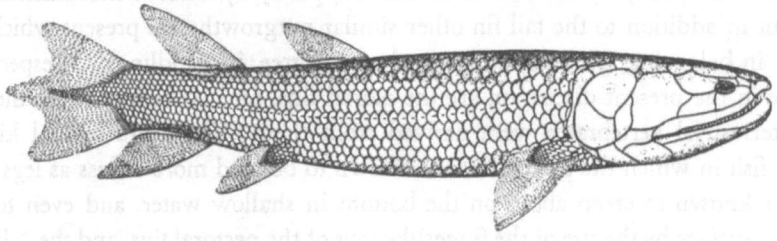

FIG. 53. The fringed-finned ganoid, *Eusthenopteron*. From W. L. Bryant.

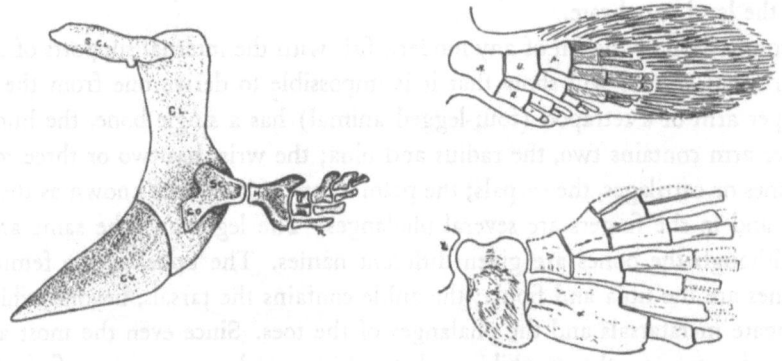

FIG. 54. At left, the fundamental elements in the arm and shoulder girdle of *Eusthenopteron*. From W. L. Bryant. At right, above, Broom's diagram of the pre-*Sauripterus* fin. Below, drawing of the fundamental cartilages of the arm of *Sauripterus*, also by Broom. S. Cl., superacleithrum; Cl., cleithrum; Cv., clavicle; Co., Sc., scapula with coracoid; H., humerus; U., ulna; R., radius.

nection, for the type-specimen presents an unusually well-preserved pectoral fin and shoulder girdle. The fins of these three fish are similar in that each has at the proximal end a single cartilage corresponding to the humerus of the arm. Articulating with its distal end are two cartilages having the position of the radius and ulna. Beyond them are several more, two or three rows, which may be compared with the wrist bones (carpals) and palmars (metacarpals) of terrestrial animals. These in turn support the rays of the fin. Although the cartilages are not much like the bones of the amphibian arm, their arrangement is fundamentally the same. Furthermore, the upper one, the humerus, is articulated at its proximal end with another which corresponds to the shoulder blade or scapula, an important element of the pectoral girdle of all tetrapods.

So much for the possibilities in the way of limbs. How did the air-breathing habit come about? Here paleontology can help less, but the rocks are not entirely silent.

Everyone has probably seen goldfish come to the surface to swallow air. Since they have no means of aerating their blood effectively by direct contact with the air, this is not a method of breathing. Fortunately, however, lungfish are still living in Africa, South America, and Australia, and from their habits it is possible to infer a reason for the origin of a lunglike organ. All fish other than sharks and skates, flounders, and some teleosts which inhabit swiftly running streams have a saclike outgrowth of the alimentary canal which is known as an air or swim bladder. Its function in most fish seems to be to counteract the effects of the various pressures under which the animals live at varying depths. In the adults of most modern teleosts it has no connection with the oesophagus, but in the ganoids and lungfish it opens directly into the gut. Since there are only three species of modern lungfish, living in widely separated regions, they may be briefly described.

Neoceratodus, the Australian lungfish, is a sluggish inhabitant of streams. Besides breathing by gills, it uses its air bladder as a lung, for at regular intervals it comes to the surface, expels foul air, and takes in a fresh supply. During the dry season the rivers in which it lives cease to flow, and the fish are confined to pools which become foul and deficient in oxygen. At such times *Neoceratodus* finds the lung of great use, for it enables the fish to survive till the next wet season. A very closely allied lungfish, placed by some in the same genus as the modern form, has been found in the Triassic of England, Germany, India, and South Africa, and in the Jurassic and Cretaceous of England, Colorado, and Wyoming.

The air bladders of the other modern lungfish, *Protopterus* (African) and *Lepidosiren* (South American), are still more lung-like, since they are double organs. *Protopterus*, an inhabitant of marshes in the vicinity of rivers, is large and eellike, some specimens reaching six feet in length. Instead of trying to follow the receding water into the rivers during the dry season, it burrows into the mud, where it forms a hardened capsule about itself by means of mucus secreted by the glands of the skin. This capsule has an aperture, the margins of which are pulled inward to form a tube which ends within the mouth. The fish lives in this condition for nearly half the year, breathing through the tube by means of the lungs and living upon the fat which it stored up during the feeding season. When thus dormant the animals may be dug out and shipped safely to aquaria in any part of the world, to be resuscitated by soaking in tepid water. *Lepidosiren* has much the same habits.

The habits of these fish and their use of the air bladder for a lung strongly suggest the method of origin of the air-breathing habit among the Amphibia. As has already been pointed out, there is a *Neoceratodus*-like fish in the Triassic rocks. When one traces the group further back, it appears that similar ones were present in the Devonian, and that the Devonian lungfish were so closely allied to the lobe-finned ganoids that it is probable that they were descended from the same ancestor. Curiously, it is known that at least one Mesozoic representative of the crossopterygians had a swim bladder. Specimens of the Jurassic *Undina* have been found so preserved that the outline of this organ can be seen. By some freak of nature the bladder happened to have its wall partially calcified, to the detriment of the individuals cursed with this pathological condition but much to the satisfaction of the paleontologist.

No calcified air bladders have been found among the remains of the Devonian representatives of the lobe fins, but Professor D. M. S. Watson has recently discovered evidence that *Osteolepis*, a Mid-Devonian representative of the group, could breathe air. He found that the anterior region of its skull is occupied by a mass of cancellous bone in which are a pair of small, nearly spherical cavities which must have been occupied by the olfactory organs. Passages extend forward from them to external narial openings, and, what is more important, posterior canals pass downward through the roof of the mouth, showing that the animal could swallow air just as an amphibian does. Evidently it must have depended in part at least upon a lung for the oxygenation of its blood. We are justified, therefore, in postulating the existence in early or Mid-Devonian times of a ganoid with fins somewhat like those of *Sauripterus* and lungs like those of the dipnoans. This was material from which the Amphibia might evolve. What gave the impetus which set the machinery in motion?

For this it is necessary to look at the physical side of the geological record. Since the oldest so-called fish are indeterminate fragments of ostracoderms, the Ordovician record is useless, and the Silurian not much better, for remains of fringed-finned ganoids of that date have not yet been discovered. In Middle Devonian strata sharks are found more commonly in marine than in fresh-water associations, but most of the ganoids and lungfish are in fresh-water deposits. It is probable that the hypothetical ancestral ganoid lived in rivers and lakes as yet undiscovered.

It will be noted that modern sharks lack the swim bladder, and show no evidence of ever having had one. Is it not possible that this structure arose only in such Lower Devonian fish as occupied fresh waters, whereas the sharks are descendants of marine animals which did not undergo the particular stresses which led to the formation of this organ?

The Devonian was a time of uplift and mountain-making in northern Europe and northeastern America. As the mountains rose, they may have cut off precipitation on the landward side, so that there ensued a time during which the rainfall on their northwestern slopes became deficient and seasonal. This would result in alternations of wet and dry seasons, with consequent raising and lowering of the water level in lakes, marshes, and rivers, and the production of conditions similar to those under which lungfish are living at the present time. Professor Joseph Barrell, who suggested this theory, remarks: "The seasonal dryness, with the shrinkage and fouling of the fresh waters, or even their complete local disappearance, was a feature of the Devonian which appears to have increased in intensity to a maximum in the epoch of the Upper Old Red Sandstone" (Upper Devonian). Under such conditions it is obvious that then, as now, the forms capable of making most use of air would stand the best chance of surviving. This was probably the time for the development of lungs, although there is, of course, no inkling of the *modus operandi*. The long continuation of the conditions which operated to start the air-breathing habit, acting upon ganoids equipped with lungs and jointed fins, may have caused them to quit the water entirely, or, more likely, may have caused the water to quit them for months at a time. Their only chance of survival was as air-breathing, terrestrial creatures.

This is, of course, only a theory, based on a series of coincidences, but it seems to have the support of a considerable number of facts. Lobe-finned ganoids, capable of swallowing air, existed during the Middle Devonian. Amphibia first appeared in the Upper Devonian. The ganoids and lungfish inhabited fresh water, and the red color of the rocks in which their remains are found is considered by many as evidence of an increasingly arid climate. Add to this the fact that the early amphibia show many ganoid-like characteristics in skull and other parts of the skeleton, and Barrell's theory appears at least plausible. It has, indeed, been widely accepted by paleontologists and geologists. But in recent years there has been an ever-increasing suspicion that the Devonian was not a time of arid or even semiarid climates. Red sandstones and shales must have been formed from materials derived from lands on which the soils were red. At the present day red soils are not produced in regions with a semiarid climate, but in moist tropical or subtropical countries where the accumulation of humus is prevented by constant bacterial activity. The writer has therefore suggested an alternative hypothesis based upon the idea that the red beds of the Devonian were accumulated rapidly during a time of warm and moist climate.

At present much of the rainfall of the northeastern part of North America is precipitated from masses of air which derive their water vapor from the Gulf of Mexico. During the greater part of the Devonian the interior of the continent was covered by a shallow sea, so that air coming from the southwest should have carried even more moisture than now. It may therefore be inferred that the rainfall on the western side of the Devonian mountains would have been greater, rather than less, than at the present time, a supposition borne out by the fact that the oldest known forests are buried in Mid-Devonian strata on the western side of the Catskills.

If the red beds of the Devonian were really accumulated under moist and warm conditions, we must change the theory somewhat, although the explanation remains the same. The important factor in Barrell's scheme was the evanescent nature of the aquatic habitat, and such a *situs* is supplied by any alluvial fan, under any climatic condition. There streams are constantly changing their courses, pools and swamps are formed in which the waters are renewed by periodic overflow from the main channels, and in these any trapped animals are subjected to the same process of gradual exhaustion of oxygen as under semiarid conditions.

There are certain advantages to be gained from the explanation according to the conditions now postulated. Where extreme fouling or actual drying takes place, as in the homes of the lungfish of the southern hemisphere, the animals hibernate for a part of the year; sloth rather than progress is encouraged. Fish inhabiting the constantly changing waters of a fan would doubtless often be routed from their homes by sudden changes in channel, to be left flopping about in shallow water as the streams spread over a flat. Those best able to travel by means of fins might win their way to deeper pools. A premium was placed on activity; the weak and unlucky were weeded out. The instinct for overland migrations to other and fresher waters may early have been developed under such precarious conditions of life. Food, too, would have been more abundant and the supply more constant under moist and warm than under semiarid conditions. Further, with a greater supply of vegetable matter, fouling by decay could occur more readily. Last and most important, amphibians have been connected with water throughout their entire history. They constantly return to it to breed, except for a few specialized groups of the race, and no theory of their origin is satisfactory which requires their complete exclusion from it.

Since it has become obvious that the crossopterygians and not the lungfish were the ancestors of the tetrapods, writers have repeatedly pointed out that the pectoral fins of the former contain the fundamental elements of the limb of terrestrial animals. Although the shoulder blade and arm bones can be identified, they are associated with numerous other cartilaginous or horny fin rays. There are so many of these extra elements that the method of their elimination has presented an obstacle to the acceptance of any known lobe fish as an ancestor. Barrell's suggestion that the fins were worn down to stumps as the animals became crawlers instead of swimmers has been accepted by many, but has seemed fantastic to others. The Lamarckian doctrine that use promotes growth is contravened by this explanation.

Professor Robert Broom has lately brought forward an idea which appears to be a logical solution of the difficulty. He believes that, as the fish came more and more into contact with the bottom, the anteroventral part of the fin came to be used in crawling, whereas the distal portion retained its normal function (Fig. 54, upper right). *Sauripterus*, according to him, had become specialized for life on the bottom, using the ventral part of the pectoral fin for scooping out troughs for itself in the sand or mud. Such use is indicated by the flattening and the anterior expansion of the humerus and radius. This removes *Sauripterus* from the ancestral line but serves to emphasize the bottom-living habits of the group to which it belongs. *Eusthenopteron* shows less of this specialization than *Sauripterus*, having a fin only slightly different from the hypothetical one shown in Broom's drawing. An animal with such a fin, forced to adopt a crawling method of existence, would in the course of time lose the distal (swimming) portion because of disuse. At the same time, the proximal crawling part would continue to increase in size and strength. This theory accounts for the loss of the superfluous elements in the same way that the absence of lateral toes in many mammals is explained.

Barrell surveyed the field for other possible causes of the emergence of the vertebrates from the water and suggested three. These are: first, enemies in the water; second, food on the lands; and third, the lure of atmospheric oxygen. Enemies of the crossopterygians were practically nonexistent. They were the largest, strongest, most formidable carnivores of their day. It is true that lack of food or overpopulation of the waters may have led them to prey upon each other, but in such a case only the weaker members of the community, not the most fit, would have been driven from the water. It is not probable that such sudden catastrophic action would lead to evolutionary changes. It is more reasonable to suppose that the modifications which prepared the fish for life on land occupied a long period. Food on the land could hardly have attracted fish unless they were aware of its existence, and it is not possible to credit them with reasoning power sufficient to make them conscious of anything outside their immediate environment. If the amphibians and their forebears had been vegetarians, it is conceivable that feeding on such terrestrial plants as hung in the water would have led them to try to climb the banks, but carnivores were exposed to no such temptations. Furthermore, so far as is known, there was no food for them on the land at that time. Even after they became semiterrestrial they must have returned to the water to feed. Finally, the lure of atmospheric oxygen cannot be classed as an active, driving force. Oxygen was plentiful in the air even in pre-Cambrian times, but not till near the close of the Silurian did any animals or plants come sufficiently into contact with it to develop any organs through which it could be utilized. It must have been necessity, not desire, which effected the change.

Since none of these causes, escape from enemies, desire for the food of the land, or lure of atmospheric oxygen, seems to have been sufficiently powerful to lead to terrestrial life, one is forced to return to the theory that it was brought about by changes in the environment. In this connection it should be pointed out that scorpions, insects, diplopods, gastropods, and plants, all of aquatic ancestry, adopted the air-breathing habit at about the same time that the fish did. The environmental changes that affected one of the groups may have influenced all, driving them to the air-breathing habit, though the result was accomplished variously.

Whatever we think of the theories outlined above, it is evident that amphibians are more closely related to crossopterygians than to any other fish that have ever existed. Although no lobe fin yet found is exactly the sort of animal which could be considered the direct ancestor of the tetrapods, it is probable that some early short-finned race not very different from *Osteolepis* or *Eusthenopteron* furnished the progenitor. This is indicated by the fundamental characteristics common to lobe fins and amphibia, unshared by the other fish.

First, and perhaps most important, the Amphibia, and through them all other tetrapods, including man, appear to have inherited directly from the crossopterygian the fundamental arrangement of the skull bones and the pineal eye. If one examines the skull of an amphibian, a reptile, or a mammal, he sees that it is made up of many bones united more or less firmly along lines which are known as sutures. The bones have an orderly arrangement, basically the same in the three groups mentioned, although the skull of a primitive amphibian has many more parts than that of a mammal. The bones are symmetrically placed on opposite sides of the median line

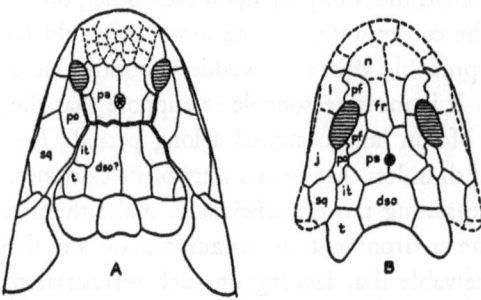

FIG. 55. Diagram to show the similarity of the bones of the top of the skull of (A) a crossopterygian, and (B) *Ichthyostegopsis*, one of the oldest known amphibians: *n*, nasal; *fr*, frontal; *pa*, parietal; *dso*, dermosupraoccipital; first *pf*, prefrontal; second *pf*, postfrontal; *po*, post-orbital; *it*, intertemporal; *t*, tabulare; *l*, lacrimal; *j*, jugal; *sq*, squamosal. Modified after T. S. Westoll.

and are easily subdivided into three regions: the dorsomedian, the marginal, and the lateral or cheek areas, the last including all between the dorsomedian and the marginal series. On the median area (Fig. 55; see also Fig. 64) there are three or four pairs of important elements, the nasals, extending backward from the nostrils, then the frontals, and back of them, the parietals. Behind the latter are the dermosupraoccipitals, called by some the postparietals. In the marginal series there are premaxillaries in front of the nasals; behind each is a long maxillary; these two pairs bear teeth. Behind each maxillary is a quadratojugal, and beneath the hinder corner in amphibia and reptiles, but not in mammals, the quadrate, the element with which the lower jaw articulates. The number of bones in the cheek series is highly variable in both amphibia and reptiles, so only a few of them need be mentioned. Most important are the jugal, below the eye, and the squamosal, above the quadrate and the quadratojugal. There may be others between the squamosal and the parietal. In front of each eye is a lacrimal and, in most cases, a prefrontal; behind it, a postorbital and a postfrontal.

THE RISE OF THE AIR-BREATHING VERTEBRATES

The similarities between amphibia and crossopterygians are marked. Though different skull bones appear in different sorts of crossopterygians the fundamental arrangement is that just outlined. Since they are the only fish which have it, it indicates that they are directly related to the amphibians. The palatal part of the skull, likewise, has the same arrangement of bones in both, as is shown by the accompanying diagrams. The lower jaws are morphologically identical in crossopterygians and early amphibia, but both possess so many elements that it is not profitable to go into a detailed account of their structure.

The distribution of the teeth is alike in lobe fins and amphibia. It appears further

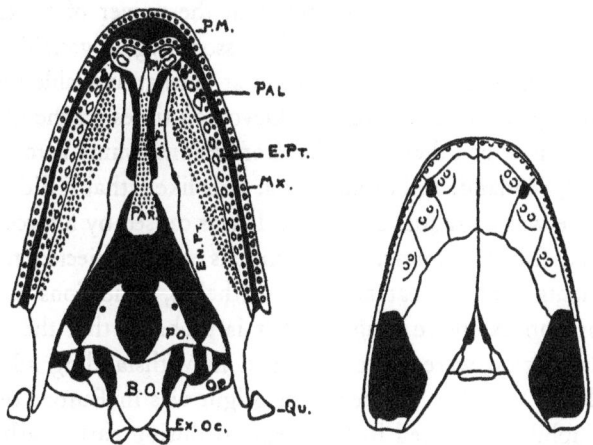

FIG. 56. At left, palatal view of *Eusthenopteron*, for comparison with that of a Mid-Carboniferous amphibian at right. P. M., premaxilla; Pal., palatine; E. Pt., ectopterygoid; Mx., maxilla; Qu., quadrate. Figure at left, from W. L. Bryant; at right, after D. M. S. Watson.

that their emplacement is the same; they are of what is called the acrodont type, fixed on the bones without sockets. In replacement new teeth are formed independently alongside the old and gradually oust the latter as their bases are resorbed. They are simple, conical, and structurally alike, for in both enamel is infolded into the dentine, producing what is known as the labyrinthodont type of tooth.

Another interesting feature, which has only recently been made known by Watson's work, is that one of the oldest amphibians whose skeleton is known, the Mid-Carboniferous *Eogyrinus*, has a fishlike shoulder girdle attached to the head, like that of the crossopterygian. This pectoral girdle consists of the primary support, the scapula, and four dermal elements. The latter are external in *Osteolepis* and have the histological structure of scales. In *Eogyrinus* they appear to have been drawn within the skin, but the arrangement is the same.

Finally, in the most ancient amphibians the dermal bones are not attached to the brain case; they are superficial in location, and the body is covered with bony ganoid scales, instead of being naked as in modern forms. These many similarities are obviously not to be ignored.

There was really remarkably little change in the anterior part of the body in the transformation of a fish into a terrestrial animal. The ornamented surfaces show that the cranial bones of the amphibians were practically external, just as in the fish. The scales beneath the jaw (gulars) were lost, as were also the large plates protecting the neck. The loss of the latter left a notch in the back of the skull over each ear, the so-called otic notch. This was probably covered by a tympanic membrane, the new arrangement permitting a great improvement in the power of hearing.

All the evidence points to the lobe-finned, crossopterygian ganoids as the ancestors of the amphibians. The geological information at present available indicates that the transformation took place in Lower or Mid-Devonian times on the alluvial fans west or north of mountains which then extended from Pennsylvania across New England, Quebec, Greenland, Scotland, and Scandinavia. It is likely that it did not happen suddenly but that there was a gradual evolution. It was caused by a force that influenced not one individual fish, but many. Like animals were subjected to like forces in similar environments over a great area. Nevertheless, conditions are never exactly identical over such an extensive range, and it is probable that the crossopterygians of some particular region were more successful in withstanding adverse conditions than those in others. Whether this center of origin was in Europe or America is as yet unknown; amphibians existed in both regions during early Carboniferous times and even earlier in an intermediate area, now Greenland.

The lobe fins were not the only chordates which experienced this period of stress during the elevation of the Devonian mountains. The cephalaspids and Antiarcha throve at first in the fresh waters but became extinct at the end of the Devonian. The chondrostean (cartilaginous-skeletoned) ganoids arose during this time, learned the trick of tiding over particularly adverse conditions by means of the air bladder, but failed to gain the land because their fins were not of such a type as to permit them to do so. They remained rare and unimportant animals until after the Devonian, but were ancestors of the bony ganoids of the Triassic and Jurassic. The latter in turn gave rise during the Jurassic to the true teleosts. The sharks, as a group, stayed in the sea. Those which essayed the evanescent streams perished; for some reason they could not develop air bladders. Some early arthrodires seemed well adapted to the fresh-water habitat, but most of them remained in the sea, where they evolved into the mighty *Dinichthys* and *Titanichthys*. The little sharklike ganoids, the acanthodians, survived till the end of the Paleozoic, but their simple fins were not such as could be converted into legs useful in terrestrial life. The lungfish, too, have survived, even to the present day, but theirs was evidently a passive resistance; theirs

THE RISE OF THE AIR–BREATHING VERTEBRATES

was not to do or die, but entrench themselves till better times came. Many a modern family has survived, with little honor, in the same way.

As the paleontologist sees it, it was fortunate that during times of adverse conditions there was a family of fish, not nice, gentlemanly creatures, but fierce, aggressive, self-assertive carnivores, which had the strength and vitality to make a struggle for existence. Forced by the conditions under which they lived to swallow air, they eventually developed an outgrowth from the alimentary canal which acted as a lung. Periodically deprived of water, they used every effort to return to it. The weaklings of each generation died. The strong survived, ever gaining strength, increasing their lung capacity and perfecting their limbs. He that hath can get more. Eventually they conquered their environment. Tetrapods came into being.

XII

AMPHIBIA, THE FIRST TETRAPODS

At table I had a very good discourse with Mr. Ashmole, wherein he did assure me that frogs and many insects do often fall from the sky, ready formed.

Pepys, *Diary*, May 23, 1661

Amphibians were born at a time when life was hard. Cast up from the waters, they still clung to them, returning always to their ancestral home to breed. The widespread marshes and swamps of the Carboniferous afforded them an ideal habitat, and they increased rapidly in numbers, variety, and size, becoming for a brief period the most important inhabitants of the earth. In the Permian they were still more abundant, but the extraordinary rise of reptiles had already reduced them to a secondary position. The Triassic witnessed a decline in variety, although it was the period during which they attained maximum size. From the Triassic onward they have been relatively lowly and unimportant creatures, showing more or less recession from the high position attained by their ancestors.

Frogs and toads are the most familiar of modern amphibians; it is their life cycle which has given the name to the class, for their existence is a double life, although not after the manner of Dr. Jekyll and Mr. Hyde. As everyone knows, the eggs are laid, and the young born, in water. The larva has a rounded head and a fishlike body, ending in a long tail. This tadpole possesses external gills, covered later in life by flaps or opercula, as internal gills are developed in the three clefts which open between the five branchial arches on either side. At the age of about two months the animal reaches a lungfish stage through the formation of a pair of lungs, so that for a time it can breathe either in air or in water. Simultaneously with the lungs, the hind legs appear, then the fore ones. The gills and tail are gradually resorbed, the gill clefts close, and the animal reaches its adult form. Certain other modern amphibians, such as the mud puppies and hellbenders, fail to complete the metamorphosis, and retain the gilled and tailed condition throughout life. They are not considered primitive, but have apparently reverted to ancestral habits and habitat by a simple process of arrested development. The frogs and toads represent the order Anura, so called because of the lack of a tail. The aquatic forms just mentioned and the terrestrial salamanders are known as the tailed amphibia, technically the Caudata or Urodela. Still another type of amphibian now living is represented by the caecilians, small, limbless, snakelike inhabitants of tropical countries.

With the exception of the last group, modern amphibians possess scaleless, rather slimy bodies and relatively large heads. The skeleton exhibits distinctive features. When viewed from above, the skull shows large openings, the result of the failure of certain elements to ossify. The short neck has only a single vertebra, which articulates with paired processes (occipital condyles) on the back of the skull. The ribs are short, failing to reach the breastbone, and the pelvis is connected with only a single vertebra of the spinal column. Because of the short neck, the shoulder girdle is close to the skull. The supports of the fore limbs are relatively complicated, with more elements than are found in the same girdle of the more familiar mammals. The pelvic girdle, on the other hand, is simple, but unusual in that the anterior (pubic) processes of many forms remain cartilaginous. Only two or three exceptions are known to the rule that amphibians of all ages have four toes on the front feet and five on the hind, provided, of course, that feet are present at all. Aquatic forms have cartilages in place of bones in the wrists and ankles, whereas in the fully terrestrial species these elements are ossified. Since cartilage is not readily preserved, many fossils have a gap between the bones of the fingers and toes and those of the lower arm or leg.

The oldest American amphibians are represented by tracks found many years ago in the Mississippian (Mauch Chunk) red sandstone near Pottsville, Pennsylvania. A specimen collected near Warren, Pennsylvania, from the Upper Devonian, has been described as an amphibian footprint, but it is a single imperfect impression of doubtful value. The tracks found near Pottsville show clearly that a five-toed animal passed that way while the sand was moist and prepared to take a good impression. Many similar tracks occur in strata of late Paleozoic and Triassic age in North America, England, and Germany. That they were made by amphibia is shown by the impressions of five toes on the hind tracks and four on the front, a broad, hand-like outline, and the absence of angular terminals such as are made by claws.

One of the most important of the many valuable discoveries made by Lauge Koch during his recent brilliantly successful expeditions to Greenland was that of Devonian amphibians. The material consists of incomplete skulls and other parts of skeletons, from which animals named *Ichthyostega* and *Ichthyostegopsis* (Fig. 55 B) have been reconstructed. The names are supposed to indicate that the animals were fishlike stegocephalians, that is, intermediate between the fish and the ancient amphibians. Unfortunately, they are typical four-footed creatures only slightly more fishlike than some of the later members of their group. Primitive conditions may be seen in the presence of a few bones not present in later amphibians, among them an unpaired rostral element at the front of the skull. The curious position of the external narial openings, submarginal, and almost ventral, is superficially like that of the crossopterygian fish, but the opening is bounded by different bones. When more fully known, these little creatures may prove to have characteristics which justify their names.

Though modern amphibians, except the caecilians, have a naked skin, the late Paleozoic members of the class show more or less protective armament. Because their heads were covered with bony plates, they are called Stegocephalia, or roof-headed. Strangely enough, the armor was thicker on the lower than on the upper surface of the body, for some which had naked skin above were covered with scales beneath. They varied in length from two inches to eight or nine feet. Most of them had two pairs of limbs, but a few were long and snakelike, the legs poorly developed or en-

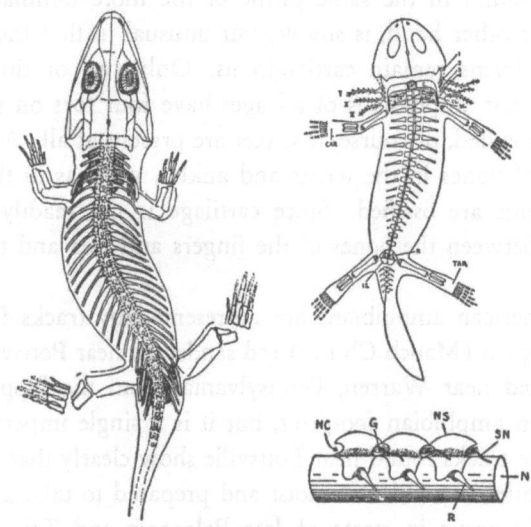

FIG. 57. At left, restoration of the primitive embolomerous amphibian, *Diplovertebron*. Two-thirds natural size. From D. M. S. Watson. At right, above, a restoration of *Branchiosaurus*, with (I, II, III) external gills. CAR., carpus, and TAR, tarsus, are blank spaces because of failure of the original cartilages to be preserved. A little larger than natural size. From Bulman and Whittard. At right, below, restoration of three rudimentary branchiosaurian vertebrae above the notochord. N, notochord, NC, neural canal, NS, neural spine, R, rib. From W. F. Whittard.

tirely absent. Although the teeth of these early Amphibia were simple cones, many have flutings at the base that indicate that the enamel is infolded, as in the lobe-finned ganoids. This type of tooth, because of the convoluted structure shown in transverse sections, is known as labyrinthodont. Such stegocephalians as possess it are united in a large group called the Labyrinthodonta.

Practically all the labyrinthodonts were large. The head was long, wide, and flat, the orbits for the eyes and the pineal opening large. The body was broad, the limbs short, with the humerus and femur approximately at right angles to the body. With one exception, which has five on both, there were four fingers on the front feet and five toes on the hind. These animals appear to have been rather numerous in the

British Isles during the Upper Carboniferous; a few representatives have been found in the American Pennsylvanian. Their remains are relatively common in the Permian of Texas and New Mexico, but their greatest development occurred during the Triassic, when they spread to Germany, Russia, India, South Africa, and Australia. Recently many of their bones have been found in Triassic strata in New Mexico.

The oldest labyrinthodonts, those found in the Lower Carboniferous of Scotland and northern England, have been described by Watson. They have a small head, a single occipital condyle, four or five cervical vertebrae, a long cylindrical trunk, double-ringed (technically embolomerous) vertebrae (Fig. 59 A–C), double-headed ribs, and probably, a long compressed tail. The dorsal surface of the head is covered with dermal bones, loosely articulated with the case about the brain. The palate is

Fig. 58. *Diplovertebron*, one of the few five-fingered amphibians. Restoration, about one-eighth natural size, by L. I. Price. A different species from that shown in Figure 57.

entirely bony. The shoulder girdle is fishlike, being attached to the skull. Several animals of this general type (Embolomeri) are now known, but unfortunately from far too few specimens and from too restricted an area to allow as full a knowledge of them as could be desired. The structure of their limbs appears to indicate that they were terrestrial during their adult lives.

They seem to have given rise to another group which existed from Mid-Carboniferous through Permian times, a few persisting into the Triassic. These differ from the Embolomeri in having rachitomous vertebrae (Fig. 59 D), each consisting of four pieces. The prongs of the neural spines saddled the upper side of the notochord; a half ring below (hypocentrum or intercentrum) bore the ventral chevrons; and the remaining parts consisted of two lateral pieces, the pleurocentra. A good example of this type (group Rachitomi or Temnospondyli) is the well-known *Eryops* (Fig. 64 A), several specimens of which have been obtained from the Permian of Texas. It is the largest of American Paleozoic amphibians, reaching a length of five feet, or if, as is possible, it had a long tail, even more. It has a large head, a heavy body, and short massive limbs. The bones of the pelvis are ossified, as in a terrestrial animal, but this may be an inheritance from more active ancestors. The position of the eyes on the top of the head indicates that it may have awaited its prey partly submerged in the water like a modern alligator or crocodile. *Cacops* is a similar animal.

Closely allied to the amphibia with the four-piece vertebrae is a group with a backbone more like that of modern fish. Labyrinthodonts (group Stereospondyli) of this sort are characteristic of the Triassic. They were the largest of all, some having

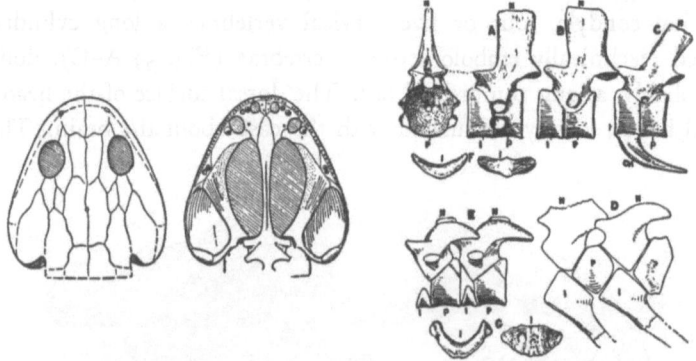

FIG. 59. At left, dorsal and ventral views of the skull of *Pelion*, a broad-headed branchiosaur, rather common in the American Carboniferous. From A. S. Romer. At right, A, B, typical embolomerous vertebrae with neural spine saddled on two centra (*i*, intercentrum, *p*, pleurocentrum). C, the same type with haemal arch attached to anterior centrum. D, rhachitomous caudal vertebrae, with haemal arches attached to the intercentra (*n*, neural arch). E, a primitive reptile, with small intercentra. F, lateral and dorsal views of an intercentrum of a primitive synapsid reptile. G, the same element in a rhachitomous amphibian. From S. W. Williston, *The Osteology of the Reptiles*.

heads as much as four feet long. Since the lower jaws articulated with backward-projecting quadrates, the gape was tremendous. Like modern crocodiles, they may have opened the mouth by lifting the upper part of the head, while the jaw rested on the mud. They were carnivores, with numerous sharp, conical teeth, two at the front of the lower jaws being so long that in some species they passed through holes in front of the nostrils of the upper jaws, actually protruding from the top of the skull when the mouth was closed. A veritable crocodile among the Amphibia! All of the late Triassic members of this group appear to have been aquatic or semiaquatic, for they have short feeble limbs.

The large labyrinthodonts were only one branch of the roof-headed amphibians. Other branches, among them the branchiosaurs (Phyllospondyli), microsaurs, and Adelospondyli, were represented by small creatures, seemingly inconspicuous and feeble, yet in the sequel more important than the labyrinthodonts, for they appear to have furnished the ancestors of the modern amphibians.

The branchiosaurs are small, primitive stegocephalians which appeared as early as the Devonian. The osseous part of the backbone is simple, composed of what are known technically as the phyllospondylous type of vertebrae. The notochord is continuous and not enclosed in bone, only the neural arches being ossified (Fig. 57, right, below). Such vertebrae are even more primitive than the double-ringed ones of the embolomerous type. The teeth are conical; some species show simple infoldings of the enamel, suggesting the convolutions of the labyrinthodonts. All members of this group have large, broad, flat heads, short bodies and tails, and feeble limbs, which indicate that they were chiefly aquatic (Fig. 57, right, above). The ventral surface of the body was covered with thin hemicycloid scales; the upper side was naked. The large short head, straight ribs, short tail, large openings in the palate, cartilaginous wrists, ankles, and pectoral girdle parallel the structure of modern

FIG. 60. A restoration of *Cacops*, one of the large Permian Rhachitomi of Texas. More terrestrial than most of its relatives, it had the rudiments of a dorsal armor, and rather efficient legs. Original drawing by L. I. Price.

frogs. Specimens of branchiosaurs have been found at Linton, Ohio (Fig. 59, left), and Mazon Creek, Illinois, in Mid-Carboniferous strata, but they are best known from the abundant material collected from the Permian near Dresden, Germany. More than a thousand specimens of *Branchiosaurus* from the latter locality were studied by Professor H. Credner. He found that many immature individuals showed external gills (Fig. 57, right, above; Fig. 61), supported by projections from the gill arches, which seem to have been at least partially calcified (Fig. 62). When the larvae reached a length of about 100 mm. (four inches), they lost these structures. This change was accompanied by a reduction in the length of the tail, which in the young had been as long as the body, an expansion of the pelvis, and an increase in the ossification of the skull and other parts of the skeleton. One sees in this a beginning of the metamorphosis which plays so conspicuous a part in the life history of a modern frog.

Somewhat like the branchiosaurs, but probably not derived from them, are the microsaurs (Lepospondyli), small animals with "lepospondylous" or hourglass-shaped vertebrae. The notochord was continuous, but constricted by the partial ossification of the centra of the vertebrae. In general appearance they were much like the branchiosaurs, but the body and tail were longer. The hind limbs were longer than the fore; the pubes were ossified, an unusual feature in amphibians other than the primitive labyrinthodonts; and in many the wrists and ankles had true bones. The

Fig. 61. Restoration of a larval branchiosaur, with external gills and a long tail. Two-thirds natural size. Original drawing by L. I. Price.

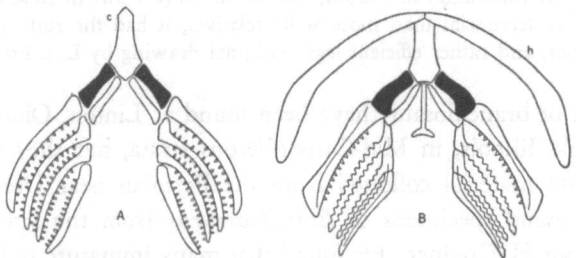

Fig. 62. A, gill-supports of a larval branchiosaur, with those of a modern axolotl, B, for comparison. The elements in black are ossified, the others cartilaginous. After H. Credner.

ventral armor consisted of oval rather than semicircular scales. Most of the group must have been fully terrestrial.

Although the history of this group is at the moment somewhat obscure, it is evident that it began rather early in the Carboniferous, for it includes a Mid-Carboniferous branch known as Aistopoda. The fact that in the microsaurs the fore limbs were shorter than the hind suggests a "degeneration" of the appendages. Various transitional forms led to limbless, snakelike creatures (Fig. 63), numerous in Carboniferous strata of Ohio and Ireland. These animals had short ribs and elongate, hourglass-shaped vertebrae with centra so nearly solid that there is only a small perforation for the notochord. Some had as many as sixty, others one hundred segments in the backbone. There is some evidence that their bodies were completely covered with scales, so they must have looked much like modern caecilians, although not closely related to them. All were small, from three to ten inches in length, and probably aquatic in their habits, for the neural and haemal spines of the posterior vertebrae are elongate, as if they supported a tail fin.

Other possible descendants of the early microsaurs are members of a small group known as the Adelospondyli. The vertebrae appear to be a modification of the hourglass type, the centra much like those of the Aistopoda. The neural spines are not attached to the centra. A well-known member of this group, *Lysorophus*, an animal found in the Permian red beds of Texas, has been much discussed by paleon-

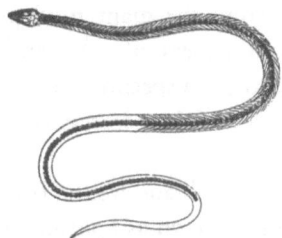

FIG. 63. A restoration of *Dolichosoma*, one of the snakelike aistopodan amphibia. One-sixth natural size. From Anton Fritsch.

tologists. Some have thought that it was a reptile; others that it was an amphibian, but probably an ancestor of the reptiles. Fortunately, a couple of nodules containing skulls and portions of skeletons of this creature were given to Professor W. J. Sollas at Oxford. Sollas has his own way of studying fossils, a method which has yielded important information in groups as widely separated as graptolites and reptiles; this is, in essence, simply the use of serial sections. The zoölogist cuts sections with a microtome; Sollas mounts fossils, still embedded in the original matrix, on a frame in such a way that he can grind them on a wheel in planes strictly parallel. Each newly revealed surface is drawn or photographed, and thus the entire structure is determined. Studying *Lysorophus* in this way, he demonstrated that it was truly an amphibian, with branchial arches like those of urodeles and no pineal opening. This seems to indicate that the modern Caudata, but not the reptiles, may be descendants of the Adelospondyli. Specimens of other small animals belonging to this group have been found in the Lower Carboniferous of Scotland and in the Mid-Carboniferous and particularly the Permian of both Europe and North America.

The modern tailed amphibians differ from all the "roof-headed" in lacking scales, in having extremely short ribs, and in having lost various parts of the skull. Several of the elements regularly fail of ossification, whereas others are bony in members of some genera, cartilaginous in others. As a result, the skull appears to have a sketchy framework, in marked contrast to that of the stegocephalians. A seemingly trivial but important characteristic is the absence of a pineal foramen. The limbs are short, and the pectoral girdle is chiefly cartilaginous, as are the pubic elements of the pelvic region. The fact that haemal arches are attached to the centra of the caudal vertebrae indicates that it is the anterior of the primitive rings which is retained.

Although most of these creatures are aquatic, they are rare as fossils. A single specimen has been found in the Mesozoic, the Cretaceous of Belgium. It is the oldest known mud puppy, but it is closely allied to the famous *Andrias scheuchzeri* of the Miocene of Oeningen. The latter is Johann Jacob Scheuchzer's *Homo diluvii testis*, whose story is told to all students of paleontology; the "man" was about three feet long and differed but slightly from the giant mud puppy (*Megalobatrachus*) still living in Japan. Several extinct genera of salamanders, differing but little from modern ones, have been described from specimens obtained in the Tertiary of France, Germany, and Bohemia. The oldest of these terrestrial caudates is of Lower Cretaceous age.

The most specialized of all amphibians are the frogs and toads. The Anura have an advanced type of vertebra, that known as opisthocoelous because the posterior end of the centrum is concave, the anterior convex. Only ten or twelve vertebrae are present in the trunk. They bear stout transverse processes, but ribs, except in one family, are absent. The caudal vertebrae are coalesced into one short piece, the coccyx. The skull shows certain peculiarities; the orbits are large, the parietal and frontal bones are fused together, and a platelike bone is interposed between the frontals and nasals. The teeth are small, bristlelike, or entirely wanting. The pectoral arch is partially ossified, the fore limbs long, with coössified radius and ulna, and ossified wrist bones, an unusual feature in the Amphibia. The hind limbs are long, the tibia and fibula united, and the ankle ossified.

The oldest frog is known from specimens found in the Triassic of Madagascar. It is so like modern ones, however, that it is obvious that the group originated much earlier. There is, as has been mentioned, a suggestion that it was derived from the Paleozoic branchiosaurs. A great many specimens have been obtained from Tertiary strata of Europe, some of them so remarkably preserved that histological studies of their tissues have been made. Phosphatized mummies have been found in France, and Miocene deposits near Bonn have produced both adults and tadpoles in abundance.

Absolutely nothing is known of the geological history of the glass snakes, the caecilians.

To recapitulate: the amphibians, arising from the lobe-finned ganoids probably in Silurian or early Devonian times, inherited from their ancestors a skull consisting of numerous bones, the dermal elements of which were attached but loosely to the brain case; labyrinthodont teeth; a functional pineal body; a continuous notochord, only partially enclosed by bone; a largely external pectoral girdle; the rudiments of two pairs of limbs; a long tail; and a covering of scales.

The general trend of evolution in the group appears to have been toward a reduction in the number of bones in the skull, the closing of the pineal opening, the constriction and elimination of the notochord through the formation of the centra of the vertebrae, the withdrawal of the dermal bones of the skull and of the pectoral girdle beneath the skin, the ossification of all the bones of the limbs, or, secondarily, their loss, a reduction of the length of the tail, and the loss of the external scales. Some of these trends had reached fulfillment early in Carboniferous times, when the real record of the amphibians begins. The limb bones were fully ossified; in fact, many of the Carboniferous and Permian forms have ossified wrist and ankle bones, even ossified pubic bones. Other trends, such as the reduction of the number of bones in the skull, perfection of the centra of the vertebrae, loss of the pineal foramen, and loss of scales, did not reach their culmination till well into the Mesozoic.

All of the Paleozoic amphibians were roof-headed stegocephalians. Their truly progressive line was that of the labyrinthodonts, which reached their culmination shortly before they became extinct in the Triassic. The collateral lines are represented by the branchiosaurs, which may lead to modern frogs, and the more terrestrial microsaurs, possibly ultimate ancestors of the salamanders. It remains, however, for paleontologists to find in late Paleozoic and early Mesozoic strata proof of the ancestry of the modern amphibians.

One of the principal lessons to be learned from the study of the history of this group is that degeneration does not necessarily result in degeneracy. The amphibians reached the peak of their form during the late Paleozoic, but their most highly specialized representatives are the modern frogs and toads. No one could call the latter degenerate in any accepted sense of the word. They lead a free, nonparasitic existence, and their insect-catching habits make them acceptable members of the modern community. Aloysius, our farmstead friend, comes to the outside faucet for his daily bath and then establishes himself beneath a grapevine to catch flies. Degenerate? Not at all. But he lacks the size, some of the skull bones, teeth, cervical vertebrae, ribs, and the tail of his Carboniferous ancestors. Structurally he is degenerate. Probably he has neither more nor less brain. If pressed, he can jump a hundred times further than any Carboniferous amphibian, but he climbs the back steps in the same way as any other four-footed animal. He has a mixture of primitive, "sub-primitive," and specialized characteristics. It is the "sub-primitive" characteristics which zoölogists commonly designate as "degenerate." If they represented reversions to an ancestral condition, they might well be called recessions. Unfortunately some of the so-called degenerate characteristics are distinctly different from any possessed by the ancestor. They represent progress, but progress backward. Perhaps "retrograde" is as good a word as any we have to apply to such lines of evolution. If this term be used, the aquatic caudates are simple-retrograde, the salamanders semi-retrograde, and the Anura specialized-retrograde.

XIII

THE FIRST REPTILES

These are our blood and bone that climb and crawl
Up from the mire through the Neanderthal.
Humbert Wolfe, *The Uncelestial City*

Many people have an inherent dislike for reptiles, a prejudice fostered by age-long tradition. This antipathy is directed chiefly against snakes, abundant and conspicuous modern representatives of the class. The paleontologist is little interested in snakes, for their history is brief; they are a specialized branch of a good old stock, their ancestors having been the principal inhabitants of the earth during the middle period of its history. From the beginning of the Permian till the end of the Mesozoic, reptiles dominated the life of land, sea, and finally air. Our own ancestry reaches back to them; so, even though their glory has departed, we must look with a certain interest upon a race whose supremacy persisted longer than that of any other single group. In retrospect it seems almost as if nature had been experimenting, trying out the possibilities of the four-footed creatures she had evolved, seeing what could be done by making various modifications of the fundamental structures. But the reptiles were handicapped by the cold blood their ancestors had brought with them from the water. A long period of life on land, eons of adversity, of pursuit of prey, of struggle to escape destruction, were necessary for the evolution of that warm blood which enabled mammals and birds to displace their sluggish forebears.

To understand the course of evolution it is, as usual, necessary to delve somewhat into details of structure which might seem technicalities important only to the specialist. The general results of the studies of paleontologists since the time of Cuvier could be summarized in a few pages, but we prefer to investigate the evidence on which the findings are based.

Reptiles are cold-blooded animals with bodies commonly encased in scales, although some are naked. Some have bony plates in the inner layer of the skin, beneath the outer epidermal scales. Incidentally it may be remarked that the reptilian scale appears not to be an inheritance from the amphibians. Paleontologists, however, are interested chiefly in the skeleton. This in most cases is fully ossified; the vertebrae usually have solid centra, for, except in the oldest and simplest reptilian forms the notochord does not persist in the adult. Since the skull is variously formed in different groups, more will be said of its construction on later pages. The teeth are simple, conical in the majority but variously modified according to the diet. All reptiles, saving a few dinosaurs, have single-rooted teeth. At the back of the skull there is a single condyle for articulation with the first of the vertebrae. It is a rounded knob beneath the opening through which the spinal cord issues from the brain case. The neck of some reptiles is long, that of others short, so that there is great variation in the number of cervicals. It will be remembered that most Amphibia have only one; most mammals have seven. The shoulder girdle is rather complicated and consists of more bones than are found in the same structures in placental mammals. The pelvic girdle is fully ossified, the upper elements (ilia) connected with at least two vertebrae, which are joined together to form a sacrum. Many reptiles have three sacrals, a few as many as eight or nine.

Primitive reptiles have five toes on both front and hind feet, but the number is reduced in specialized groups. Oddly enough, one of the most constant reptilian characteristics is the number of bones in the digits. Although the term "phalangeal formula" seems technical, its meaning is simple; it is merely an easy way of expressing the number of bones in fingers and toes. One can readily learn his own formula by counting the segments of his fingers. He should remember, however, that in walking in the normal mammalian position, the big toe and thumb are inside, the little toe and little finger outside. Digit number one, the thumb, has two bones; all the others have three each. Hence our phalangeal formula is 2,3,3,3,3, which, incidentally, is the typical formula for primitive mammals. If one investigates the bones in the fingers and toes of reptiles, he finds that the thumb has two, that there is increase by one each to the fourth finger, then a decrease of one segment; therefore the formula for the hand is 2,3,4,5,3. That of the foot is 2,3,4,5,4, the little toe having one more bone than the little finger. Not all reptiles have exactly this arrangement of bones, but since the majority, including the oldest and most simple, do, these are the primitive formulae.

The oldest reptilian remains are found in rocks of Mid-Carboniferous age. For a brief period, many years ago now, the writer, a student of invertebrate fossils, enjoyed the enviable reputation of being the discoverer of the oldest known reptiles, found in Mid-Carboniferous strata east of Pittsburgh. But glory is transient. As soon as their description was published, the vertebrate paleontologists began to rummage among their stores and soon discovered that Cope many years earlier had described as a reptile a partial skeleton found in a coal mine at Cannelton, Ohio. This coal was formed a few thousand years earlier than the Mid-Carboniferous strata which held the bones found near Pittsburgh; the specimen from Cannelton, for which Williston suggested the name *Eosauravus*, is therefore older, and till the present date, 1937, remains the oldest reptile known (Fig. 69). The skeleton is a small one without skull or fore limbs. Although Cope had correctly identified it, later students concluded that it was an amphibian. Once the fact was demonstrated that reptiles did exist as early as Mid-Carboniferous, restudy showed that the great naturalist was right. The only obvious characteristic which shows that it is a reptile is the phalangeal formula of the hind foot, 2,3,4,5,4.

THE FIRST REPTILES

Little is yet known about Carboniferous reptiles, but Permian strata, particularly the red beds of Texas, continue to furnish specimens, and rocks of the same age in Europe, Africa, and South America have yielded their contributions. The most primitive fossils from these beds belong to a group known as the cotylosaurs. The present evidence suggests that these primitive reptiles originated in the north, perhaps in that ancient Eria which bounded the northern Atlantic. They spread in Carboniferous and early Permian time to North America and Europe, and during the middle and late Permian to South Africa. They disappeared from North America at the end of the Paleozoic, but remains of them are found in Lower and Middle Triassic rocks in South Africa and Europe. Nowhere did they survive later.

That some sort of stegocephalian was the ancestor of the reptiles is amply proved by the skeletal structure of the cotylosaurs. The skulls are depressed and completely roofed with bones; otic notches are present in some of them. Detailed comparison reveals that one or another shows every paired bone which was present in the skulls of the labyrinthodonts; in general, however, there are fewer bones in the skulls of primitive reptiles than in those of amphibians (Fig. 64). The backbone also is amphibian-like, for the individual centra are biconcave (amphicoelous), and the fact that the centrum is perforated in most shows that the notochord, although much constricted, was continuous.

The outward appearance of the cotylosaurs must have been much like that of the contemporary amphibians. Like them, they were semiaquatic, living in swamps and marshes, crawling about with their plump, rounded bodies almost dragging on the ground, their short legs being inadequate for rapid or energetic movement. The skin is practically unknown, but horny scales probably covered the body. A few had bony plates serving as a partial protection. The teeth were conical, or modified cones, in some forms blunt, adapted for crushing. All appear to have been flesh-eaters but not fierce carnivores. It seems, rather, that some of them fed on crustaceans or insects. The great cockroaches of the day, some four inches long, were doubtless appetizing and succulent morsels. Other reptiles were probably fond of snails and mussels, their teeth being well adapted for crushing shells. In Texas they lived on a lowland facing the sea but got their food on the land and in the fresh waters. Most of these animals were small, from six inches to four feet long, although a few reached the length of ten feet.

A few of the cotylosaurs are so well known, at least among paleontologists, that they deserve special description.

Seymouria (Fig. 64), an animal about two feet long, represented by nearly complete specimens from the Permian of Texas, has been widely heralded as the most primitive known reptile. It was a short-legged, crawling creature, its widely spaced, sharp, conical teeth especially fitted for catching insects. The short neck, massive shoulder girdle, and numerous bones of the skull suggest its near relationship to the amphibians whose habitat it shared. So amphibian-like is it that one specialist on early reptiles, Professor Broom, still insists that it is merely an amphibian which has certain reptilian characteristics; on the other hand, American students believe that it is a true reptile. When doctors disagree, the layman is not particularly interested in whose opinion is correct but in the reason for the quarrel. From our standpoint the significant fact is that here again we find evidence that late Paleozoic reptiles were closely allied to amphibians.

The most important characteristics of *Seymouria* are those of the backbone. The vertebrae suggest the two-ringed, embolomerous type of the early labyrinthodonts, but there is a marked difference, for in *Seymouria* it is the anterior part, the inter-

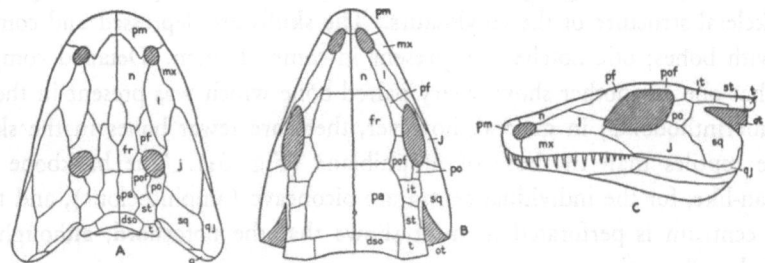

Fig. 64. Diagrams to permit comparison of skull bones of (A) *Eryops*, a Permian amphibian, and (B, C) *Seymouria*, a primitive Permian cotylosaur: *pm*, premaxilla; *n*, nasal; *fr*, frontal; *pa*, parietal; *dso*, dermosupraoccipital; *l*, lacrimal; *pf*, prefrontal; *pof*, postfrontal; *st*, supratemporal; *t*, tabulare; *mx*, maxilla; *j*, jugal; *sq*, squamosal; *qj*, quadrato-jugal; *q*, quadrate; *po*, postorbital; *it*, intertemporal; *ot*, otic notch. Note the unpaired bone on the median line in A. A, after R. Broom; B, C, after S. W. Williston, with modifications.

centrum, which is reduced, whereas the posterior ring (pleurocentrum) is complete and supports the neural arch (Fig. 59 E). This indicates the workings of a process exactly opposite to that in the amphibians, where the posterior ring tended to disappear. All the earliest reptiles appear to have had vestiges of the anterior part of the primitive double-ringed vertebra, but in most lineages they disappeared during the early or middle Permian, although they have persisted till the present in the "living fossil," *Sphenodon*, and in the geckos among the lizards. Evidence of their former presence is shown by the position of the haemal arches of the tails of many reptiles. These appear to alternate with the vertebrae; that is, they are intercentral in position, souvenirs of the lost anterior halves of the centra.

It is unlikely that the discrete pleurocentra of the rhachitomous vertebra would grow together again to form a ring; therefore it is not at all likely that the vertebrae of the early reptiles could have been derived from a four-piece type. The vertebra of a cotylosaur seems to have been formed directly from an embolomerous type by

THE FIRST REPTILES

the loss of the upper part of the anterior of the two primitive rings. The ancestor of the reptiles must therefore be sought among the Embolomeri of the Lower Carboniferous. Only four or five genera are yet known. One had five fingers, but unfortunately not enough phalanges. Further discoveries in the older coal measures of Scotland or northern England will probably reveal the form with just the proper characteristics. If *Diplovertebron* (Fig. 57, 58) had a phalangeal formula of 2,3,4,5,3, instead of 2,3,3,3,4, it would doubtless now be heralded as the most important of all amphibians, the ancestor of the reptiles. By what narrow margins is glory missed! Three little useless bones in the hand!

Pareiasaurus, represented by many skeletons from the Upper Permian of South Africa and northern Russia, is a good example of an Old World cotylosaur. It has a massive skeleton, six to eight feet long, short but sturdy legs, strong, clawed toes, and a broad heavy skull. This animal, like some other reptiles inhabiting the same regions, stood higher than its American cousins. It was one of the first to begin to solve the

Fig. 65. A, the skull of the Triassic cotylosaur, *Elginia*, compared with (B) that of the modern horned toad, *Phrynosoma*.

problem of getting the legs beneath rather than beside the body. This was accomplished by a change in the axial direction of the upper arm and leg bones which brought them into a vertical rather than a horizontal position. If some of the restorations of *Pareiasaurus* may be trusted, this change was a necessity, since otherwise the animal could hardly have moved its exceedingly corpulent body. A more fundamental explanation of the change may be seen, however, in the altered positions of the various bones of the pectoral and pelvic girdles which resulted in changed directions for the muscles. The fusion of the pelvic bones greatly strengthened the support for the limbs. *Pareiasaurus* is supposed to have been a herbivore.

The last cotylosaur to be mentioned is *Elginia* (Fig. 65), remains of which are found in the Triassic sandstone near Elgin, Scotland. Only the skull is known, pieced together from plastic casts made from natural molds in sandstone. Its chief interest lies in the fact that its head instead of being smooth, like those of most of its relatives, is covered with spinelike protuberances, those along the posterior margin being elongated into conspicuous spines. Its skull was in some respects like that of the modern horned toad, which is a reptile, not a toad. The late Professor C. E. Beecher of Yale, who was especially interested in spinescence, investigated its development and history in many groups of animals and plants, with the view of learning its significance. His studies led him to the conclusion that in general the production of spiny outgrowths is confined to the later periods of the history of any race, occurring usually only shortly before the disappearance of a particular group.

Although the cotylosaurs themselves disappeared with the closing of the Mid-Triassic, they left numerous descendants, the least modified of which are the turtles, discussed briefly in a later chapter on aquatic reptiles. If one wishes to envisage the fully roofed skull of the primitive reptile, he should inspect that of a large marine turtle. Possibly it is because most turtles, like their cotylosaurian ancestors, are semiaquatic that some have retained a primitive arrangement of the skull bones. All other living reptiles show one or two pairs of openings (vacuities) behind the eyes. These fenestrae are considered so important that the division of the group into five subclasses is based upon their distribution. This grouping was first suggested by E. D. Cope, first applied by H. F. Osborn, and fully amplified into a definite scheme by S. W. Williston, a paleontologist with a wide knowledge of fossil and recent reptiles. In no other group, perhaps, is a knowledge of fossils more important in proposing a classification. No living reptile is truly primitive; most are highly specialized, and there are many more groups of extinct than of living members of the class.

> All that tread
> The globe are but a handful to the tribes
> That slumber in its bosom.

All classifications published before that proposed by Williston are unsatisfactory because they fail to recognize that those reptiles are most primitive which are most like the stegocephalians from which they were derived.

The cotylosaurs are obviously the parent stock, for they are the oldest, and but slightly different from the contemporaneous amphibians. They show no openings in the skull behind the orbits (Fig. 66 A). Like the early amphibians, they are roof-headed. How were the reptiles with one or two pairs of temporal openings derived from them?

The mechanics of the explanation are simple. It must not be supposed that because the stegocephalians, the cotylosaurs, and the turtles had big, broad heads they had big brains occupying the space beneath the roof of the skull. On the contrary, their brains were small, enclosed in a narrow, bony box beneath the median portion of the parietals and frontals, bones adjoining the mid-line of the skull. The great area of the inner surface served merely for the attachment of muscles, particularly those working the jaws. The operation of these muscles exerted a considerable stress upon the roof bones, pulling them downward and inward. They would naturally be most apt to give way along their margins, the sutural union being inadequate. It will be remembered from the analysis of the skull bones of the amphibians (Chapter XI) that three series are readily recognized: a median series, two marginal ones, and lateral ones between them. If the bones were pulled apart along the sutures between the median and the lateral series, a pair of dorsal openings would be formed: if between the laterals and marginals, the vacuities would be at the sides. If they were severed along both lines of weakness, both dorsal and lateral fenestrae would appear.

All these things have happened, and the arrangement of the openings is characteristic of the subclasses. The reptiles with none, the primitive cotylosaurs and the somewhat specialized turtles, form a group known as the Anapsida. Those with a single pair of lateral ones, the Synapsida (Fig. 66 B), appear next in sequence. These

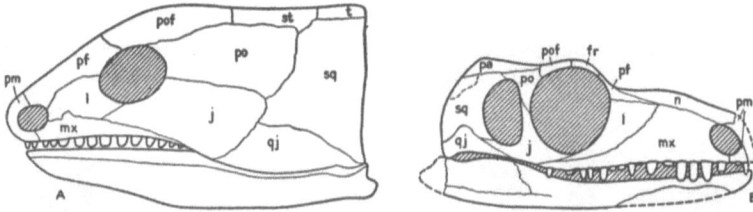

FIG. 66. Diagrams to show lateral views of the skull of (A) an anapsid and (B) a synapsid reptile. Lettering as in Fig. 64. After S. W. Williston.

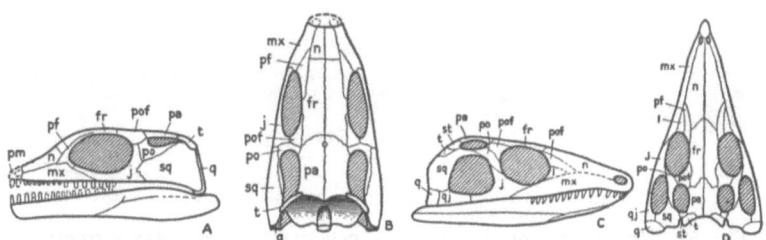

FIG. 67. Diagrams to show lateral and dorsal views of skulls of (A, B) a parapsid and (C, D) a diapsid reptile. Lettering as in Fig. 64. After S. W. Williston.

are all extinct but of great importance, for among them are the ancestors of the mammals. Next are the forms with a pair of dorsal openings, divided by Williston into two groups, the Parapsida (Fig. 67 A, B), a subclass which includes the ichthyosaurs, and the Synaptosauria, the plesiosaurs and their allies. Last, and largest, is the subclass known as the Diapsida (Fig. 67 C, D), with dorsal and lateral temporal vacuities which tend to reduce the posterior part of the skull to a sketchy framework. Here belong many of the living reptiles, the lizards, the snakes, the primitive *Sphenodon*, the alligators and crocodiles, and their first cousins, the extinct dinosaurs; here also are the most specialized of all reptiles, the flying pterosaurs, and early members of the subclass are supposed to have been ancestors of the birds.

Broom has made the ingenious suggestion that the synapsids were derived from cotylosaurs with broad heads, and the parapsids from narrow-headed forms, the more direct downward pull in the latter resulting in dorsal rather than lateral openings. The diapsids, according to him, probably were derived from some parapsid which had become secondarily broad-headed. Since the lateral temporal openings seem to have appeared before the dorsal, it is possible that the diapsids were derived from narrow-headed descendants of the synapsids. This would be more in accordance with what is now known of the geological record, for synapsids were common in Permian times, whereas the parapsids were chiefly Mesozoic. The remains of the oldest diapsid, *Youngina*, are found in Permian strata in South Africa.

FIG. 68. Restorations of the primitive pelycosaur, *Ophiacodon retroversus* (Cope). Original, seven and a half feet long. Note the compressed heads, as compared with the depressed ones of cotylosaurs. Original drawing by L. I. Price.

Not all paleontologists agree that the temporal fenestrae were formed by muscular stresses. Some think the apertures were produced by the resorption of the bones to which the muscles were attached. This seems to the uninitiated a bit like the action of a man in a tree who saws off the limb on which he is seated, but resorption of bone and shell is undoubtedly a much greater factor in change of skeletal form than has been generally recognized. R. T. Jackson has for years advocated more study of this process.

Most Paleozoic reptiles belong to one or the other of the two primitive subclasses, the Anapsida and the Synapsida. The anapsids, the cotylosaurs, have already been discussed. Their companions in the American swamps were the pelycosaurs, readily recognizable as synapsids by the presence of lateral temporal openings.

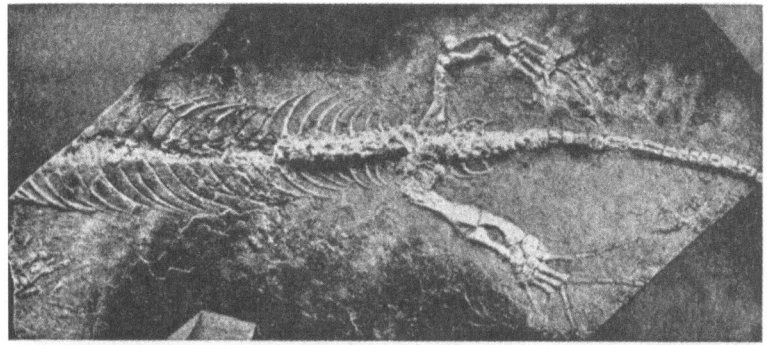

Fig. 69. The oldest known reptile, *Eosauravus*. From Roy L. Moodie.

Fig. 70. *Edaphosaurus*, the "telegraph-pole" lizard. A microcephalous, disharmonic, unadaptable type of pelycosaur. Specimen mounted in the Museum of Comparative Zoölogy, Harvard University, under the direction of A. S. Romer. Mount and photograph by George Nelson.

FIG. 70A. The oldest known egg, possibly that of an *Ophiacodon*, a Permian reptile. Courtesy of Alfred S. Romer.

THE FIRST REPTILES

Most of the pelycosaurs were externally similar to the cotylosaurs, their immediate forebears. Their skulls were, on the whole, somewhat narrower and higher, compressed rather than depressed (Fig. 68). Although they were short-legged, they succeeded in carrying their bodies off the ground. In these respects they show further departure from the form and habits of their amphibian ancestors than the cotylosaurs. Probably they were all carnivorous, for most had long, sharp, conical teeth. The exceptions are those with a dental apparatus consisting of numerous small, blunt, marginal, and palatal cones, adapted for crushing thin-shelled mollusks and crustaceans; only a few have been suspected of being herbivorous, and the case has not been proved against them. Most were small, inconspicuous creatures, two or three feet long, unimportant members of their community. They received no plaudits in their day, nor do they now, as one looks hastily at their skeletons or reconstructions. Yet in the fullness of time it was their stock which gave rise to the mammals, and their blood flows in our own veins. Man ignores the John Smiths in his ancestry.

Yet the group is not entirely neglected. Museum visitors pause before skeletons of *Dimetrodon* and *Edaphosaurus* (Fig. 70), the pelycosaurian freaks, ancestral to nothing. Tremendously elongated neural spines on their vertebrae gave these extraordinary reptiles a bizarre appearance. It is easy to distinguish the genera, for *Dimetrodon* has simple spines and a large, long-toothed skull. He was a predaceous carnivore. *Edaphosaurus*, the "telegraph-pole lizard," was more specialized, for it had cross bars on the spines. But the skull is small, and the teeth indicate a diet of small invertebrates. The two genera obviously result from two independent lines of pelycosaurian evolution. The function of the elongated spines is still unknown. They were hardly strong enough to stand alone, so it is inferred that they were inside a continuous membrane which formed a dorsal fin. But why should a terrestrial animal have such a fin? It was too weak to be protective. Cope long ago suggested that the animals were aquatic and used the fin as a sail. The relatively feeble limbs of *Edaphosaurus* may support this hypothesis, but neither skeleton shows any really aquatic adaptations. Lacking other suggestions, paleontologists have fallen back upon the orthogenetic idea that, once started, the growth of neural spines continued to a climax in which they were overdeveloped and led to the extinction of their owners. This is in line with Beecher's ideas about spinescence in general. Nevertheless, this explanation somehow fails to explain.

When the writer looks at the restorations of *Edaphosaurus*, and realizes how microcephalous that snail-eating creature really was, he sometimes regrets that Professor E. C. Case "honored" him by naming the oldest (Mid-Carboniferous) species of the group *E. raymondi*.

XIV

THE TERRIBLE LIZARDS

> Upon my head they plac'd a fruitless crown,
> And put a barren sceptre in my gripe,
> Thence to be wrench'd with an unlineal hand,
> No son of mine succeeding.
>
> *Macbeth*, Act III, scene 1

Dinosaurs have long been the spectacular assets of the paleontologist. Many a student has elected a course in "Paleo" because he or she heard that it was "all about dinosaurs," and much money has been raised for scientific purposes by using these gigantic creatures as bait. Although they loom so large in the public eye, their place in the great scheme of evolution is a comparatively unimportant one. It was their fate to make a triumphant, awe-inspiring march up a blind path, only to perish at the height of what seemed to be a career of ever-increasing prosperity. As we learn their history we see that the dinosaurs were just another branch of the reptiles, interesting because they were the chief terrestrial animals of the Mesozoic.

The greater part of our knowledge of them has been obtained from the wonderful skeletons collected in the western parts of the United States and Canada during the past fifty years. Footprints and scattered bones in the Weald of southern England attracted attention as early as the middle of the eighteenth century and were described by G. A. Mantell in his *Medals of Creation*. A great English paleontologist, Sir Richard Owen, was the first to point out that these fragmentary traces represented a group of enormous reptiles, named by him Deinosauria, the terrible lizards. The first specimens, however, were too incomplete to allow a full understanding of the creatures. Since Cuvier's law of "correlation of parts" could be applied with only partial success, the early attempts at making restorations resulted in what now seem to be mere caricatures.

In America, as in England, attention was first attracted to dinosaurs by the discovery of tracks, which are abundant on the surfaces of certain layers of the Upper Triassic red sandstone in the Connecticut valley. Such impressions have been known for a long time, the earliest record being that of a specimen plowed up by Pliny Moody on his farm at South Hadley, Massachusetts, in 1802, but it was not until 1836 that the tracks came to the attention of the scientific world. In that year Dr. James Deane of Greenfield pointed out to Professor Edward Hitchcock of Amherst the birdlike footprints to be seen on the flagstones which were being used for sidewalks in Greenfield. Hitchcock soon convinced himself that the impressions had been made by birds. He not only described them as such in the *American Journal of Science* but at once set about the task of obtaining all possible specimens, thus building up the unrivaled collection now to be seen in the museum of Amherst College (Fig. 1, at right). Hitchcock founded a new science, ichnology, the interpretation of tracks, and was even able to persuade the state to publish two excellently illustrated books on the subject, the sole occasion, so far as I know, on which the Commonwealth of Massachusetts has taken cognizance of the existence of such an abstruse study as paleontology. Dr. Deane also published a book about the tracks, likewise in the belief that they had been made by birds.

Although it is now known that there is little, if any, possibility that the footprints were left by winged animals, it is not surprising that they should have been so interpreted. The three-toed tracks are so obviously birdlike that a popular explanation of them as the footprints of Noah's raven was current before they were studied by scientists. The likeness is, in fact, more than superficial. Not only do most of the tracks have three toes, but many show the imprint of the tip of a fourth, behind the others. Moreover, well-defined ones show the segmentation of the foot, which is like that of a bird. Another resemblance lies in the fact that the footprints of one of these trails follow each other in a single line, instead of being in two lines as is customary with four-footed reptiles, whose legs are far apart at the sides of the body.

How, then, has it been determined that the tracks were made by dinosaurs rather than birds? It is chiefly an inference, based upon increased knowledge of the dinosaurs and the fact that so far no remains of birds have been found in any strata older than the Upper Jurassic. The study of the tracks themselves, however, has furnished some evidence against the original theory. Some trails are accompanied by a furrow that could have been made only by a dragging tail, which suggests a reptile rather than a bird. Others show an association of four- or five-toed impressions with the three-toed tracks, which is likewise indicative of a tetrapod. Moreover, many searches and a rather careful watch of the quarries in the Connecticut valley have resulted in the discovery of a few skeletons, all of reptilian creatures, most of them dinosaurs. Many dinosaurs walked on their hind legs in a birdlike fashion, and many of them had but three functional toes on the hind legs. Since the few skeletons which have been found are of this type, one naturally infers that the tracks were made by such creatures rather than by birds.

If attention be turned from tracks to skeletal remains, we find that these animals constitute two great groups, now generally recognized as distinct orders of reptiles and distinguished from each other principally by their teeth and by the arrangement of the pelvic bones. There are, however, certain characteristics common to all which have been united under the general term "dinosaur." All are diapsid reptiles; that is, the skull has two pairs of openings in the temporal region, one pair dorsal, one lateral. This same condition is well shown in modern alligators and crocodiles, the living animals most nearly related. Dinosaurs differ strikingly from crocodiles, however, in that many of them were bipedal and that all of them walked with the legs beneath rather than alongside the body, in a mammal-like fashion. Furthermore, few had as much defensive armor as a crocodile, the skin of the majority being naked or covered with thin scales.

The first group of dinosaurs includes those with sharp, simple teeth, which indicate that they were carnivorous or descended from carnivores. To these the name Saurischia — reptile-hipped — has been given, because theirs is a normal reptilian pelvis, each half made up of three bones (Fig. 71, at left). The ilia are attached to from three to eight united sacral vertebrae; a pair of convergent ischia extend downward and backward; and simple pubic bones, directed forward and inward, unite at their distal ends in a broad symphysis. This group contains two suborders, the true carnivores, or Theropoda, and the great four-footed amphibious dinosaurs, the Sauropoda.

The second order is known as the Ornithischia, since its members had somewhat more birdlike hips (Fig. 71, at right). The pelvis differs from that of the Saurischia in that it appears to be made up of four pairs of elements, instead of three. It is sometimes spoken of as tetraradiate, in contrast to the more common triradiate reptilian type. There are really only three pairs of bones, but each pubis has two branches, a prepubis extending forward, and a postpubis projecting backward parallel to and beneath the ischium. There is some, although a remote, similarity between this sort of pelvis and that of a bird. All the Ornithischia were herbivorous animals with rather long teeth adapted for cutting coarse vegetation. All of them exhibit an unusual characteristic in that the lower jaw has a beaklike bone in front of the tooth-bearing ramus. This is called a predentary; hence the older and more appropriate name of Predentata for this group. The anterior portion of the upper jaw is also toothless, the muzzle probably sheathed in horn, forming a turtlelike beak. There are three rather distinct suborders. The first are the Ornithopoda, the bird-footed kinds, herbivores which walked on their hind legs, as did the Theropoda. Next are the four-footed armored creatures, Stegosauria, best protected of all dinosaurs but rather short-lived, geologically. Last are the Ceratopsia, the big-skulled, four-footed race which was latest to appear and in some respects most specialized of all.

We shall follow the histories of the five groups, trying to find their relationships to one another and the particular adaptations of each.

The true carnivorous dinosaurs, the theropods, represent the most vital stock of the saurischians. Most of the tracks in the Connecticut valley were made by members of this group, and almost all the skeletons so far found in Triassic rocks belong to it. As would be expected, their chief characteristics are their long, sharp, conical, or bladelike teeth, and the sharp claws on their toes, the impressions of which have been of great assistance in the interpretation of tracks. The skull is small, with large openings behind and in front of the eyes. The vertebrae are lightly constructed, the fore limbs definitely shorter than the hind. The number of toes varies considerably, the early forms having five on both front and hind feet, the latest only three or, more rarely, two fingers and four toes. No species is known with five functional toes on all the feet, at least one, and generally two, being short and more or less useless. Even the oldest known members of the group appear to have walked habitually on the hind legs in a birdlike fashion, the long, stout, and somewhat rigid tail serving to balance the anterior part of the body. The numerous footprints found in the Connecticut valley do not in any instance show evidence that the carnivores ever brought the fore feet to the ground. Although many skeletons of theropods have been described, enough to give at least a glimpse of the general lines of evolution in this group, it is not yet possible to outline any satisfactory phylogeny.

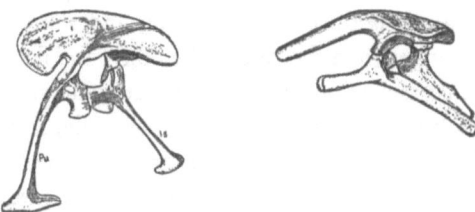

FIG. 71. At left, pelvis of a saurischian dinosaur. *I*, ilium; *Pu*, pubis; *Is*, ischium. At right, pelvis of an ornithischian dinosaur, with a posterior branch of the pubis below the ischium. From O. C. Marsh.

Plateosaurus (Fig. 75) seems to be the most primitive carnivorous dinosaur so far discovered. It was a relatively large animal for its time (Upper Triassic), the skeletons varying from fifteen to nineteen feet in length. It has five digits on both hands and feet, but both the big and little toes of the foot are short, so that the median ones bear most of the weight. The arms are about two-thirds as long as the legs. The pubic bones are joined mesially and are greatly expanded and flattened at the union. Ventral ribs are present. Remains of this animal have been found in the Upper Triassic of Germany and France. An allied American genus is the somewhat smaller *Anchisaurus*, represented by some of the few remains which have been found in the Triassic of the Connecticut valley.

Although *Plateosaurus* and *Anchisaurus* are themselves slender, agile animals, a contemporary of the latter in Triassic days in the Connecticut valley is of even lighter build. This is *Podokesaurus holyokensis*, described by Dr. Mignon Talbot from an almost complete skeleton, the most recent of the finds in our American "New Red Sandstone." It is a little creature, only about four feet long, nearly half the length supplied by the slender tail. The arms are short, the hands retaining but three fingers. All the bones are slender and of extremely light construction, indicating an active animal with a birdlike gait. This is one of the oldest of the slender theropods known as coelurosaurs.

Other Triassic carnivores which should be mentioned are two small, slim forms somewhat allied to *Podokesaurus*. One is *Hallopus*, one-time resident of Colorado, and the other *Procompsognathus*, whose remains were found in Württemberg. Their chief claim to fame is that the calcaneum, or heel bone, is so much elongated as to suggest muscles in the back of the leg strong enough for leaping. That some dinosaurs had this kangaroo-like habit would be a natural deduction, but study of tracks and trails gives no evidence to support the idea that they ever departed from a sedate and even-balanced gait.

Several theropods are known from the Jurassic, the most interesting, because best known, being *Allosaurus*, represented by a wonderful specimen from Wyoming now mounted in the American Museum of Natural History. Larger than any known Triassic dinosaur, *Allosaurus* is thirty-four feet long and a little over eight feet high at the hips. The tail is longer than the rest of the body; the flat articulations between the vertebrae indicate that it was not very flexible. It could not have been curled up or thrown over the back but probably stuck out rather stiffly, serving as a counterpoise to the anterior part of the body. The hind limbs are long, the bones straight, and the knees bent forward as in the mammals, not outward as in most reptiles. The arms are short, not long enough to be used in walking, but each hand has three long, sharp, recurved claws, effective for grasping and rending flesh. The skull is large and wide, with a long lower jaw, articulated at the extreme posterior end, so as to give the mouth a large gape. The teeth are of medium size, conical, but with sharp, somewhat compressed, cutting crowns.

That *Allosaurus* preyed upon the big amphibious reptiles is shown by fragmentary skeletons of the latter found in the same strata. Vertebrae, particularly those of the tail, appear to have had their spines bitten off; others are scratched and scored. When a jaw of *Allosaurus* was compared with these marks, it was found that they agreed exactly, the spacing of the teeth corresponding with that of the scratches. Moreover, while the vertebrae of a *Brontosaurus* were being removed from the rock, numerous broken teeth of *Allosaurus* were found but no other traces of their skeletons, an indication that the teeth had been fractured while the animal was feeding. The specimens mounted in the American Museum combine the fragmentary and scored *Brontosaurus* skeleton with that of *Allosaurus*, posed above it in a feeding position. Whether *Allosaurus* successfully attacked such a huge animal, or whether it was a carrion feeder, devouring only individuals that had died through accident or from old age, is still a debated question. The sauropods were, however, so unprotected and dull of wit that it seems as though they must have fallen easy prey to large and active carnivores.

Allosaurus was the largest and most spectacular of Jurassic theropods, but, curiously, strata of about the same age in Germany have furnished the remains of the smallest known dinosaur, the diminutive *Compsognathus*, no larger than a cat. So small is this skeleton, and so curved its attitude on its slab of Solenhofen limestone, that the suggestion has been made that it represents an embryo. Viviparous habits are not unknown among modern reptiles, and the inference that the dinosaurs may have brought forth their young in a well-developed condition is not illogical. The discovery that representatives of one of the most specialized dinosaur groups, the ceratopsians, were oviparous, seems, however, to indicate that this theory is untenable. There is, in fact, no evidence that the single known specimen of *Compsognathus* was not that of an adult animal. It is a slender, light-boned creature, possibly a resident of the Jurassic upland forests, where food may have been scarce enough to limit its growth. On the other hand, like many modern carnivores of small size, it may have been one of the creatures which exist upon the scraps which fall from the rich man's table.

The theropods continued to prosper throughout the greater part of the Cretaceous, reaching their culmination in *Tyrannosaurus* only a short time before the dawn of the age of mammals. Parts of three skeletons of this greatest of carnivores have been found, two in central Montana and one in South Dakota. The largest, as mounted in the American Museum in New York, is nearly forty-eight feet long, the top of the head nineteen feet above the floor. The skull, although of light construction, is imposing, being about four feet in length. Armed with large, sharp, recurved teeth, the longest of which project five inches from the jaw, this is perhaps the most savage mouth the surface of the earth has ever known. The upper part of the skull has numerous openings bounded by narrow bones, and a brain case so disproportionately small that the low mentality of the great brute is readily apparent. The neck vertebrae are large and closely articulated, indicating great strength and little flexibility. The shoulder girdle and ribs are of light construction, but the pelvis is large. The heavy pubic bones are greatly expanded where they unite at the distal ends and furnished excellent support for the viscera, whereas the long slender ilia attached to several vertebrae welded together in the sacrum formed the basis of a pelvis well adapted to support the great animal in its semierect position. The bones of the arms are short and slender; only three fingers were functional, but they are armed with great recurved claws which show that the fore limbs, although reduced, were by no means useless. The hind limbs are exceedingly strong, the articulation of the bones unusually close-knit for a reptile. The head of the femur has a large knob (trochanter), which fits into the acetabulum of the pelvis nearly at right angles to the shaft. The shaft bears various knobs and rugosities which indicate the places of attachment of large, strong muscles. The feet have three large digits, each tipped with a powerful recurved claw, and there is a fourth toe, which, although it did not reach the ground, is armed with a sharp spur evidently not without its function. Not only the teeth but the hind limbs of *Tyrannosaurus* could have been used as weapons of offense. To see two such animals locked in a firm embrace, using the hind legs after the manner of cats, roweling each other, would have been a stupefying spectacle. There can be no doubt that the tyrannosaurs were complete masters of their world; yet they existed for only a short time and probably were not numerous.

A Canadian contemporary of this the greatest of all terrestrial carnivores was *Gorgosaurus* (Fig. 76), a little creature only twenty-nine feet long, but worth separate mention because it was even more specialized than *Tyrannosaurus*. Its diminutive arms have only two functional fingers, the smallest number yet found in any dinosaur.

Several other large and truly terrible lizards have been found in Cretaceous strata, but the only other theropods to be mentioned here are small, slender ones somewhat like the Triassic *Podokesaurus* and classified in the same group, the coelurosaurs. The best known of these is *Struthiomimus*, the "ostrich mimic," from Albertan Upper Cretaceous. It is of medium size, thirteen feet long, with slender bones, long neck and tail; it differs from most other known dinosaurs in lacking teeth of any sort, a birdlike horny beak probably having taken their place, although it is not preserved. The four-fingered hand is elongate, adapted for grasping and pulling down branches but useless in seizing active prey. The skull, posture, and hind limbs are remarkably like those of running birds, such as the ostriches, but the long tail proclaims *Struthiomimus* a dinosaur. A very similar creature, armed, however, with slender conical teeth, is the Upper Jurassic *Ornitholestes* of Colorado.

Glancing back for the moment over the theropods, we find that they existed from the Upper Triassic almost to the end of the Upper Cretaceous. The first of them still retained the typical five toes of hands and feet, although not all digits were fully developed. Throughout their history there was a constant tendency toward decrease in size and length of the arms, accompanied by reduction in the number of functional fingers to three, at least, and in one instance to two. The hind limbs became progressively longer and stronger. The outer (fifth) toe disappeared in some; the first was rotated so that it was in opposition to the rest, as in the birds, and became so short that only rarely was it long enough to reach the ground. The theropods were all digitigrade; that is, they walked on the toes. The upper bones of the foot (metatarsals) came to lie closer and closer together, so that they acted more or less as a single unit, although they were not joined together. There seem to have been three or more distinct phyletic lines, but the collected skeletons are not yet sufficiently abundant to establish them. *Plateosaurus*, which retains five digits, appears to be near the stem from which all were derived, and also to be ancestral to or closely related to the ancestor of the next group, the Sauropoda.

The sauropods, the most gigantic of all dinosaurs, were in some respects rather primitive. Not only did they walk humbly on all fours but they retained the five toes of the early reptiles. All had small heads, long necks and tails, large bodies, and stocky legs.

THE TERRIBLE LIZARDS

Brontosaurus was a heavy-boned animal, sixty to seventy feet long, fifteen feet high at the hips, of such massive proportions that its weight has been estimated at as much as ninety tons, although a more reasonable figure is about one-third as great. The bones of the limbs are massive and solid, and indicate great strength. The feet are short; the distal bone in each of the outer toes is stubby, suggesting the presence of a hoof. One or more of the inner toes bore claws. In contrast to the massive limbs and feet, the backbone, although strong, is of light construction. The centra of the cervical and anterior dorsal vertebrae are opisthocoelous; that is, each is convex at the front, concave behind. The remainder of the centra, except the distal caudals, have flat ends. The cervicals bear short neural spines but seem complicated because the hatchet-shaped ribs are more or less fused to them. The trunk vertebrae are most extraordinary, the complicated neural spines being tall and provided with various pairs of processes by which adjacent ones articulate with one another. Although some

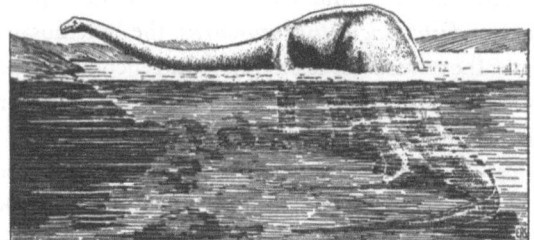

Fig. 72. *Diplodocus*, one of the amphibious sauropods. Original drawing by E. R. Schmitz.

of the spines are as much as three feet high, they are nevertheless light, for all their elements are in the form of thin bony plates. Even the centra of the vertebrae are not solid but are deeply excavated at the sides. As someone has said, the sauropods were the first animals to discover the principle of angle-iron construction. All the cervical and trunk vertebrae are built in this way, but the caudals are more solid, without lateral cavities. As with many other vertebrates, the tail retains a primitive type of vertebra. The skull is small, of light construction, with small openings back of the eyes and larger ones in front of them. The teeth are long and slender, confined to the anterior parts of the jaws. Their tips are somewhat spatulate, with poorly developed anterior and posterior cutting edges. For an animal of such size they seem utterly inadequate.

Diplodocus (Fig. 72) was in many respects similar to *Brontosaurus*, although a longer and more slender animal. The best skeleton, that of *Diplodocus carnegiei* in Pittsburgh, is eighty-seven feet long, the greatest length known from any complete specimen of a dinosaur. All the bones are more slender than the corresponding ones of *Brontosaurus*, and the height at the hips is less by a foot or more. Both neck and tail are longer and slimmer, one of the most remarkable features of *Diplodocus* being the long whiplash at the distal end of the tail. The appearance of one of these animals partially submerged in water must have been uncannily snakelike.

The hind limbs of both *Brontosaurus* and *Diplodocus* were stouter than the fore, and the same condition prevailed in several other genera of sauropods. There is, however, one group in this suborder in which the fore limbs were actually longer and stronger than the hind. These are the brachiosaurs, best known from specimens collected in eastern Africa but also represented by various bones found in the western United States. Though no complete specimen has been recovered, the present opinion seems to be that some of these animals reached a length of ninety feet. If, as paleontologists seem to believe, the brachiosaur could raise its head to the limit of its tremendous neck, it must have been the giraffe of dinosaurs, rearing to heights not even remotely attainable by any other creature which kept its feet on the ground.

The oldest sauropods now known are those found in the Mid-Jurassic of England. They reached their maximum abundance in the Upper Jurassic, the age graced by the presence of *Diplodocus*, *Brontosaurus*, the brachiosaurs, and other huge animals. At that time they were especially well intrenched in the region which is now Colorado, Wyoming, and Utah, and in East Africa. Few remains of sauropods, a single species from Maryland being the most important, have been found in the North American Cretaceous, but recently several large forms have been described from Upper Cretaceous strata in South America, and others have been reported from Africa, India, and Australia. Were it not that C. W. Gilmore has found a shoulder blade nearly five and a half feet long in the Upper Cretaceous of New Mexico, we should be inclined to believe that the sauropods were banished to the southern hemisphere after Jurassic times.

Although skeletons of ornithischian dinosaurs are much more common in museums than those of sauropods and theropods, their evolution is less fully known, for Triassic representatives of the order are so few as to afford but meager information about them. That the connecting links between the three known groups are missing does not surprise one, for nearly all the material has been derived from strata of Upper Jurassic and Cretaceous age.

The least specialized of the Ornithischia are the Ornithopoda, or bird-footed dinosaurs. Misnomers seem to reach their culmination in the name of this group, for the Ornithischia do not have a really birdlike pelvis and the foot of the ornithopod is less birdlike than that of a theropod. The ornithopods were actually birdlike only in their habit of walking on their hind legs, a characteristic shared by the theropods. The oldest evidence of the existence of this group consists of tracks on the Upper Triassic sandstones in the Connecticut valley. They are identified as belonging to the ornithopods rather than the theropods because the digits terminate in blunt, hooflike toes, not in claws. These oldest members of the group had four functional toes and a vestigial one (the outer one) on the hind feet and five fingers on the hand, the two outer ones reduced in size. Unlike the Triassic theropods, they were not at that time entirely bipedal but upon occasion brought the fore feet to the ground. Skeletal material from the Triassic is highly unsatisfactory. The small and incomplete remains of *Nanosaurus*, long known from the supposed Triassic at Canyon City, Colorado, show few primitive characteristics. In all likelihood the few remains found in the Upper Triassic of South Africa will eventually prove to be more useful in tracing the ancestry of the group.

Neither are Jurassic ornithopods any too well known. *Camptosaurus* (Fig. 77), an animal from ten to seventeen feet long, possessing a small head, short neck, rather long tail, short arms, and long, slender legs, is represented by fairly complete material from the Morrison beds of Colorado and neighboring states. The hands retain five fingers, and the sacral bones are not coössified, both primitive characteristics. The feet have four toes, the fifth absent, the first reduced, so that the functional ones are the second, third, and fourth. The general appearance of *Camptosaurus* was much the same as that of the contemporaneous carnivores. According to Gilmore, the front limbs, although much reduced, were probably still useful in locomotion; he has therefore mounted the specimen in the National Museum at Washington in a quadrupedal position.

The Cretaceous was the heyday of the ornithopods; at least, so it seems to us because of the numerous remains which have been found, but every skeleton represents an individual tragedy. Perhaps for the ornithopods it was a time of "war, pestilence, and famine," when the strong and ingenuous perished and the weak and cunning survived.

A striking ornithopod is *Iguanodon*, one of the few European dinosaurs represented by complete material. Trapped in fissures in the Carboniferous rocks of Belgium while they roamed that country in Lower Cretaceous times, the skeletons of these creatures were buried by the sands of the encroaching sea. Miners, following coal seams, have penetrated the filled crevasses and recovered the remains of many. Several are now mounted in the museum at Brussels. From twenty to thirty feet long, *Iguanodon* was a powerful animal, with a heavy, stocky skeleton. Locomotion appears to have been largely bipedal, although the front limbs were half as long as the hind, more powerful than those of most bipedal dinosaurs. Five fingers were present, the third and fourth slender, the first reduced to a single spikelike segment. This thumb, which extends outward at right angles to the axis of the hand, was a powerful spurlike instrument, probably used as a weapon of offense. A none too gentle nudge with it may have warned friend or foe to keep a respectful distance. The skull was laterally compressed, with deep, powerful jaws armed with closely set, serrated teeth arranged in a single row on each side. The feet were similar to those of *Camptosaurus* but both first and fifth digits were absent.

Ornithopods have also been found in southern England and the Isle of Wight.

The latter locality has furnished the skeleton of the slender *Hypsilophodon*. It is of interest because it shows that there were light, slender herbivores as well as carnivores. Furthermore, it is one of the few with teeth in the premaxillaries. The anterior teeth are simple, with narrow, sharp crowns. The fact that they do not reach the median line, but leave room for a narrow beak, indicates that the ancestor had a full set like those of the carnivores. There appears to have been in all the ornithischians a general simplification of the teeth in the modification for a vegetable diet, for although much more numerous than those of the theropods they are more loosely attached to the jaws, in grooves rather than in sockets. Finally, *Hypsilophodon* is of interest because it may have been capable of climbing about in trees. If this is proved, it will be the only known instance of arboreal habits among the dinosaurs. Othenio Abel has pointed out that the long slender hands and feet, and especially the long arms, may have been an adaptation to climbing. The animal was small, only a little more than three feet long.

One more family of ornithopods, the Upper Cretaceous trachodonts or duck-bills, remains to be discussed in some detail. Numerous skeletons of *Trachodon* have been collected, the most nearly complete of them in Wyoming, Montana, and South Dakota. About thirty feet long, the animals reached a height of nearly seventeen feet when standing on their hind legs. The arms were only about one-sixth the length of the legs but were capable of supporting the individual in a quadrupedal position. There are three functional fingers, the first being absent and the fifth small. The feet have three toes, each ending in a broad, hooflike expansion. The skull is depressed, constricted in the middle but broadened at the front into a bill-like expansion which suggested the popular name for the family. Although the teeth were restricted to the posterior part of the jaws, they were extraordinarily numerous, many of them being in use at the same time, while replacement teeth were held in reserve beneath the gums. In many respects the dental succession reminds one of that which obtains in sharks and skates. Barnum Brown is authority for the statement that each jaw is provided with from forty-five to sixty vertical and ten to fourteen horizontal rows of teeth, making a total of more than two thousand altogether.

Trachodon is perhaps most generally known because of the "mummies" found during the past twenty-five years. They are the remains of individuals which dried up instead of decaying or being devoured after death. After being throughly dried, they were buried, probably as the result of being caught by the rising waters of a flood. In the process of desiccation the skin shrank about the bones of the limbs and chest, collapsing over the visceral area. When it decayed, as it did later, it left its impression on the matrix in which the animal was entombed. The skin was thin, covered with small scales, apparently horny in nature but unlike the familiar overlapping ones of the snakes or the large horny and bony plates of the crocodiles. It is obvious that they were of no defensive value. Their reduced condition suggests an aquatic adaptation of the trachodons, for scales tend to be lost by reptiles which live in the water. Several characteristics of the skeleton indicate that *Trachodon* was a good swimmer. The laterally compressed tail must have afforded an effective organ of locomotion, and the broad processes of bone on the inner posterior face of the femur indicate powerful tail-muscles similar to those of the crocodile. The general opinion is that *Trachodon* was amphibious in its habits, a dweller on the seashore, equally at home on land and in the water. Many remains are found in marine deposits, a circumstance which is unusual for dinosaurs.

Strata of Upper Cretaceous age in Alberta have produced wonderfully preserved skeletons of duckbills allied to *Trachodon*, most of them smaller but nearly all more peculiar. *Corythosaurus*, with a remarkable bony crest reminding one of a cock's comb, is one of the most striking. A specimen showing part of the skin, tendons, and

FIG. 73. At left, a restoration of the ornithopod, *Iguanodon*, after W. E. Swinton and Vernon Edwards. At right, the spiny-frilled ceratopsian, *Styracosaurus*. Restoration from various sources.

even impressions of some of the muscles is to be seen in the American Museum. It has many features that suggest aquatic habits. The Royal Ontario Museum in Toronto also possesses a splendid representative of this animal and of *Saurolophus*, a duckbill with a tremendous recurved crest.

Taken as a whole, the evolution of the ornithopods parallels that of the theropods to a remarkable degree. Both appear first in the Upper Triassic and reach their culmination in the Upper Cretaceous. In both there was reduction in the fore limbs, loss of fingers and toes, and increase in the size of the largest individuals. In all these changes, however, the theropods went further than the ornithopods.

Theropods, sauropods, and ornithopods were all, so far as is known, without defensive protection. The theropods carried the war into the enemy's country, their teeth and activity being their best defense. The sauropods must have relied largely upon their bulk and their aquatic habitat for protection. The ornithopods, so far as can be judged, flourished most after they had learned to depend upon swimming as a method of escape. Throughout Jurassic and Cretaceous times, however, there existed a group of sluggish, small-brained, but somewhat heavily armed ornithischians known as the stegosaurians. Representatives of this group are neither common nor particularly large.

The oldest of the stegosaurs, *Scelidosaurus*, was about thirteen feet long. Like all the armored dinosaurs it walked on all fours. Its small head was held rather low, because the fore limbs were short. Its armor, which was not greatly developed, consisted of small dermal ossicles and tubercles in longitudinal rows. The skeleton shows a curious mixture of primitive and specialized characteristics. The hand had four fingers, the foot three functional toes; yet the centra of the vertebrae are amphicoelous, as in the earliest reptiles. Another curious feature is the absence of any enlargement for the spinal cord in the sacral region, a most un-dinosaur-like characteristic. Were

Fig. 74. *Stegosaurus*. A restoration, after an outline sketch by R. S. Lull.

it not, indeed, for the presence of the predentary bone, one would doubt if this were really an ornithischian. The specimens were found in the oldest Jurassic rocks at Charmouth in Dorsetshire, England.

Stegosaurus (Fig. 74), the most bizarre of all dinosaurs, is well known from complete skeletons about twenty feet long, found many years ago in the Upper Jurassic of Colorado and Wyoming. The tiny skull, short front legs, and extraordinary bony plates embedded in the skin of the back combine to give this quadruped an aspect unlike that of any other animal. The broad plates, alternated in position in two rows, one on either side of the middle of the back, had their thickened bases embedded in what must have been a tough hide. Carried erect along the body, the thin edges upward, they seem to have been of little worth as a protection. Perhaps, like the bristling fur of an angry cat, they served rather to impress the enemy than to ward off his attack. Near the end of the short tail the broad plates give place to heavy clublike spines, two feet or more in length. These were obviously not designed

FIG. 75. The Upper Triassic theropod, *Plateosaurus*, one of the most primitive dinosaurs known. Photograph by George Nelson of the specimen he mounted in the Museum of Comparative Zoölogy, Harvard University. It is about sixteen and a half feet long.

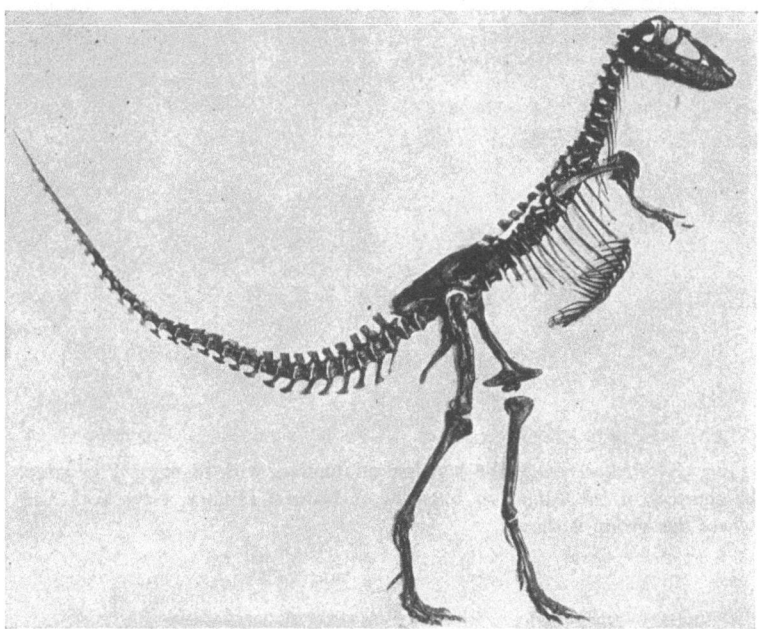

FIG. 76. *Gorgosaurus*, the most specialized of the heavy-boned theropods. Skeleton about twenty-nine feet long. Photograph by courtesy of the American Museum of Natural History, New York City.

FIG. 77. Two specimens of *Camptosaurus*, a primitive Jurassic ornithopod. The larger is seventeen feet long, the smaller almost ten feet. Photograph through courtesy of Charles W. Gilmore and the United States National Museum, Washington, D. C.

FIG. 78. *Protoceratops*, the hornless ceratopsian, with its eggs. Photograph by courtesy of the American Museum of Natural History, New York City, where this group is shown.

for mere passive resistance. The swishing tail of a cornered and enraged stegosaur must have been a powerful deterrent to the advances of even *Allosaurus*, the mighty carnivore domiciled in the same region. *Stegosaurus* had a small head and the most diminutive of brains but was not an animal to be attacked with impunity if he saw his opponent in time to turn his back.

Small-headed armored creatures, more or less allied to *Stegosaurus*, are found in rocks of Upper Cretaceous age, the Belly River formation, chiefly in Alberta, and also in the Lance, the latest Cretaceous strata in Montana. Several genera are known, one of which, *Palaeoscincus*, has been dubbed, aptly, the "animated citadel." Most completely armored of all known dinosaurs, it was broad-headed, flat-backed, short-legged, dull-witted, slow-moving, almost completely encased in bone. Bands of stout thick plates alternating with irregularly arranged bony bosses covered the sides, particularly above the limbs, protecting the more vulnerable ventral surface. Seemingly these huge creatures were safe even from the gigantic *Tyrannosaurus*, but like all dinosaurs, armored and unarmored, they were on the verge of extinction at the very zenith of their powers.

Most dinosaurs had great bulk, massive hips and legs, long necks and tails, but practically all that have been mentioned so far had small heads. It is the distinction of the ceratopsians, the latest of the predentates in order of appearance, that their heads were large — not merely proportionally large, but actually the largest heads possessed by any terrestrial animal. Many skulls and some nearly complete skeletons have been collected from Middle and Upper Cretaceous strata in western United States and Canada. Until recently it had been thought that the ceratopsians were strictly American animals, but the discovery of a member of the group in the Gobi Desert has dispelled this misapprehension. The swampy lowlands east of the rising Rockies were, however, their chief habitat.

Most of the ceratopsians had horns over the eyes, the feature which suggested their name, and a horn on the mid-line, above the nose. Pictures, models, and mounted skeletons have made *Triceratops* familiar to almost everyone. Between twenty and twenty-five feet long, with an estimated weight of about ten tons, it was an exceedingly compact, bulky animal. The neck and tail were short, a large part of the former covered by the great frill of bone at the back of the skull, which in some individuals occupied fully a third of the total length. Both fore and hind limbs were straight and massive, the five short toes at the front and the four at the back terminating in broad, hooflike, ungual phalanges. The fore limbs were considerably shorter than the hind, so that the greatest height was at the hips, as in nearly all quadrupedal dinosaurs. The impressions of blood vessels on the surface of the solid horns show that they were really horn cores, like those of modern cattle; in life, they must have had a thick covering, probably of horny material, which may have increased the length six inches to a foot.

Triceratops, from the Lance formation of the uppermost Cretaceous, is one of the last of the dinosaurs. It is therefore instructive to look at *Monoclonius,* one of the ceratopsians from the Belly River formation (Mid-Upper Cretaceous) of Alberta. A splendid complete skeleton, seventeen feet long, is mounted in the American Museum of Natural History. The skull, about five feet in length, has a long horn above the nose but only a very short one over each eye, an arrangement that gives the animal much more the appearance of a rhinoceros than of a *Triceratops.* The frill of bone extending back over the neck has a scalloped margin. It is not solid, but has two large lateral openings. The partially preserved skin of this specimen shows rather large polygonal plates surrounded by areas of smaller ones, an indication that the body of the ceratopsians was covered with a thick hide in which were embedded protective ossicles of considerable strength.

Associated with *Monoclonius* are the remains of *Styracosaurus* (Fig. 73, at right), described by Laurence M. Lambe. *Styracosaurus,* although one of the most ancient ceratopsians, was one of the most spinescent. Not only did it have the long nasal and two short orbital horns of *Monoclonius,* but the margin of the frill bore long conical spines. The head was almost as spinose as that of the horned toad, a modern lizard of our southwestern states.

Other genera are known, more or less closely allied to those which have been mentioned. Nearly all the earlier forms, from the Mid-Upper Cretaceous, have a long nasal horn, whereas the horns above the orbits are rudimentary or short. On the other hand, all of the ceratopsians from the uppermost Cretaceous have long orbital horns and a short nasal one. In *Diceratops,* in fact, the latter has almost disappeared. This transformation suggests a change in the habits of the animals during the late Cretaceous times.

The only ceratopsian yet known outside North America is the small hornless creature discovered by members of the third Mongolian expedition of the American Museum of Natural History, and named *Protoceratops andrewsi* by Walter Granger and W. K. Gregory. It is the female of this species which is supposed to have laid the famous dinosaur eggs, many of which were collected during the expeditions led by Dr. Roy Chapman Andrews (Fig. 78). More primitive than any American ceratopsian in its hornless skull, short frill, and small size, the adult being only about seven feet long, it was at first hailed as the ancestor of the whole group. Unfortunately, intermediate forms between the Mongolian and American representatives have not yet been found, and the late Cretaceous age of the strata from which the Asiatic specimens were obtained makes it seem doubtful that *Protoceratops* is actually ancestral. It is, however, very like the theoretical ancestor.

A few more words should be added about the peculiarities of the ceratopsians. They differ from the other predentates in having a predentary bone in the upper as well as the lower jaw. They are also the only known reptiles with double-rooted teeth. Although the teeth are not particularly numerous, being confined to a single row in each side of the jaws, they have broad crowns, well adapted for crushing the coarse vegetation plucked by the turtlelike beaks. The curious bony frill at the back of the skull is not made up of bones especially secreted in this group but is a backward extension of ordinary skull bones, the parietals on the median line and the squamosals at the sides. Although secondarily a protection for the neck, its primary function is to furnish areas of attachment for the huge muscles needed to support so heavy a skull.

Although vegetarians, the ceratopsians were by no means pacific in their habits. Broken horn cores and healed wounds in the frill above the neck attest the presence of the fighting instinct. It may be that this was dominant among the males, and that the various mutilated skulls which have been found represent honorable scars received in knightly jousts for "ladyes faire." To the credit of ceratopsians it may be said that they at least engaged head-on, not turning their backs as did the pusillanimous stegosaurs.

XV

MORE ABOUT DINOSAURS

> How did the jolly dinosaur
> Improve each shining era?
> By getting large and specialized,
> Extinction drawing nearer.
>
> Radcliffe "Blue Book," June 1926

So overwhelmingly do the strange, massive dinosaurs bulk in Mesozoic history that they compel us to spend more time with them than they really deserve. They thrust themselves upon our attention so insistently that we are not satisfied with a mere catalogue of their characteristics and structural organization. What do we know of their evolution, their relationships to modern reptiles, their habits, and the cause of the extinction of a ruling class which had no sooner reached its maximum prosperity than its doom was sealed?

If skeletons of modern reptiles be compared with those of dinosaurs, the crocodiles and alligators appear to be the most closely allied. They agree in having two pairs of temporal openings, backward projecting quadrates which give the jaws an enormous gape, teeth restricted to the marginal bones, and similar vertebrae and limbs. The chief differences are in the bony palate of the living forms and the pelvic structure. The palate of the crocodile, like that of the mammal, is formed by the growing together of the inner edges of the premaxillaries and maxillaries over the roof of the mouth. These plates, joined to the palate bones, push the inner narial openings so far back into the throat that, the external nostrils being at the tip of the skull, the crocodile can drown its prey in its mouth, while itself still able to breathe. Crocodiles and alligators with similar palates have been found in Tertiary strata, but as the line is traced back through the Mesozoic the ancestors show less and less of the bony roof of the mouth, till, in the Triassic, creatures are found which lack this specialization entirely. The other important difference between crocodiles and dinosaurs is that in the latter the acetabulum — that is, the socket into which the head of the femur is received — is a depression shared by all three bones of each side of the pelvis; the ischium, ilium, and pubis all reach the acetabulum. The same socket in the crocodiles, on the other hand, is in the ilium and ischium only, the pubis being in front of it. Here again, as we trace the history of the crocodiles, we find that the ancestral forms had a dinosaur-like pelvis. The earliest (Triassic) crocodiles, called parasuchians, differed but slightly from the dinosaurs, and among them we may seek the common forefathers.

There are various sorts of parasuchians. Some, such as the phytosaurs and longirostrines, had long skulls, with tremendously elongated premaxillary bones. They were obviously specialized. Other short-headed, more primitive forms, known as pseudosuchians, whose remains are found in the Upper Triassic of Germany, Scotland, and New Mexico, and in the Lower Triassic of South Africa, are the probable ancestors of the dinosaurs and the crocodiles. Broom, who has studied this question in detail, points out that the skulls of *Euparkeria* and *Ornithosuchus* (the latter from the Triassic of the Elgin district in Scotland) are practically like those of dinosaurs, and that there is nothing in the skeleton that is incompatible with our ideas of the theoretical dinosaurian ancestor. The simplest pseudosuchians have short necks, with eight to ten cervicals, slightly concave or flat-ended trunk vertebrae, three sacrals, and moderately long tails. Their gait was quadrupedal, and all walked with the legs beneath the body, an adaptation to terrestrial life. Each foot had five digits, but the inner and outer ones were smaller than the others, the axis of the foot being through the third, rather than the fourth toe. The fact that the metatarsal bones were somewhat elongated suggests that, when startled, these animals may have adopted a bipedal gait, as some modern lizards do. All seem to have had more or less protective armor, consisting of bony plates and scales.

The carnivorous dinosaurs were probably direct descendants of such early Triassic pseudosuchians. Some students believe that the former diverged from other reptiles as early as the Permian, but this seems unlikely, since only one reptile with two pairs of temporal openings has been found in strata of so great an age. That dinosaurs arose in South Africa, the cradle of progressive reptilian evolution, is probable. From this birthplace they appear to have moved northward into central Europe, where the remains of *Plateosaurus*, the most primitive of the theropods, are found in strata of Upper Triassic age. During that period evolution and dispersal were apparently rapid, for near relatives of *Plateosaurus* reached North America at the same time.

Although the origin of the predentate group is as yet unknown and the direct ancestor has not been found, these animals share so many characteristics with the theropods that it seems probable they too were derived from pseudosuchians. One group of the ancestral animals seems to have continued to live upon flesh; the other, outstripped in the competition for food, resorted to a vegetable diet. The oldest known predentate, *Nanosaurus*, has teeth much like those of the carnivores. In any case, the herbivorous dinosaurs were distinct from the carnivores as early as late Triassic times, the oldest evidence of their existence being found in North America. Whether they originated on this continent or were early immigrants is not known.

The early carnivores quickly differentiated into a slender, swift-moving, light-boned stock, the coelurosaurs; a heavier, more sluggish, but more powerful group, the megalosaurs; and a still more sedentary, stupid, and conservative race, the sauropods. Both the megalosaurs and coelurosaurs adopted a completely bipedal mode of existence; both were active, aggressive carnivores; both became more and more specialized as bipedal beasts of prey, losing digits but, through the elongation of the hind legs, gaining in mobility and adaptation to their predatory habits of life. The sauropods, on the other hand, as we have seen, although undoubtedly descended from bipedal creatures similar to *Plateosaurus*, retained the primitive five digits, reverted to the ancestral quadrupedal mode of progression, and waxed fat by the simple process of eating anything they chanced upon. They were probably omnivorous; anything was grist which came to their mill. So heavy that even their massive bones could not support their bodies on land, they lived an amphibious existence, denizens of the swamps along the coastal plains. Half submerged in water, they were able to crawl about, feeding upon real amphibians, fish, water weeds, and the ancient "sago palms," which the powerful claw on the first finger enabled them to rip apart. A happy, luxurious life, while it lasted. In spite of their sluggishness, they were able to migrate, for their remains are found in the Upper Jurassic strata of East Africa, Madagascar, central Asia, and western United States, and in the Cretaceous of New Mexico and South America.

The somewhat clumsy, heavily built iguanodonts represent the central stock of the Predentata (Ornithischia). Some of their descendants, such as the trachodonts, seem to have become semiaquatic, but did not lose the bipedal method of locomotion. Such animals are found in North America and Asia. Others became more fully armored, the weight of their protective plates bringing them down upon all fours. One group, the Stegosauria, appears to have had its inception in central Europe but reached its culmination in America. The skull, like that of its ancestors, remained small, the protecting plates lying chiefly along the back. In another line, that of the ceratopsians, the head became enlarged, with great horns, useful in both offense and defense. This group was, as we have seen, chiefly American but may possibly have originated in Asia, since *Protoceratops*, the only ceratopsian known outside North America, is so simple that, although perhaps not the real ancestor, it may be one of its slightly modified descendants. That the ceratopsians were the offspring of bipedal ancestors is shown by their short forelegs. In this case, the increasingly massive skull was in itself enough to necessitate the resumption of a quadrupedal mode of locomotion. The retention of five toes on the front feet and four on the rear, a relatively primitive condition, shows that they must have separated rather early from the parent stock. Some authorities have held, with assurance, that they were derived from the trachodonts; others are not so positive. Further collections from Asia or Africa may greatly modify our ideas of their relationships. That they were highly specialized predentates is shown not only by the huge skull but by the extra bone in front of the premaxillaries, the double-rooted teeth, and the great spinelike horns.

As has been said, both primitive carnivorous and herbivorous dinosaurs were bipedal in their locomotion. How did this come about? The habits of the smaller modern lizards furnish a clue. Ordinarily quadrupedal, some of them, when alarmed, scuttle away, rising, as their speed increases, upon their hind legs until their hands are entirely free from the ground. The tail, held rigidly behind, serves to balance them in this unusual bipedal posture. Thus they achieve their greatest speed. Since these lizards inhabit semiarid or arid regions, it has been argued that the conditions under which they live account in large measure for their customs; when water holes or places of refuge are remote from one another, rapidity of movement is at a premium.

The red color of the Upper Triassic strata of eastern North America, Colorado, and Great Britain has been interpreted as indicating that they were deposited under semiarid conditions. Combining these observations, Professor R. S. Lull and others have suggested that the bipedal gait of the early dinosaurs resulted from the climatic conditions under which they lived. Overlooking the fact that the red beds are of late Triassic age only, and of relatively restricted distribution, proponents of this idea have assumed that there was a universally arid climate throughout the whole of the Triassic. A suggestion of this sort, however, ignores generally accepted ideas regarding climates. So long as there are oceans, water will be evaporated from their surfaces, and some of it will fall as rain somewhere on the land. It is unlikely that the dinosaurs should have sought out the unfriendly semiarid localities of the earth as the theater of their first development. Animals are naturally attracted to regions of abundant food, not to arid wastes. If they invade the latter, it is in retreat from powerful competitors; and the carnivorous dinosaurs had none. Furthermore, as was mentioned in the discussion of the origin of terrestrial vertebrates, the red color of the strata probably signifies warm and moist rather than semiarid conditions. Let us quote what Lull himself quotes from a quotation which Schuchert makes from Knowlton's inferences regarding the climate in eastern North America during late Triassic times: "The trees show no annual rings, from which is inferred a uniformity of warm subtropical climate, without change of seasons." "Without change of seasons" certainly rules out a semiarid climate, in which periods of rainfall alternate with long dry intervals.

As a matter of fact, bipedality probably had its inception in early Triassic times. The first reptiles, living with their amphibian forebears in swamps, got along very well with a sprawling gait. As soon as the draining of the swamps at the end of the Paleozoic caused them to leave their ancestral habitat, it was necessary for them to get their bellies off the ground, get their legs under them, and learn to walk. A human baby learns how in a few months; the reptiles probably accomplished it in a few hundred thousand years.

There may, perhaps, have been a climatic incentive. Ellsworth Huntington tells those of us who live in the north temperate zone that the changeable climate we detest is our real salvation. It saves us from stagnation, spurring us, in exasperation perhaps, to continuous endeavor. The reptiles were subjected to similar but harsher times during the Permian. Glacial conditions supervened, with effects that did not wholly cease to be felt till the end of the Paleozoic. Although for a time so severe as to drive all terrestrial life from some regions, this season of refrigeration did not oust all reptiles permanently. Some sorts perished; others appear to have been driven out only temporarily, for their descendants returned when the time of stress was over. The changeable seasons which preceded and succeeded the actual glacial era seem to have affected favorably such races as survived. Forced to march, they marched. Exercise strengthens muscles; strength overcomes mechanical difficulties. The reptiles of South Africa learned to walk with their bellies off the ground. The fore legs reach ahead, but the hind limbs really push the body forward; therefore the hind limbs of practically all terrestrial vertebrates are the stronger ones. As the population increased, internecine strife began. The carnivores preyed upon friends and strangers alike. Survival was the only desideratum. To the strongest and swiftest belonged the food. Desire for food and terror of attack were the dominant influences. Terrestrial animals simply got on their hind legs and ran.

We are still uncertain about the cause or causes for the extinction of the dinosaurs. It is probable that there were many contributory factors, some the result of their own organization, others purely external, due to geographic and climatic changes.

Their low order of intelligence doubtless had much to do with their downfall. *Stegosaurus* is generally considered to have been about the stupidest of all animals. So small is the cavity within which it was lodged that the total weight of the brain is estimated at only two and a half ounces, whereas that of the body must have reached a couple of tons. The great sauropods, *Diplodocus, Brontosaurus*, and their relatives, were certainly no better, if as well, supplied with gray matter. At most, they had only one or two pounds, whereas the weight of their bodies was from thirty to thirty-five tons. It is perhaps unfair to use man as a standard in judging the brains of other animals, since he is so highly specialized in that respect, but when we realize that man has on the average about two pounds of brain for each hundredweight of body, it helps to emphasize the almost incredible stupidity of the dinosaurs. Bipedal dinosaurs, both herbivores and carnivores, had proportionately larger brains than the quadrupedal forms. The carnivores had the better ones, but even the Jovian *Tyrannosaurus*, greatest of all hunters, had a remarkably small brain case. Much has been written about the importance in the evolution of the primates of the liberation of the fore limbs from their sordid use in supporting the body, to become the servants of the head. Their use in grasping and handling foreign objects is supposed to have quickened the perceptions, giving the animals increasing stores of knowledge concerning their environment and thus inducing greater development of the brain through greater use. This process doubtless had its influence, a greater one, perhaps, in the warm-blooded mammals than in their more phlegmatic cousins. However that may be, bipedal dinosaurs walked on their hind legs and used their anterior ones as arms and hands over a period of several million years without any great cerebral improvement.

In the pelvic region of many dinosaurs there is a striking enlargement of the cavity for the spinal cord, which has led to the idea of a "sacral brain" at the hips. As a matter of fact, some did not have it, and it is questionable if, in all cases, the enlargement was due to an increase in nervous tissue; fatty matter, forming a protection for the ganglia in that region, may have occupied much of the space. The enlargement is particularly great in some of the quadrupedal dinosaurs, notably in *Stegosaurus* and the sauropods, so that the mass of the pelvic ganglia has been estimated at as much as twenty times that of the brain in the skull. Certain German paleontologists have suggested that the "sacral brain" acted independently of that in the head and took care of the functions of nourishment, digestion, and procreation. This idea originated in an attempt to explain how animals with such small heads and great bodies as the sauropods could possibly have got enough to eat; a relatively small amount of food might conceivably do, if entirely digested, and digestion might be especially complete under the control of a special nervous arrangement.

Professor Lull has examined this suggestion in so far as it relates to digestion, and finds nothing to support it. The tenth cranial nerves, the vagi, are the principal transmitters of the stimuli which initiate digestion. Certain of their branches are distributed to the oesophagus and stomach, terminating at the beginning of the intestine and the pancreas. Casts of the brains of *Tyrannosaurus, Stegosaurus,* and *Morosaurus* (a sauropod), representatives of three groups of dinosaurs, all show that the exits for the tenth nerves as well as those immediately in front and behind are relatively larger than in the crocodile. This indicates that the vagi were at least as well developed as in the modern reptiles; consequently, there is every reason to believe they had the same function. Certain nerves of the recent reptiles do extend from the sacral region to the hinder portion of the alimentary canal, but beyond the region of the glands secreting the digestive juices. They appear to regulate the elimination of waste products. Lull also thinks that the stomach of the dinosaur, like that of the crocodile, and much like the gizzard of birds, must have had thick muscular walls. Like birds, crocodiles occasionally swallow stones which remain in the thick-walled stomach and aid in the trituration of food. There is some evidence that dinosaurs had the same habit, for smooth stones, "gastroliths," have been found associated with their skeletons. Like the crocodiles, they may have had a powerful digestive juice which dissolved even the bones and other more resistant parts of the prey.

Such studies as have been made of the central nervous system of the dinosaurs suggest that its various parts had the same relationships and functions as in modern reptiles. Intelligence was low, but reflex action good. A blow upon the head might merely maze the brute, but a step upon the tail would doubtless have been followed by instantaneous and convulsive writhings. Obviously, animals with such small brains, which showed little improvement during the ages of the Mesozoic, were not particularly well fitted for survival.

As has been indicated in the preceding chapter, different sorts of dinosaurs had somewhat diverse habitats and foods. Those best equipped to survive were the more sizable carnivores, their strength and their large daggerlike teeth enabling them to prey upon any animals of their time. The more slender, lighter-boned coelurosaurs were carnivorous also, but because of their size must have been content with a diet of small amphibians, reptiles, fish, and, in the later Mesozoic, possibly birds. Their long, slender, grasping hands connote various specialized feeding habits, though they still afford an interesting subject for speculation. *Ornithomimus* and *Struthiomimus*, Upper Cretaceous coelurosaurs, were toothless. Since loss of teeth is a gradual process, it is probable that their condition was due to their ancestors' long persistence in the use of some food which was easily picked up with a bill-like mouth, and which did not try to escape by wriggling out of the jaws. Baron Nopcsa has suggested as such a food the eggs of dinosaurs, other reptiles, and birds, for the hand, composed of three fingers of equal length, with an opposable thumb, seems especially well adapted for gathering them. Of this possibility A. S. Romer says, "As if in confirmation of this suggestion was the subsequent discovery in Mongolia of a crushed skull of an ostrich-like dinosaur in a nest of fossil eggs belonging to a dinosaur of a different group. It would almost seem as if this reptile had actually been caught in the act of egg-stealing." If a sufficient supply of eggs was not obtainable, small crustaceans, fruits, and seeds may have supplemented the diet. In short, birdlike food habits seem to be indicated.

The other carnivorous dinosaurs, so far as one can learn from their structure, did not have specialized food habits. If their extinction was due to lack of food, it must have resulted from a wholesale destruction of all the animal life within their locality. The sauropods, on the other hand, appear always to have been menaced by extermination from lack of nourishment. One marvels that they existed as long as they did. In spite of their cool blood and sluggish nature the ponderous creatures must have required huge quantities of food, and they were not adapted to travel far in search of it. The slightest unfavorable geographical or climatic changes would affect them seriously. Like the sauropods, the true stegosaurs had feeble teeth, probably adapted for gathering some of the more tender plants from the marshes in which they seem to have lived, rather than the coarser and tougher ones which nourished the ornithopods and ceratopsians. Animals of both the latter groups were so well equipped with beaks and crunching teeth that they must have been capable of feeding upon almost any sort of vegetation. It is true, however, that the later ornithopods, the duckbills, appear to have become somewhat dependent upon a particular environment, for many of them seem to have adopted a semiaquatic mode of life.

Food habits alone, then, would account for the extinction of only one group, the sauropods. Some more general cause for the wholesale destruction of dinosaurs must be sought. It has been suggested that their disappearance may have been due to the increasing numbers of mammals. These, although all small, were becoming fairly numerous throughout Jurassic and Cretaceous times, and it may be that they had a fondness for the eggs of reptiles. That such small creatures should have been able to annihilate the gigantic dinosaurs by robbing their nests is not impossible, but it seems rather improbable. Racial senescence also has been urged as a possible cause. This hypothesis assumes that races, like individuals, lose their vitality in the course of time; perhaps through failure of the reproductive processes, the group disappears. Dr. R. L. Moodie's studies suggest also that old races are more susceptible to disease than young ones, and some have thought that epidemics of disease swept away the dinosaurs.

Most students agree, however, that the basic cause for the extinction of this remarkable group must be connected in some way with the geographic changes which took place during the later part of the Mesozoic. For lack of sufficient detailed evidence from other parts of the world, let us focus our attention upon happenings in North America. Similar changes took place at the same time in South America. Just what occurred in the other continents is not known. Late in the Jurassic, mountains arose parallel to and relatively near the Pacific coast. East of them was a huge low area, now occupied by the Basin Ranges and the Rocky Mountains. This was the happy hunting ground of the late Jurassic and Cretaceous dinosaurs, the ones about which we know most. This region, with that for a long distance to the east of it, appears to have been sinking throughout most of the Cretaceous. During the Lower Cretaceous the waters of the Gulf of Mexico spread as far north as the center of Kansas, and the Arctic flooded part of northern Canada. After a short recovery, more extensive flooding took place in the Mid-Upper Cretaceous, when a great sea covered the region east of the Jurassic mountains from Mexico to the Arctic. This was the first critical period for the dinosaurs. The swampy homes of the sauropods and the stegosaurs were completely inundated. A few of the sauropods managed to survive, but most were blotted out long before the sea actually reached the places of their former habitations, the swamps having apparently been drained during some of the preliminary movements. During Upper Cretaceous times the whole region west of the Mississippi began to rise, the interior sea was drained, and the formation of the Rocky Mountains began. It was during this period that dinosaurs became most abundant. The sauropods and true stegosaurs were gone, but the ceratopsians came on the scene, the carnivores and the ornithopods flourished, and various armored types replaced the stegosaurs. As the Rockies continued to rise, however, the plains of the old sea floor to the east of them were correspondingly uplifted, the swamps were drained, and the climate became colder in proportion to the increasing altitude. Soon the new mountains cut off the moist sea winds. In that climate, chill and semiarid, the huge, cold-blooded creatures could hardly find food enough, or warmth enough, for survival. Like the protagonists of many a prince-to-pauper tragedy of our own times, they who had been monarchs of the earth perished miserably and obscurely of hunger and cold.

XVI

MARINE REPTILES

> Beyond the shadow of the ship,
> I watch'd the water-snakes:
> They moved in tracks of shining white,
> And when they rear'd, the elfish light
> Fell off in hoary flakes.
> Coleridge, The Rime of the Ancient Mariner

Born of the water, amphibians have never strayed far from it. Born of the amphibians, reptiles remained, during the earlier eons of their existence, in the environment of their ancestors. Relatively few of these cold-blooded creatures became fully terrestrial in their habits during the Paleozoic era. Living as they did in swamps, along rivers, and by the borders of the sea, many sought their food in fresh or salt waters. Comparatively few became vegetarians, even in the Mesozoic. The waters, even at the present day, afford a greater supply of animal food than the lands; it was natural, then, even in the early days of the reptiles, that many should return to the rivers or seas in search of their prey.

We are molded by our habits. Those animals, even though they be air breathers, which adopt a fishlike life naturally come to be more or less fishlike in appearance. A terrestrial animal which returns to the aquatic environment must, if it is to survive, become a skillful swimmer. Youth is the plastic age, in great groups as well as in individuals. Reptiles which returned to the water early in their racial history, before they were fully adapted to terrestrial life, came more completely into accord with their new environment than those which essayed it later. To this class doubtless belonged the ancestors of one of the most remarkable Mesozoic reptiles, one commonly known as *Ichthyosaurus*, the fish-lizard. I say commonly known, for purists tell us that the correct name is *Proteosaurus*; still worse, a recent investigator of the group has subdivided it until no trace of the original genus remains. But what's in a name anyway?

Remains of ichthyosaurs are most abundant in the Jurassic strata in some parts of Europe; some have been found in the Triassic of Europe and western United States; and fragmentary remains have been collected from the Lower Cretaceous of various parts of the world. The best-preserved specimens are found in quarries in the Lower Jurassic (Liassic) strata at Boll and Holzmaden in Württemberg, although complete individuals have been obtained from rocks of the same age at Lyme Regis in Dorsetshire. Most of the specimens are between four and eight feet long; exceptional ones may reach thirty feet.

The body of the ichthyosaur was remarkably fishlike (Fig. 79), a mimic indeed of such fish as the mackerel, which are best adapted of all for travel through the water. The body was rounded, the head tapered forward, there was no neck, and the tail was a powerful fin, obviously the chief organ of propulsion. The limbs were paddle-shaped and were used as fins, the anterior pair larger than the posterior ones. As in fish, the pectoral appendages were well supported, whereas the pelvic arch had lost any firm connection with the backbone. A large dorsal fin, on the middle of the back, is known only from impressions on the rock, being entirely unsupported by any bony or cartilaginous processes. The sharp, recurved, conical teeth, although they are set in grooves, not sockets, are reminiscent of those of fish, some actually showing the infolding of the enamel characteristic of the lobe-finned ganoids and labyrinthodont amphibians. They are well adapted for seizing and holding slippery prey; hence it is not surprising to learn that alimentary tracts and coprolites of ichthyosaurs show that their food consisted chiefly of fish and the Mesozoic representatives of the squids, the belemnites. One individual, preserved at the museum at Stuttgart, has in its stomach the remains of over two hundred of the latter.

Another interesting observation is based on the discovery of several individuals in which small skeletons are associated with the remains of an adult in a position indicating that they must have been within the body at the time of death. As many as seven have been found between the ribs of a large ichthyosaur. When such specimens were first found, it was supposed that the small ones had been swallowed as food, for it is not at all likely that such animals were squeamish as to what they ate. More recently the general opinion has been that the small skeletons represent unborn young. It is argued, for example, that an animal would not be likely to select for a single meal seven young, all of its own species and all of the same size. Furthermore, some of the small individuals have been found in such a position as to suggest that they were still enclosed in their egg-coverings at the time of death. Most modern reptiles lay eggs, but some are viviparous. Consequently, it would not be at all surprising if a reptile highly specialized for life at sea did carry the young until they were well developed. The limbs of the ichthyosaurs were so fully adapted for use as fins that they must have been incapable of supporting the body on land; even visits to beaches to deposit eggs would have been dangerous to an animal entirely helpless out of water, more particularly since the skin appears to have been practically scaleless and the body entirely unprotected.

Although so fishlike in its general characteristics, the ichthyosaur has a typically reptilian skeleton. The skull is tremendously elongated, chiefly because of the great overdevelopment of the premaxillaries, which make up the larger part of the dolphin-like snout. The orbits are huge, and contain a ring of unusually large but thin sclerotic plates. Evidently the eyes were large, with powerful vision, capable of being focused by the muscles attached to plates surrounding the eyeball. The narial

FIG. 79. An unusually well-preserved ichthyosaur, retaining the outline of the skin of the dorsal and caudal fins. Photograph by courtesy of the American Museum of Natural History, New York City.

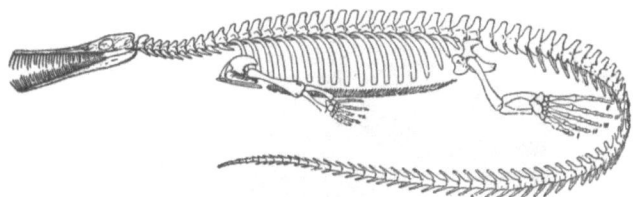

FIG. 80. *Mesosaurus*, the Permian swimming reptile of South America and South Africa. One-fifth natural size. From S. W. Williston, *The Osteology of the Reptiles*, after McGregor.

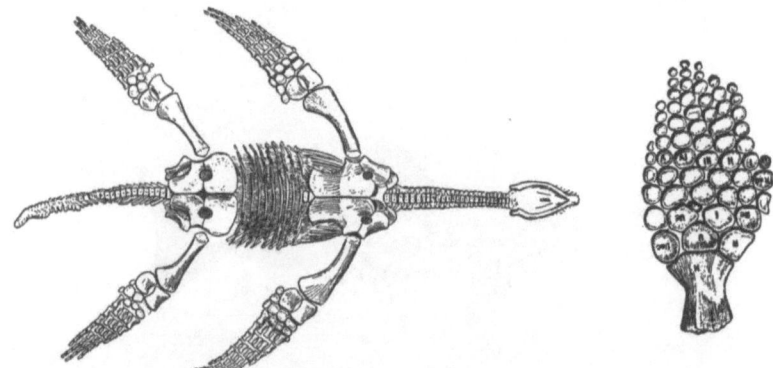

FIG. 81. At left, a plesiosaur, in ventral aspect, to show the broad plate-like bones of the pectoral and pelvic girdles, and the ventral ribs. At right, the arm of an ichthyosaur, showing the supernumerary fingers. Both from S. W. Williston, *The Osteology of the Reptiles*.

FIG. 82. Restoration of a plesiosaur, redrawn after E. Fraas.

openings were far back, close to the eyes; they also are large, for these animals had no gills but remained breathers of air, like their terrestrial ancestors. A curious feature is the position of the pineal foramen on the median line at the juncture of the frontals and parietals. The vertebrae are simple biconcave disks, similar to those of fish. The neural arches are not united to the centra and articulate but feebly with one another, a reductive adaptation to the aquatic environment. Ichthyosaurian vertebrae of the trunk region are easily recognized by the pair of tubercles or depressions low on the sides, for the attachment of the double-headed ribs; the anterior caudals have single-headed ribs.

The most remarkable parts of the skeleton are the limbs (Fig. 81, at right). They contain all the segments of the legs of terrestrial reptiles, being in no respect comparable in their osteology to fins of fish; yet they are curiously modified for their finlike function. The humerus, radius, and ulna of the arm, the femur, tibia, and fibula of the leg, are all much shortened and flattened. Two rows of carpal and tarsal bones are present, all similarly flattened, and all so modified into polygonal shapes that they fit together like blocks in a mosaic. The phalanges, likewise, are flattened, the fingers and toes held close together in such a way that there is little or no chance for movement between adjacent digits. Most remarkable of all, some species show extra rows of fingers and toes; there are, indeed, extra bones along the sides of all the elements from upper arm to fingers. There may be as many as a hundred phalanges, for the typical formula, 2,3,4,5,3, or 2,3,4,5,4, is ignored by these specialized reptiles. As many as eight or nine fingers or toes may be present. This is the really noteworthy feature of the ichthyosaur, for no other reptile, no amphibian, no bird, nor any mammal, barring such freaks as six-toed cats, has more than five digits. Reduction is common in various groups, but increase occurs here only. Whether it came about by bifurcation, as in freaks, or whether it is from the formation of bones *de novo*, is not known. A possible explanation is that each appendage originally had five digits but that a cartilaginous border was formed on either side as the limbs came to be used as paddles. Such cartilages may have provided centers of ossification which led to the formation of false fingers and toes.

Although the ichthyosaurs probably had terrestrial ancestors, it has not been possible to discover them. The oldest members of the group are those found in the Middle and Upper Triassic of Germany, Lombardy, Nevada, and California. These are somewhat less specialized than the true ichthyosaurs, since the legs are longer and the bones less flattened. The pelvis is larger, the skull relatively shorter, the teeth less numerous and set in sockets, like those of terrestrial reptiles. The vertebrae are longer, less like those of fish, more firmly articulated by processes of the neural arches. The posterior portion of the tail is not bent downward so sharply, a fact that indicates the presence of only a small tail fin. In some species, at least, it was not a caudal appendage comparable to that of the true ichthyosaurs but a small dorsal fin behind the pelvis. All these features indicate that the Triassic forms, commonly called mixosaurs, were less fully adapted for aquatic life than the Jurassic and Cretaceous ichthyosaurs, and hence more like the terrestrial ancestor.

It has been suggested that the Lower Permian *Mesosaurus* (Fig. 80) of South Africa and South America may be one of the initial members of the line, but proof is lacking. *Mesosaurus* is interesting both because of its geographical distribution and because it is the oldest known reptile to show definitely aquatic adaptations. It is a small, elongate, slender animal, with a long skull containing numerous long, slender teeth on the margins of the jaws and smaller ones on the bones of the palate. The tail is long and laterally compressed, evidently an organ of locomotion, suggestive of that of some of the mixosaurs. The fact that the lower limb bones are short connotes aquatic life, but the chief proof of aquatic habits is seen in the greatly elongated little toe, which in the South African species has an extra phalanx. The first or "big" toe is the smallest, and the fifth or "little" toe is the big one, an indication, according to Williston, that the feet were webbed. In contradistinction to those of the ichthyosaurs, the hind legs are larger and more powerful than the anterior ones, and the neck has several vertebrae. These characteristics suggest the plesiosaurs rather than the ichthyosaurs. On the other hand, Dr. J. C. Merriam has shown that the locomotion of the Triassic mixosaurs was accomplished largely by the aid of the limbs and not by the tail. He found that their hind limbs were nearly as large as the anterior ones; consequently, he believes that the mixosaurs are in a sense intermediate between the mesosaurs and the ichthyosaurs. As a matter of fact, however, it is not possible to say that this suggested line really indicates the ancestry of the ichthyosaurs.

The plesiosaurs (Fig. 81) were fully as remarkable in their aquatic adaptations as the ichthyosaurs, a comparison of the two groups showing particularly well the different results attained when similar animals use unlike methods in solving the same problem. The ancestors of both ichthyosaurs and plesiosaurs were terrestrial reptiles which were attracted to aquatic life by the food in the sea. Both ultimately came to feed upon fish, belemnites, and other swiftly moving animals, as well as the less elusive clams, so that both had to become good swimmers. The ichthyosaurs used the tail as an organ of locomotion; the plesiosaurs probably swam in the manner of a sea turtle.

The outward form of a plesiosaur is but little like that of an ichthyosaur. The head in most cases is relatively small, and there is a distinct neck, short in some species but abnormally long in others. The body is broad, somewhat flattened, the tail short, without a fin. As in the ichthyosaurs, the limbs are paddlelike, the fingers closely appressed and enclosed in a thick integument. But there are no extra rows of digits, and the hind limbs are longer and stronger than the anterior ones. The long-necked forms were perhaps the most remarkable, for no other animals show so great an increase in the number of cervical vertebrae. Giraffes and camels have long necks, but both have the typical seven cervicals of the mammals; the great length is due to the elongation of the individual bones, not to the introduction of new ones. The length of the necks of some birds is due both to an increase in the number and to the elongation of the vertebrae, although no bird has more than twenty-one cervicals. But some plesiosaurs have the astonishing number of seventy-six bones in this region. One had a head two feet long, a neck twenty-three feet long, the body nine, and the tail seven, the neck being more than half the total length. In spite of this great expanse, plesiosaurs (Fig. 82) apparently did not have the gracefully curved, swanlike forms usually shown in restorations. The cervical vertebrae articulated by nearly flat surfaces, allowing for little curvature. The head must have been carried rather stiffly.

Although they were well provided with teeth, plesiosaurs appear to have had the birdlike habit of swallowing their food whole. This placed the burden of mastication upon the stomach, which, like the gizzard of the bird, seems to have been a thick-walled organ containing gravel and pebbles which aided in the trituration of food. Specimens have been found with as much as a peck of pebbles within the ribs, fragments foreign to the strata in which the fossils are found. Modern crocodiles have a similar habit of picking up stones to assist in grinding their food.

Looking for a moment at the osteology, we find that the skull is in many respects similar to that of the ichthyosaur. The elongate, recurved, conical teeth are not so numerous, and are set in sockets, not in grooves. There is a single pair of dorsal temporal openings adjacent to the parietals, and the narial openings are close to the orbits. The vertebrae are similar to those of the ichthyosaur, but the centra are only slightly biconcave and thus less fishlike. The pectoral and pelvic girdles are both large and flattened, with platelike elements that gave firm support to the paddles. A rather striking feature is the presence of numerous ventral ribs, which made a sort of basket for the support of the organs between the shoulder and pelvic girdles. All the bones of the appendages are somewhat flattened, but the humerus and femur are long. The lower arm and leg bones are short, however, as are the elements of the wrist and ankle. All the fingers are elongate, the individual phalanges being long and rounded, not flattened as in the ichthyosaurs. The phalangeal formula is not constant, but each digit has supernumerary bones, the third and fourth being the longest, with, in some cases, as many as nine segments.

The oldest plesiosaurs are represented by fragments from the Rhaetic, the youngest Triassic deposits. All the good specimens are from Jurassic and Cretaceous strata. The best of them are found in the lower Jurassic of southern and eastern England and Württemberg, Germany, at all these localities associated with ichthyosaurs, and in the Cretaceous of various parts of the world. Small ones are from eight to ten feet long; the largest, from Kansas and Australia, are as much as fifty feet.

Although the ancestry cannot yet be traced in detail, there seems to be no doubt that the plesiosaurs were closely related to the Triassic nothosaurs. Skeletons of the latter are found in Germany, Switzerland, and northern Italy, chiefly in marine strata of Mid-Triassic age. They appear to have been relatively small creatures living on the shores of the great inland sea which at that time covered a large portion of central Europe. They resemble plesiosaurs in the structure of the skull, the elongate neck, the stout pectoral and pelvic girdles, and slender limbs, with short lower arm and leg bones. The legs are not modified as paddles, nor are there extra phalanges. The nothosaurs appear to have been amphibious in their habits, showing a tendency toward aquatic adaptation chiefly in the shortening of the bones of the lower parts of the limbs. No species which could have been directly ancestral to any plesiosaur has yet been found.

From time immemorial sea serpents have held a large place in the folklore of seafaring nations. Even now scientists dare not deny absolutely that large marine snakes still exist in the oceans. There is no tangible proof that they do; it is not even probable; yet a half century ago scientists would have said that the existence in modern seas of squids fifty feet long was not probable. Someone may yet catch a sea serpent. It is known, however, that sea serpents were common in the Upper Cretaceous oceans, for their remains have been found in some abundance in Holland, Belgium, and Kansas, and there are scattered records from various states of North America from New Jersey to Alabama and from North Dakota to Texas, from France, Germany, and far-off New Zealand. These, however, were sea serpents with legs, not true snakes, although closely allied to them. The best specimens are found in Kansas, but the group takes its name, mosasaurs, from the river Meuse (Maes, Maas).

The mosasaurs (Fig. 83) are elongate, round-bodied, short-legged reptiles, fully adapted for marine life. About twenty-five species, representing several genera, are known. The shortest of them is eight feet long, the longest complete specimen thirty feet, although larger incomplete remains indicate a maximum length of more than forty feet. The body is covered with small overlapping scales, in contrast to the ichthyosaurs and plesiosaurs, which probably were naked.

The snakelike characteristics are seen particularly in the structure of the skull. One of the most striking features of the snake's skull — shared, however, by that of lizards — is the mobility of the quadrate bone, which is held to the skull by ligaments only. This gives great freedom of movement at the back of the jaw and assists in the swallowing of large objects. The quadrate of the mosasaur, although larger than that of a snake, was equally free to move. As in the snakes, the two rami of the lower jaws were united at the front by extensible ligaments, not by coössification. This was another help in engorging huge mouthfuls, but not of so great assistance as was an elbowlike joint about the middle of each ramus, a feature unknown in other sorts of animals. The jaws were set with recurved teeth, two double rows above and two single rows below, exactly as in snakes. One can well imagine a mosasaur catching a huge fish, turning the elbows of the jaws inward to spread the front apart and allow a wide bite, then turning them outward to pull the prey in, till constant repetition of the movements forced the unwilling victim within the power of the swallowing muscles of the throat.

The centra of the vertebrae are procoelous, that is, concave at the anterior end, but they lack the complex processes characteristic of snakes. The pectoral girdle is strong, but that of the pelvis weak, not attached to the backbone. Evidently the limbs were used as fins, not as organs of propulsion, the long flattened tail taking over that function. So far as is known, the tail had no elaborate fin, although in some species the neural spines are elongated in the caudal region. A wriggling, eel-like method of swimming seems to be indicated. The upper and lower arm and leg

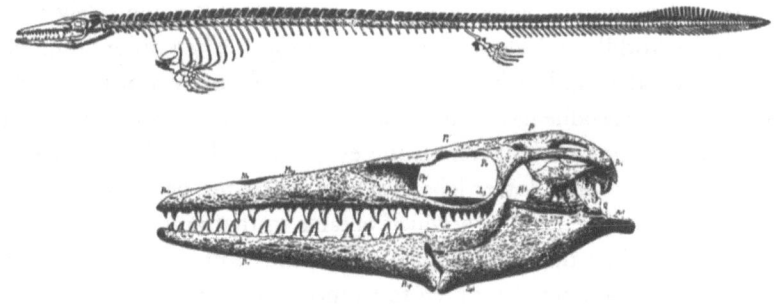

Fig. 83. Skeleton and skull of the Kansan Cretaceous mosasaur, *Clidastes*. The skeleton is 11.3 feet long, the skull 2.5 feet. From S. W. Williston.

bones are short and flat. All five digits are present, but there is a variable number of phalanges. Instead of holding the fingers and toes close together, these animals spread them apart, producing a flexible fin like that of the fish rather than a stiff paddle. All the digits were doubtless enclosed in a continuous membrane, although this has not been seen.

As with the groups described above, the ancestry of the mosasaurs is still in doubt. Some believe them to be descended from the aigialosaurs, whose remains have so far been found only in the Lower Cretaceous strata of Dalmatia. In their characteristics these animals are almost exactly intermediate between the mosasaurs and terrestrial lizards like the modern monitors. The skull resembles that of a mosasaur, including so striking a feature as the hinge in the lower jaw. The pelvis, however, is firmly attached to the backbone, and the legs are long and lizardlike, although the feet must have been webbed, as is indicated by the clawless toes, a sign of at least partially aquatic habits.

The modern monitors are elongate lizards belonging to the genus *Varanus*. Inhabitants of Africa, Asia, and Australia, they have recently been rather widely kept in zoölogical gardens in Europe and the United States. Their heads are long and pointed, and their jaws show a suggestion of the mosasauroid hinge. Most species are wholly terrestrial, but others are excellent swimmers, progressing by means of the compressed tail. Although fossils to tell us their geological history before the Upper Cretaceous are wanting, it is probable that animals much like the monitors existed in Jurassic times. Some of the ancestral forms appear to have been terrestrial, others semi-aquatic. From the latter may have arisen the Lower Cretaceous aigialosaurs, and from them, the Upper Cretaceous mosasaurs.

The ichthyosaurs, plesiosaurs, and mosasaurs were the most conspicuous marine reptiles of the Mesozoic, but associated with them were members of two other groups which should be mentioned briefly. These are the Thalattosuchia, commonly called marine crocodiles, and the marine turtles.

The first of these groups contains only a few members, known solely from skeletons found in the Middle and Upper Jurassic of Europe. They differ from crocodiles in lacking a bony palate and in having biconcave vertebrae, two primitive characteristics. The head is crocodile-like, with a long, slender snout and numerous teeth. A wide ring of sclerotic plates surrounds the eye. The body is slender, the tail long, ending in a caudal fin whose lower lobe is supported by the backbone, as in the ichthyosaurs. Contrary to the condition which obtains in that animal, the hind legs are longer and stronger than the anterior pair. The limbs form narrow paddles, somewhat like those of plesiosaurs. From ten to twenty feet long, these creatures in some respects suggest overgrown tadpoles, the hind legs seeming rather superfluous. Their short geological range and limited geographical distribution mark them as unsuccessful competitors of ichthyosaurs and plesiosaurs, some of whose characteristics mingle in them in mongrel fashion. The absence of any of the bony armor so well developed in the true crocodiles deprived them of a needed protection.

The marine turtles have the distinction of being the only "old salts" to survive from the Mesozoic to the present day. The carapace is incompletely ossified, the limbs are paddlelike. One of the oldest, the most completely specialized for marine life, and certainly the largest, is *Archelon* (Fig. 84), whose magnificent skeleton is one of the ornaments of the museum at Yale. Since the carapace is reduced to a row of marginals, there are numerous vacuities between its long, slender ribs. The massive plastron, on the other hand, is composed of great plates margined by long, fingerlike prongs. The hands and feet are large, the phalanges elongate, clawless, apparently good organs for use in swimming. The skull is three feet long, the neck short, not retractile, and the carapace six feet in length. Including the tail, the whole skeleton is about twelve feet long; the width across the outspread flippers is a little greater. Dr. G. R. Wieland, who discovered this huge "leatherback" in the Upper Cretaceous rocks of South Dakota, estimates its weight when alive as three tons. Its parrotlike beak suggests that it may have fed on shellfish, in which the Cretaceous seas bounded. Whether *Archelon* was in the ancestral line of the modern leatherback turtles is a disputed question. At any rate, it and its Cretaceous relatives, the modern leatherback (*Dermochelys*), and the true sea turtles (Chelonidae) all show much the same sort of adaptation to marine life that we observed in the mosasaurs, that is, a lengthening and spreading of the digits.

Marine and terrestrial reptiles alike reached their maxima in the Mesozoic. They received equally severe checks at the end of the era. What controlled their destiny is still a mystery. It would be interesting to know in what ways Mesozoic lands and seas were particularly favorable to reptiles and what could have happened at the end of the Mesozoic to affect all environments equally.

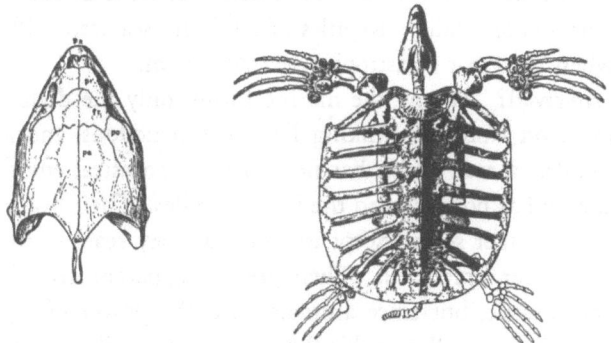

Fig. 84. At left, the skull of a large modern marine turtle showing the simple cotylosaurian arrangement of the bones. From S. W. Williston, *The Osteology of the Reptiles*. At right, *Archelon*, a huge Cretaceous turtle. From G. R. Wieland.

One theory in explanation of the decline and fall of the reptiles at the close of the Mesozoic, suggested by the all too prevalent habit of judging all other phenomena in the light of man's experience, compares racial history with the life story of the individual. Thus reptiles as a group were in a youthful stage at the beginning of the Mesozoic, endowed with abundant energy and fecundity. Having no competitors on the land, they increased and multiplied and peopled the earth. Though they were carnivorous at first, the abundant food led some of them to vegetarian habits. Before the end of the Mesozoic all reptilian phyla were ages old, and it may be that, like old men, they had lost their youthful vitality and fertility, and were no longer resistant to disease. Energy had run down; old age had overtaken the race. In their doddering senility some had lost part or all of their teeth; a few had actually grown spines, considered by some students a sure sign of approaching extinction.

A plausible theory, but poorly supported by facts. Among other animals, both carnivorous and herbivorous dinosaurs increased constantly in size and numbers till the time of their disappearance. Even the sauropods, ousted from their ancestral homes, survived in the southern regions where they found refuge. It is true that some of the spinose forms, notably *Stegosaurus*, fell by the wayside, but since this happened millions of years before the extermination of the group it can hardly be accepted as a portent. Many stegosaurians continued to flourish. The ceratopsians are the only other dinosaurs which developed spines, but they appeared early in the history of that line, and were evanescent features, the last and largest of the animals having the fewest horns. When we turn to the marine reptiles, we find that their history is parallel, except that there is no development of spinosity. Throughout their era there was constant increase in size, numbers, and diversity. Strangely enough, a large group of aquatic forms, the mosasaurs, appeared during the Upper Cretaceous at about the same time as the terrestrial ceratopsians. There is no suggestion whatever of racial old age or loss of vitality. Reptiles of land and sea were still advancing and differentiating when the great catastrophe overtook them.

What sorts survived? Among the marine forms, only the doddering old toothless turtles; of those on land, the slinking lizards, sphenodons, and snakes, and the semiaquatic crocodiles, a race which has never left its ancestral environment.

So far this survey has not included the flying reptiles. Their history is not exactly parallel to that of the other sorts, except in so far as they reached their culmination in size during the Upper Cretaceous. Their greatest apparent diversity was attained during the Upper Jurassic, but since animals with the power of flight are seldom preserved as fossils it is not at all probable that their true distribution is known. There is every reason to believe that their story is the same as that of the other groups.

In our speculations upon possible causes of the extinction of dinosaurs, various factors were considered, such as epidemic diseases, the devouring of eggs by mammals, and alteration of the vegetation through changes in climate resulting from mountain-building. Although any one, or all, of these happenings may have been important in the case of the dinosaurs, it is obvious that no one of them could originate a world-wide catastrophe affecting reptiles of land, sea, and air. Racial senescence having been ruled out, there remains only one guess — that is, worldwide reduction in temperature. As is well known, cold-blooded reptiles thrive in tropical regions. The further northward or southward one proceeds from the present equatorial belt, the smaller and the less numerous are the reptiles, till one reaches a region where there are none. There is a definite northern and southern limit to reptilian distribution at the present day, but that limit undoubtedly is considerably further from the poles now than it was during Mesozoic times. The evidence for this is not wholly dependent upon the known occurrences of Mesozoic reptiles; it is supported also by what is known of the contemporaneous plants and corals.

It is generally recognized that the end of the Mesozoic was a time of an unusually high stand of continents and restriction of shallow seas. Such conditions, at the end of the Proterozoic, in the Paleozoic, and in the Pleistocene, were accompanied by worldwide refrigeration and glaciation. It must be admitted, however, that there is little evidence for such a happening at the end of the Mesozoic. There were, it is true, mountain glaciers in the San Juans of Colorado during the Eocene, but these ice streams seem to have been local and relatively unimportant. In short, the evidence is confusing, for Eocene floras in general indicate that the tropical and subtropical belts were then wider than at the present day. Perhaps we can only finish rather weakly with a prediction that it will eventually be shown that sometime late in the Cretaceous or early in the Tertiary there was a period of general refrigeration during which the flying and the large terrestrial reptiles, all the marine reptiles except the turtles, all the ammonites, and the large sessile pelecypods were extinguished.

XVII

FLYING REPTILES

He seems to be a man sprung from himself.
Tiberius

A third of a century ago there was published in London a fascinating book, *Dragons of the Air*, H. G. Seeley's summary of his lifelong studies of those most extraordinary reptiles, the Pterosauria. The popular conception of reptiles is that they are sluggish creatures, prone to spend their lives basking in the sun, moving only when disturbed. On the other hand, flying suggests the acme of activity, constant motion, supreme vivacity. These estimates are, in the main, justified, but anyone who has witnessed the rushes of an alligator, the scamperings of a lizard, or the angry lashings of an infuriated snake realizes that reptiles may be extremely active. Flying reptiles are not, then, entirely anomalous, although since none exists at the present moment they seem a bit out of place in the general scheme of nature.

One small group of the modern lizards, the little "dragons" (*Draco*) of Java, make some pretensions to consideration as members of the aerial fraternity. They are arboreal, as most flying vertebrates are, but they make no effort to fly, confining their activities to gliding from higher to lower branches. In this they are assisted by lateral extensions of the skin of the body, supported by flexible prolongations of the ribs. Such makeshifts, however, do not lead to true power of flight. This is accomplished only if the fore limbs are provided with a patagium, that is, an expansion of the skin between bones or a series of feathers which may be moved as occasion demands. The patagium of the pterosaurs consisted of a thin but firm outgrowth of the skin, supported in front by the greatly elongated fourth finger of the hand, stretching thence to the tail and the proximal (femoral) segments of the legs. This wing lacked both avian feathers and reptilian scales; hence it closely resembled the wings of bats, though whether so "leathery" as one would infer from the common textbook statements is not clear. The impression, as preserved in the few known specimens, appears to be that of an exceedingly thin and featureless epidermis.

Paleontologists are still amazed by the extraordinary history of the pterosaurs. Their sudden appearance, so far as the fossil record is concerned, in the upper Triassic of Europe is comparable only to their sudden extinction after a brief visit to North America in Mid-Upper Cretaceous times.

Most of the best-preserved Jurassic specimens are small creatures, about the size of common bats; a few reached a length of eighteen inches. During the Cretaceous, however, there seems to have been considerable increase in size. The largest complete specimens are the pteranodons of the Kansan Mid-Upper Cretaceous, but fragments of similarly large individuals have been found at many places. In Upper Cretaceous strata in the vicinity of Cambridge, England, is a greensand which for a short period was extensively quarried for fertilizer. Although only a foot in thickness, it contained enormous numbers of coprolites, with a phosphatic composition which made them valuable. Associated with them were numerous fragmentary bones of pterosaurs, some thousands of which were saved and preserved in various English museums. All are relatively large, and Seeley found among them some which seem to have belonged to animals with a wingspread of at least twenty feet.

All pterosaurs had long, lightly built skulls, with large orbital, preorbital, and narial openings, the latter well back. The temporal vacuities of most are small, as is necessitated by the shortness of the region behind the eyes. The cranial cavity is small, enclosing a brain with birdlike arrangement of lobes but of diminutive size. The articulation of the jaws of some is peculiar, for the quadrates extend forward, bringing the back of the jaw below the eyes. The long, slender, curved teeth are of the grasping type, but are directed forward instead of backward. They were not set in continuous series but were widely separated, the anterior ones longer than those further back in the mouth. This condition suggests that reduction in number and size was already in progress in early Jurassic times; one is therefore not surprised to find that some of the later Cretaceous representatives of the group are toothless.

Numerous skeletons appear to indicate that the head was carried in a birdlike position, approximately at right angles to the vertebral column. Although the evidence for this deduction is so ample that such a posture is accepted as the normal one by most paleontologists, one should accept the statement with some caution. It is true that the heads of flightless birds are held at an angle of 90° or less with the axis of the neck, and volant birds when perching or walking show the same attitude. But birds in flight, with outstretched neck, raise the head, bringing it more or less into the axial trend of the body. If one looks critically at the specimens of pterosaurs in which the skull appears to be at right angles to the cervical vertebrae, one notices that most of them have broken necks, as indicated by an abrupt change in the axial direction of the cervicals near the skull. The rounded knob of the single occipital condyle must have allowed great freedom of movement of the head, but no more than that enjoyed by most other reptiles. The position of the skull was probably more dinosaurian than avian; if angles must be mentioned, it is probable that the axis of the skull was more nearly 55° than 90° off the direction of the body.

The neck is relatively short, with seven large vertebrae whose broad neural spines afforded places for the attachment of strong muscles. The trunk also is short, with ribs on nearly all vertebrae. The sacrum contains from two to five vertebrae in addition to the typical reptilian pair; hence it is more nearly comparable to the dinosaurian than to the avian condition. The ischium is attached to the ilium, which is narrow but elongate. These bones share the acetabulum for the head of the femur, as in the crocodiles. The pubes are free bones without sutural connection with the other elements of the pelvis. The significance of this unusual condition has not yet been determined, but it should not be interpreted as indicative of the descent of pterosaurs from crocodiles.

The limbs are of light construction, the bones hollow, with dense walls, much like those of birds. The pectoral girdle also appears at first glance to be avian, for there is a large sternum or breastbone, a portion of the skeleton which is ossified in but few reptiles. Some pterosaurs have, as an anterior extension of this bone, a process like the keel of a bird of flight. Situated as it is, with its flattened surface at right angles to the body of the sternum, this anterior process must have afforded support to the pectoral muscles; hence it was functionally, if not morphologically, a keel. Its position, chiefly in front of the breastbone, is doubtless to be correlated with the structure of the wing.

There are two possible interpretations of the bones of the hand. The presence of four fingers is obvious. Three of these are of normal reptilian type, each provided with a terminal claw. The outermost is greatly elongated, many times as long as any of the inner ones. This finger is now considered to be the fourth, but it was for many years identified as the fifth, or little finger. Why the difference of opinion on this subject?

If one examines any well-preserved specimen, one finds on the inner side of each arm a backward-pointing bone which seems to articulate with the carpals. The obvious interpretation is that it is a metacarpal, and the vestige of a thumb. If so, the clawed fingers were the second, third, and fourth, and the elongate one the fifth. It was, however, long ago suggested that the reflexed bone was not a metacarpal but the remnant of an imperfectly ossified tendon. If so, this "pteroid" bone has no homologue in the normal reptilian hand, and the first clawed finger is the thumb. According to Williston, the latter interpretation is the proper one, the evidence lying in the number of phalanges in each digit. Ignoring the inner, questionable element, the first clawed digit has two, the second, three, and the third, four phalanges. The greatly elongated finger also has four. The formula may therefore be written 0,2,3,4,4, or 2,3,4,4,0. The typical reptilian formula for the hand is 2,3,4,5,3; hence it is obvious that the interpretation advocated by Williston implies less modification than the other, and so is the more plausible. It should not, however, be stated dogmatically that the interpretation now widely accepted is the correct one; it is merely the more probable one. Incidentally, it is that of Cuvier, the first paleontologist to describe a pterosaur.

There are two types of pterosaurs, the short-tailed or Pterodactyloidea, commonly called pterodactyls, and the long-tailed or Rhamphorhynchoidea.

Rhamphorhynchus (Fig. 86) is the best known of the long-tailed tribe. Several good specimens have been found at Solenhofen, the most complete of them the one now in the Peabody Museum at Yale. This individual retains the impression of a large part of each wing and of a small rudder or stabilizer at the tip of the tail. The rhamphorhynchoids are considered somewhat less specialized than other pterosaurs,

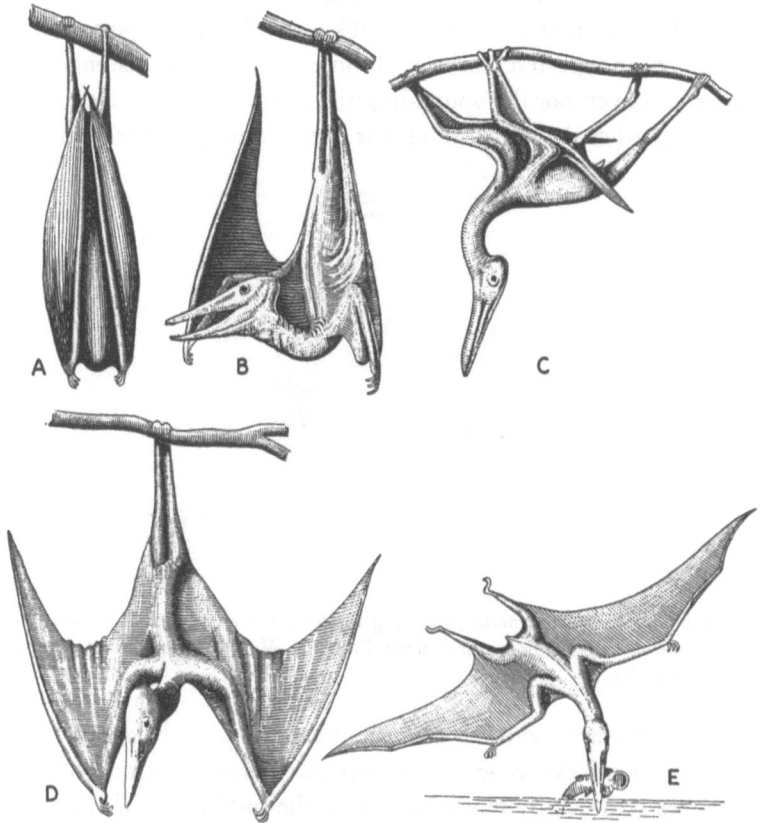

FIG. 85. Sketches showing possible habits of pterodactyls, according to Abel. A, sleeping; B, awakening; C, walking; D, ready to glide; E, catching a fish. All redrawn, with slight modifications, from O. Abel.

since an elongate tail is primitive, and the little toe, although shorter than the others, is less modified than that of any of the pterodactyls. *Rhamphorhynchus* and its allies are found only in the Jurassic of Europe.

The history of the short-tailed pterodactyls is also chiefly European, entirely so during the Jurassic. Seeley stated that the group shows no evolution, but it is a far cry from the toothed *Pterodactylus* of sparrowlike proportions of the Upper Jurassic to the huge toothless *Pteranodon* of the Upper Cretaceous. What he had in mind seems to have been that the steps in the evolutionary change have not yet been traced.

The small pterodactyls of Solenhofen need no particular description. They differ from *Rhamphorhynchus* chiefly in having a short tail, a short fifth toe, a rather elongate skull and lower jaw, and smaller teeth, which are directed forward less conspicuously. The Cretaceous *Pteranodon* (Fig. 87), however, deserves at least a paragraph, partly because it is the only well-known American pterosaur and partly because it is one of the greatest freaks of all time. Neither the anatomical knowledge of a Cuvier or an Owen nor the opium-inspired imagination of a De Quincey would have sufficed to predict the structure of this extraordinary reptile. The skull is the

Fig. 86. *Rhamphorhynchus*, showing the long tail and portions of the wing membranes. Two-ninths natural size. From S. W. Williston, *The Osteology of the Reptiles*.

most remarkable feature, with its elongate toothless jaws balanced by an almost equally long backward extension of the posterior bones. Not one of *Pteranodon's* Jurassic ancestors showed any tendency toward the production of such a crest. The eye seems totally out of place, far behind the posterior ends of the lower jaws and above the lateral temporal openings. The skull is nearly twice as long as the vertebral column, the neck as long as the inadequate and feeble body. The hind legs are long but slender, obviously ill-adapted for walking or perching. The fourth finger and its metacarpal are so greatly elongated that the wingspread must have been from twenty to twenty-five feet. Despite its great size, *Pteranodon* probably weighed no more than twenty-five pounds, according to Dr. George Eaton, who arrived at this result after weighing the exceedingly thin-walled bones of a nearly complete skeleton.

There has been a great deal of discussion of the habits and possible habitat of the pterosaurs. Their general similarity to birds, their large wings, hollow, air-filled bones, and keeled sternum, together with other minor characteristics which have not been mentioned, suggest that they had considerable powers of flight. On the other hand, the wings are too large and awkward for efficient flapping, and the elongate anterior support makes them at once much less mobile than those of birds or bats and much less under control. Recent studies of the soaring action of birds and experimentation with motorless gliders suggest that the action of the flying reptiles was more in the nature of volplaning than of true flying. With light bodies and a great expanse of wing, animals could accomplish long flights by taking advantage of various currents in the atmosphere.

Seeley thought that on land the pterosaurs were chiefly quadrupedal, as is shown by his numerous restorations, in which the animals are shown on all fours, the wings

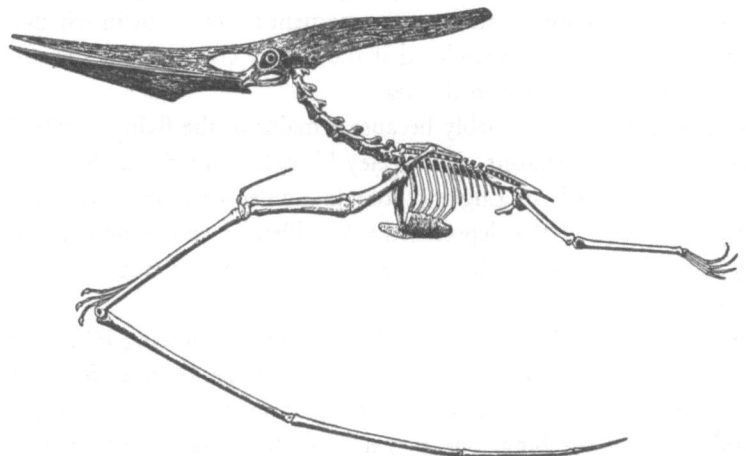

FIG. 87. Skeleton of *Pteranodon*, the most specialized of the pterosaurs. From George F. Eaton.

folded up at the sides. This interpretation was doubtless inspired by the walking attitude of the barbastelle bat which he figured. It is a question, however, just how accurate this figure is. Most members of this group of flying mammals have the knee twisted about so that it points backward rather than forward and hence are no great pedestrians. Seeley also considered that bipedal locomotion was possible in the pterosaurs, but Williston argued against this on the ground of the weakness of the hind legs and the slenderness of the toes. It should be pointed out, however, that the vestigial nature or absence of the fibula indicates that either the pterosaurs or their ancestors must have made good use of the hind legs. This bone does not tend to disappear in sluggish quadrupeds or bipeds.

Abel, whose studies of the habits of extinct animals have been based upon wide knowledge of comparative anatomy, came to the conclusion that the pterosaurs were arboreal. Their clawed front and hind feet enabled them to climb, not perhaps with the agility of a squirrel but sufficiently well to get about. There is no evidence of the rotation of a first toe which would enable them to perch like a bird, but the long slender toes would serve to hold them in an inverted, batlike position. Folding their wings about them and pulling in their heads, they may have passed slumberous hours swaying in the breeze. Tired of one position, they had only to spread their wings and glide to another. A favorable breeze might tempt them to a long soaring flight. If food became scarce, a whole colony might glide away to settle elsewhere. Seeley deduced from the abundance of their bones in the Cambridgeshire greensand that they were gregarious.

What was their food? Seeley and many others have said fish, and there are numerous pictorial restorations showing a pterosaur swooping down with a fish, held crosswise, in its mouth. I cannot at the moment think of one in which the animal is swooping up. I have often wondered if the kindly creature was not on the point of putting a stranded fish back in the sea.

Why the fish-motive? Probably because remains of the flying reptiles have been found chiefly in marine deposits. But Seeley himself pointed out that practically all the pterosaur-bearing beds in England contain numerous remains of terrestrial animals and plants. One of the deposits, the Wealden, is chiefly or entirely of freshwater origin. Seeley interprets this as meaning that pterosaurs fished in rivers or in near-by seas.

But why fish at all? More natural is the restoration showing *Rhamphorhynchus* pursuing a dragonfly, for some fairly large ones rest beside it at Solenhofen. It is doubtful if any animal ever climbed a tree primarily to catch fish. There are two good reasons for tree-climbing, one to get food, the other to escape enemies, and possibly a third, intellectual curiosity.

> *Zaccheus*, he
> Did climb the tree,
> His Lord to see.

But it is doubtful if this sort of curiosity was rampant in Mesozoic times. It is probable that the ancestors of pterosaurs and birds climbed trees originally because of one or both of the incentives first mentioned. Perhaps they were chased up and, having found food there, returned again; perhaps, having tasted the fruit upon branches within their reach, they pursued it onto higher levels. Where flowers and fruit are present, insects abound — an excellent mixed diet, much enjoyed for a time by our own ancestors. If persisted in, however, it leads to the swallowing of small insects, seeds, and fruits whole. There is no particular need for mastication. Through disuse, the teeth become feeble, and some or all of them disappear, as they did in the pterosaurs.

It is interesting to compare the history of the loss of teeth by the pterosaurs with the same process in birds, although not much is known about it in either line. Apparently it began at the back of the jaws in the flying reptiles, at the front in the birds. The ancestral condition of neither group is known, but it is supposed that each was derived from a carnivore with small, closely set conical teeth on the margins of the upper and lower jaws. The oldest known representatives of both have rather widely spaced teeth, an indication that reduction had been going on for some time. As to the flying reptiles, let us quote from Seeley. "A Pterodactyle's teeth vary a good deal in appearance. The few large teeth in the front of the jaw of *Dimorphodon* (Lowest Jurassic), associated with the many small vertical teeth placed further backward, suggest that the taking of food may have been a process requiring leisure, since the hinder teeth adapted to mincing the animal's meat are extremely small." If we are to trust Seeley's figures, which are the best known, the hinder teeth were not at all adapted for mincing meat. They were few and far between, and although they might have helped to break up a fruit, or husk a seed pulled off by the longer anterior teeth, they would have been capable of nothing more than putting dents in "meat." Seeley continues, "In *Pterodactylus* (Upper Jurassic) they are short and broad and few, placed for the most part toward the front of the jaws. Their lancet-shaped form indicates a shear-like action adapted to dividing flesh." Such carnivores as bother to masticate their food chew with the posterior (cheek) teeth, not the anterior ones, which are used for grasping and holding.

Not to prolong the argument, it may be said that there are two cogent reasons for believing that pterosaurs were not primarily piscatorialists. The first is the nature of the teeth. The real fish-eaters do not chew their food. Either they swallow it whole or cut it into pieces which will pass down their gullets. Those whose food was primitively fish, and some in whom the habit is secondary, have sharp recurved teeth which help to hold and ingest the prey. The forwardly inclined teeth of the pterosaurs would not have helped to retain active, struggling animals but would have served well enough in the pulling of fruits and seeds. The second is that, although it is conceivable that the flying reptiles could have swooped down, snatched a fish which happened to be at or near the surface, and then have risen again, the action would have been extremely risky. The slightest mischance would have involved a forced landing, and they were about as well adapted for getting off again as a land plane. Doubtless they would have floated well, but how could they get any lifting power from those long, ungainly wings? A bat can rise from the ground, although there is a general belief to the contrary, but it seems improbable that a pterosaur could have got off the water.

The relative abundance of the remains of pterosaurs in the quarries at Solenhofen is natural enough, for they contain many terrestrial insects and a couple of birds. The deposit was formed near shore, as is shown by the presence of the skeleton of a small dinosaur. The Kansan "chalk" from which remains of *Pteranodon* have been recovered was deposited much farther from the coast. However, remains of *Pteranodon* are extremely rare there; they probably represent a few unfortunate individuals, blown offshore and forced down in storms. Their bodies may have drifted long before the flesh had sufficiently decayed from their light, air-filled bones to allow their corpses to sink. They were too awkward, and carried too much sail, to be successful when emergencies arose. It is not at all surprising that the group became extinct.

XVIII

FROM SCALES TO FEATHERS

One never rises so high as when one does not know where one is going.
                                                    Cromwell to M. Bellièvre

The chief glory of the reptiles lies not in what they themselves accomplished but in the fact that they gave rise to the two most important groups of modern vertebrates, the warm-blooded feathered birds and the warm-blooded hairy mammals. Like many obscure mothers, they achieved their distinction through their offspring.

As has been pointed out repeatedly, the similarities between birds and reptiles are so numerous and so obvious that the appellation "feathered reptile" can be used with propriety. Modern mammals, except the monotremes, are not so obviously like the reptiles, and hence are commonly considered to be more widely removed from the ancestral stock, more specialized. If one stops to reflect, however, it will be realized that the wing of the bird is more highly modified than the limb of any mammal, and that other peculiarities, such as pneumatic bones, the curious vertebrae, and the synsacrum, combine to produce a skeleton much more specialized than that of the "average" mammal. Furthermore, in Triassic times the mammals were so like reptiles that it is possible to trace them to a particular group. On the other hand, the numerous reptilian characteristics of the birds do not suffice to lead one to the parental stock. They are the most specialized vertebrates and on anatomical grounds may be placed at the summit of the animal kingdom.

Remains of birds, in strata older than the Pleistocene, are among the rarest of fossils; well-preserved specimens are the greatest of paleontological treasures. Until 1861 none was known from strata older than the Tertiary, and, since those of the Tertiary are similar to modern ones, not much had been learned of bird ancestry up to that time. Then came, in a space of some sixteen years, a series of remarkable finds in the Jurassic and Cretaceous which served to bridge, in part at least, the gap between feathered and scaled animals.

The oldest known bird is *Archaeopteryx*, whose name, meaning ancient feather, was given to the first specimen found. To describe a modern genus if one knew nothing more of it than the imprint of a single feather on a bit of hardened mud would be considered a very foolish proceeding, but it must be remembered that when this one was found in quarries in the Upper Jurassic limestone at Solenhofen it was the first trace of a bird Mesozoic strata had produced. Curiously enough, only a month after it had been described as *Archaeopteryx lithographica*, a nearly entire individual was discovered at the same locality. It was secured for the British Museum (Natural History). Although at first supposed to be of the same species as that which lost the original feather, it was later christened *Archaeopteryx macrura*. It lacks the skull but is otherwise nearly complete. In 1877 another specimen, the most nearly complete yet known, turned up at Solenhofen and was purchased for the Natural History Museum at Berlin. Long supposed to be an *Archaeopteryx*, recent study has convinced paleontologists that it belongs to another genus; hence it is now known as *Archaeornis siemensi*. Since 1877 nothing avian has been found at this locality, so the catalogue of Jurassic birds is a brief one, two relatively complete specimens and a feather. O. C. Marsh, it is true, described a "bird" from the Upper Jurassic of Wyoming, but the fragments are so poorly preserved that it cannot be proved that they are avian.

As Alexander Wetmore points out, there are many reasons why birds should be rare as fossils. Epicurean man is not alone in his appreciation of a "warm bird," although it must be said to the credit of other animals that he alone desires the "cold bottle." Many birds succumb to the attacks of carnivorous animals; even those which die a natural death are eaten, unless speedily buried. Since the limb bones are hollow, carnivores crush them easily; only the thicker portions at the joints escape comminution. Small birds are entirely consumed, even the bones being digested. The brain is a choice tidbit, easily obtainable by crushing the thin-walled skull, and, the breast and viscera being equally desirable, little is left after the predator has lunched. It is not, then, surprising that most of the remains to be found in paleontological collections are ends of limbs, toe bones, and other scattered scraps.

The skeleton differs in many respects from that of either a reptile or a mammal. The skull, which has large orbits, shows little trace of sutures between the constituent bones. The quadrate is movably articulated with the skull, not incorporated in it, and thus is "free," like that of mosasaurs and snakes. One of the most characteristic features is the bar which extends along the lower boundary from the quadrate to the maxillary. It is a slender bone made up of the quadratojugal, the jugal, and a posterior projection from the maxillary. A bone of such form is present in no other animal. On the other hand, the single occipital condyle is an obviously reptilian feature. The neck is extraordinarily flexible, as must have been observed by anyone who has watched the snaky writhings of geese or swans. This mobility is made possible by the saddle-shaped articulations between the vertebrae and by the numerous processes to which muscles are attached. If you are unfortunate enough to be served with the neck of a fowl at dinner, your struggles to remove meat from it will convince you of the complexity of the bones.

The rigidity of the trunk vertebrae is in striking contrast to the mobility of the cervicals. The vertebral column, from shoulder to sacrum, is a solid support for wings and legs. The dorsal vertebrae are more or less coalesced, so that little movement is possible between them, and the posterior ones are united with the long, broad ilia to form what is known as the synsacrum. The true sacrum has only two vertebrae, as in the reptiles, but from nine to twenty-one more are joined with them, the whole united with the pelvis to provide an unusually strong support for a body held in a semierect position. The ilia are long and deep, the ischia in most cases fused with them posteriorly, whereas the pubes are small bones lying below the ischia. They are vestigial, apparently turned far back of the normal position for such bones. The tail, which is short, is composed of few vertebrae, most of them fused together. The flat, slender ribs are characterized by prongs directed upward and backward. These "uncinate processes" are characteristic, but, strangely enough, similar outgrowths are present on the ribs of some stegocephalian amphibians. The pectoral girdle is so constituted as to afford unusual support for the anterior appendages. Although it is not actually attached to the vertebral column, the muscular union is close. The shoulder blades project backward, parallel to the dorsal vertebrae. The collarbones are united in front, forming the furcula ("wishbone"), and the coracoids join the anterior end of the breastbone. The latter is large, although thin. It is smooth in birds which have lost the power of flight but, as every carver knows, has a deep median keel in those with functional wings. The most powerful of the wing muscles are those connected with the sternum, and the presence of a keel adds greatly to the area to which they may be attached.

The fore limbs have fewer bones than do those of most tetrapods. No vestiges of the fourth and fifth fingers remain, the wrist is incomplete, and the number of phalanges much reduced from the typical reptilian formula. If we knew nothing of modern birds other than the skeleton, we should probably apply the term "degenerate" to limbs with so few bones; knowing as we do how powerful these organs are when equipped with feathers, we realize that they are really highly specialized. The hind legs are less modified, but still they have characteristics in which they differ from those of all other animals. The femur is short; the tibia and fibula are long. The latter bone is vestigial, in some cases joined to the tibia. The ankle joint is between the two rows of tarsals, and no one of these is present as a free bone. Since those of the upper row are fused to the tibia, this combination is termed a tibiotarsus rather than a tibia. The lower tarsals are ankylosed to the upper ends of the metatarsals, which consequently are called the tarsometatarsals. The second, third, and fourth metatarsals are coössified into a single rather long bone. The first is vestigial, only the lower end remaining, and the fifth is absent. Most birds have four toes, the fifth (the little one) having been lost. The perching birds have the first toe behind, in opposition to the others. Many flightless birds have only three toes, the African ostrich but two.

Since the Jurassic birds are the oldest known, they are of the utmost importance, and their remains have been repeatedly studied and described. Nevertheless, much is still to be learned about them, and more and better specimens are urgently needed if the problems which they raise are to be solved.

An *Archaeopteryx* (Fig. 91) or an *Archaeornis* flattened in a slab of limestone covers a considerable area, but in life neither could have been much larger than a domestic pigeon. As preserved, the fossils retain impressions of the larger feathers of the wings, hind limbs, and tail, and the substance of most of the bones of the skeleton, although there are serious lacunae. Only the single specimen of *Archaeornis* retains the skull, which is flattened, badly crushed, and imperfect in the highly important posterolateral and temporal regions. In spite of these imperfections, it displays a large birdlike orbit, with remains of a ring of sclerotic plates, and a rank of thirteen small conical teeth, implanted in sockets, in the upper and presumably in the lower jaws. There is no beak.

The neck appears to be relatively short, but since cervical and dorsal vertebrae are alike, no exact cervical formula can be stated. Most students have described the vertebrae as amphicoelous, which may be true, but the statement is an assumption, for the specimens have been considered too precious to permit sufficient dissection to show the true nature. Externally the vertebrae are simple, and it is certain that the articulations are not saddle-shaped, so the neck could hardly have been as flexible as that of a modern bird. Neither could the dorsals have been rigid. The sacrum is badly preserved, but the pelvic bones of *Archaeopteryx* have lately been revealed by delicate manipulation. The ilium proves to be so short that not more than four vertebrae could have been attached to it, a marked contrast to the number involved in the synsacrum of modern birds. The tail is long, with a pair of large feathers for each vertebra. Totally unlike the modern fan-shaped tail, it was constructed on the same plan as that of a lizard. The bones of the pectoral region are imperfectly preserved. The scapula appears to have been slender, and a bone in front of it may represent the furcula. There is no trace of a sternum, although probably one was present. It may have been incompletely ossified or entirely cartilaginous.

The pelvis is peculiar (Fig. 93 B). The ilium is short; the pubes are deflected downward and turned backward on the ventral side, the opposite ones in contact, although not actually united. The ischia, bones with irregular outlines, project downward and backward. Their conformation shows that the egg must have been small, with contents not more than one quarter the bulk of that of a modern bird of the same size. The wings, although feathered, are not much like modern ones (Fig. 88). The humerus is longer than the radius and ulna, and though there are only three digits, each is complete, tipped by a long, strong claw. Apparently the fingers functioned both as climbing and as flying organs.

Until recently it was supposed that the hind limbs were like those of modern birds, but the removal of matrix which had partially concealed the bones shows that the fibula is as long as the tibia, although only a part of it is preserved, and that at least one of the tarsals, the fibulare, which is the fourth in the first row, is a free bone, not fused with the end of the tibia. Moreover, the median metatarsals are not joined to one another, although they lie very close together. Each probably has joined to its proximal end the corresponding tarsal of the lower row. The first metatarsal is behind the others and supports a short toe, opposed to three others, as in the perching birds. The outer, fifth, metatarsal is vestigial and has no toe.

From this detailed description of the skeletal features of the Upper Jurassic specimens it is evident that they have more reptilian than avian characteristics. They were, as Gerhard Heilmann has said, "warm-blooded reptiles disguised as birds." Almost their only strictly avian characteristic is the presence of feathers. The list of reptilian features, however, is formidable. It includes both positive and negative

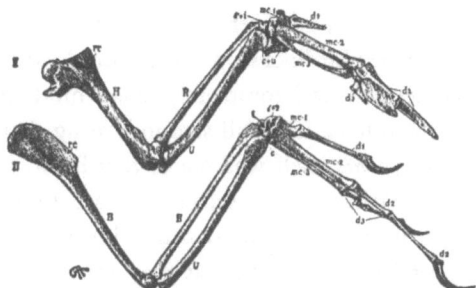

FIG. 88. Wing of a modern pigeon, above, compared with that of *Archaeornis*, below. H, humerus; R, radius; U, ulna; *c*, carpals; *mc*, metacarpals; *d1, d2, d3*, first, second, and third fingers. From Gerhard Heilmann's *Origin of Birds*, by permission of the D. Appleton-Century Company.

features, such as the lack of a bill, the presence of teeth, a fixed rather than a free quadrate, simple vertebrae, lack of differentiation between cervicals and dorsals, few sacral vertebrae, a long tail, the absence of uncinate processes from the ribs (this may be due to poor preservation), the presence of ventral ribs, a humerus longer than the forearm, fingers with claws, a weak, possibly unossified sternum, a short ilium, a long fibula, at least one discrete tarsal, and separate metatarsals.

Lower Cretaceous deposits have as yet yielded only scanty and unsatisfactory traces of birds, but considerable dispersal must have been taking place during this time, for avian fossils are widespread in Upper Cretaceous strata. Since most of the remains are exceedingly fragmentary, little is known about these latest Mesozoic representatives of the group. Only two genera are really well known. They are *Ichthyornis* and *Hesperornis*, both described by Marsh from remarkable specimens found in the yellow chalky limestone of western Kansas.

Ichthyornis (Fig. 92, at left) appears to have been about as large as a domestic pigeon. It was a true bird of flight, as is shown by the large, deeply keeled breastbone and well-developed wing. The feathers have not been found, but the ulna shows tubercles for the attachment of secondary wing feathers, and the ankylosed metacarpals afforded a firm basis for the primaries, so it may be inferred that the wings had the same general arrangement of quill feathers as modern ones. Despite numerous primitive characteristics, the skeleton is surprisingly modern. The chief primitive features are teeth in sockets along the maxillaries of the upper jaw and the whole length of the dentaries of the lower; the two rami of the jaw not fused at the symphysis; the vertebrae slightly biconcave; the three components of the pelvis not fused with one another at the posterior ends, and the small brain. In contrast to these are the following: the presence of a bill on the upper jaw, as indicated by the absence of teeth from the premaxillae; somewhat complex lateral and dorsal processes on the vertebrae, the third cervical foreshadowing saddle-shaped articulations; distinctly modern wings, with radius and ulna longer than the humerus; coössified metacarpals and reduced, clawless phalanges; birdlike scapulae, coracoids, and sternum; an elongate pelvis, attached to a synsacrum; and completely avian hind limbs, with a vestigial first toe in opposition to the others, its small size indicating that, although descended from a perching type, *Ichthyornis* itself was not of that habit.

The characteristics appearing in the second list are so much more important than those in the first that it may fairly be said that birds were about as highly specialized for flight in Upper Cretaceous times as they are at present. They must have passed through the more important stages in their evolution between Upper Jurassic and Upper Cretaceous times. This has been cited as an instance of rapid evolution, but as a matter of fact the time that elapsed between the Upper Jurassic and the Upper Cretaceous was probably as long as that from the Upper Cretaceous to the present.

In the same strata with *Ichthyornis* are found the remains of *Hesperornis* (Fig. 89), a large wingless creature fully equipped for aquatic life. Specialized as the birds of flight are, this aquatic animal superimposes specialization upon specialization. Crowded off the earth by overpopulation or by enemies, or lured therefrom by certain kinds of food, birds learned to fly and became adapted to life in the air. But no sooner had they mastered the new form of locomotion than some members of the group went back to live on land; others, still more adventurous, essayed life at sea. In the new environment, the wings, the perfection of which had been the crowning accomplishment of the race, became more or less worthless and through lack of use tended to atrophy. Everyone is familiar with the fact that the wings of many modern ground and aquatic birds are inadequate for sustained flight. Yet wings are present even in the flightless birds. Few if any of the modern Aves show so great a reduction of wing as *Hesperornis*.

Comment has just been made on the suggestion that the perfection of the wing between Upper Jurassic and Upper Cretaceous times indicated rapid evolution, and a reason for discounting it has been shown. But *Hesperornis* tells a different story. Almost completely wingless, it was nevertheless descended from a bird of flight, indicating, if one may so express it, twice as much evolution as *Ichthyornis*. That is, full adaptation to life in the air was accomplished; then, secondarily, there was a complete change to a life of swimming and diving. Unfortunately, there are no data on rates of change in evolution. Despite certain opinions, based upon Biblical training, it is not at all certain that it was slow on the narrow upward path and rapid on the broad downward way. In fact, it is unlikely that there is any definite rate; progress one way or another depends upon many circumstances, partly inherent in the organism, partly external.

FIG. 89. Restoration of *Hesperornis*, redrawn after G. Heilmann.

Specimens of *Hesperornis*, spread out on slabs of chalk, reach a length of five feet from tip to tip. Because of the peculiar construction of the legs it is difficult to say what its height was, but it was probably about as large as a swan, although the elongate head is un-swanlike. The arrangement of the teeth seems to be the same as in *Ichthyornis*, except that they were set in grooves instead of having separate sockets. The elongate cervical vertebrae are relatively simple but have saddle-shaped articulations. The ilia are long and connected with the many vertebrae of the synsacrum, an avian feature, but the long slender ischia and pubes are separate, forming a pelvis which as a whole is primitive. The ribs have uncinate processes; the sternum is long and well developed, although without a keel: the clavicles are not united to form a wishbone. The coracoid is short and wide, forming, with the long birdlike scapula, a firm support for a wing; but the wing is not there. The slender humerus dwindles to a point at the distal end, showing conclusively that it is only a vestige. It must have been without function. The hind limbs are well-developed and strong, larger, in fact, than those of a perching bird. The femur is short, the tibiotarsus long, stout, and hollow, the tarsometatarsals long and strong but curiously articulated so as to turn outward at right angles to the body. As Lull has shown, this proves that the feet could not be brought beneath the body; hence *Hesperornis* could never have walked well on land but was doomed forever to a paddling existence. For such a life it was, however, well fitted, for it not only had a thick coat of downy feathers but the outer (fourth) toe is tremendously developed and fully twice as long as the third. The first does not turn backward as in the perching birds but forward, contributing its mite to the formation of a powerful, webbed foot; yet it is far from deserving the title of big toe, for it is vestigial, much smaller than the others.

As may be seen by the sketchy description outlined above, the skeleton of *Hesperornis* presents a queer combination of primitive (reptilian), advanced (bird of flight), retrograde (wingless), and aquatic characteristics. Those of the first two categories suggest the ancestry, those of the last two the habits of the creature. It was apparently a strong swimmer, a good diver, a fish-eater. Entirely efficient at sea, it could only have flopped about on land, if it could make any progress at all. Possibly it laid its eggs in nests built of drifting vegetation, as the modern grebe does, although it is also possible that they may have been deposited in the sand, near high-tide mark. Wetmore suggests that *Hesperornis* could have floundered about on shore, progressing, as the seal does, on its breast. This does not seem likely, however, for the pectoral appendages were of much less use than those of seals.

So much for Cretaceous birds. In some respects their characteristics are intermediate between those of Jurassic and modern ones, but so few genera and species are well known that it is impossible to trace family lines. It is not probable that any one of them is directly ancestral to any living animal. Yet these few representatives of the ancient avian fauna do help to reconstruct, partially at least, the chain of structural changes through which birds must have passed in the course of their evolution.

Cenozoic birds are much more like modern ones than those of the Mesozoic are. Most of them can be referred to families represented in the modern fauna, although there are some peculiar creatures of doubtful affinities, probably representatives of extinct groups. Only the large flightless ones will find a place here.

The most spectacular of all Eocene birds was one of the unexpected discoveries made by a party from the American Museum of Natural History in 1916. While collecting mammals in the Big Horn basin of northern Wyoming, they came upon the nearly complete skeleton of a huge bird which apparently belongs to the same genus as one previously known from fragmentary specimens from New Mexico, to which the name *Diatryma* (Fig. 90) had been given. About seven feet in height, it enjoys equal distinction as the largest North American bird and the oldest of all ground birds. It has no more remnants of wings than *Hesperornis*, no vestige being visible externally. The hind legs are long, strong, provided with three large toes. The neck is short, with large vertebrae, to which were attached the powerful muscles necessary for holding the most massive head ever seen on a bird. The skull is about seventeen inches long and six and one-half inches deep in front of the eyes. There are no teeth; nevertheless, one is inclined to believe that it was a carnivore, preying upon small animals of all sorts. Although without any particular organs of defense other than the strong hind legs and the powerful beak, *Diatryma*, when on the alert, could not have been in any great danger from the rather small contemporary carnivorous mammals. The general opinion seems to be that it was related to the cranes and rails, although not ancestral to them.

FIG. 90. An Eocene flightless bird, *Diatryma*, and a contemporary horse, drawn to the same scale. The horse, about one foot high at the shoulder, is redrawn after Charles R. Knight, the bird after Matthew and Granger.

The Miocene strata of Patagonia have produced some large and strange ground birds, larger even than *Diatryma*. One of the most remarkable is *Phororhacos*. Imagine an ostrich with a head and neck as large as those of a horse, the skull bearing a great curved beak as sharp as an ice pick, and you have some idea of this peculiar creature. The largest known skull is about two feet long, exceedingly massive, but not so deep as that of *Diatryma*. The cervical vertebrae are enormous, five inches across, larger than those of any other known bird. Vestigial wings, totally inadequate as organs of flight, may have been of some assistance as offensive or defensive weapons. The legs are large, with three strong toes on each foot. One suspects that *Diatryma* was a carnivore: one is sure that *Phororhacos* was — a bold aggressive carnivore, with little to fear and itself an object of terror to all inhabitants of the land. The huge legs suggest power in running, scratching, and fighting, so that, although it is considered to be a relative of the cranes, it is difficult to think of it as a wading bird. *Phororhacos* and *Diatryma* were related, doubtless through a common ancestor.

The occidental world provided the first of the ground birds, but their greatest diversity was in the Orient during Pleistocene times. These eastern feathered creatures are not in any way related to those just described.

The largest and most spectacular of all birds are the moas of the Pleistocene of New Zealand. Almost a hundred years ago missionaries to that southern Commonwealth heard from the natives stories of a monstrous human-headed creature which lived, guarded by two huge lizards, on a mountainside far in the interior. No man had seen this monster and lived to tell the tale, for on the approach of human beings it rushed forth and trampled them to death. Although none had actually seen it, the huge bones were well known. Instigated by these reports, travelers and naturalists set out in quest of what was locally called the moa. Although they never found it alive, they came upon its bones in the muck of swamps, in old lake beds, and in caves. Specimens were found retaining bits of skin and dried tendons, and old camp sites yielded charred bones and broken eggshells. Such evidence, coupled with tradition, indicates that the moas existed until recent times. It is probable, indeed, that the aborigines may have delivered the *coup de grâce* to this tribe. Remains of many were sent to Europe in the middle of the last century, and from twenty to twenty-five species have been described, grouped under four or five generic names. Not all were giants, but all were relatively large. Many of them passed through the hands of the great anatomist, Sir Richard Owen, who fortunately had a relative in New Zealand to supply him with specimens which are now in the British Museum (Natural History).

The larger moas (*Dinornis* and *Palapteryx*) are of extraordinarily massive construction. The head is relatively small, but broad, with a wide, short, pointed beak. The vertebrae are large, especially the dorsals; the breastbone large, convex or flat, but without any trace of a keel. A scapula is present, but there is no glenoid cavity at the junction with the coracoid, hence no articulation for a humerus, and no vestige of a wing. The pelvis is curiously primitive, the single archaic feature of these remarkable birds. The ischia are not united with the ilia, nor are the pubes in contact with the ischia. In this respect the hipbones are reminiscent of those of *Ichthyornis* and *Hesperornis*. Whether this is a truly primitive characteristic or a reversion to an ancestral status cannot be determined in the absence of representatives of the group older than the Pliocene. The question is of considerable importance, for if the condition is due to secondary adaptation it constitutes an important exception to Dollo's "law of the irreversibility of evolution." The hind legs are "perfectly enormous." The femur is relatively long for a bird but not more than half as long as the tibiotarsus; both are of large diameter, greatly enlarged at the joints. Although vestigial at the lower end, the fibula is a big bone, mostly behind the tibia. Three large spreading toes are supported by the massive tarsometatarsal.

The largest of the moas, *Dinornis maximus*, appears to have been at least ten

Fig. 91. The specimen of *Archaeopteryx* in the British Museum (Natural History). From B. Petronievics.

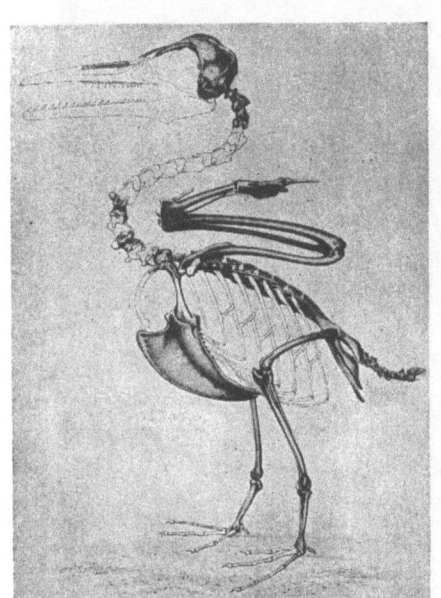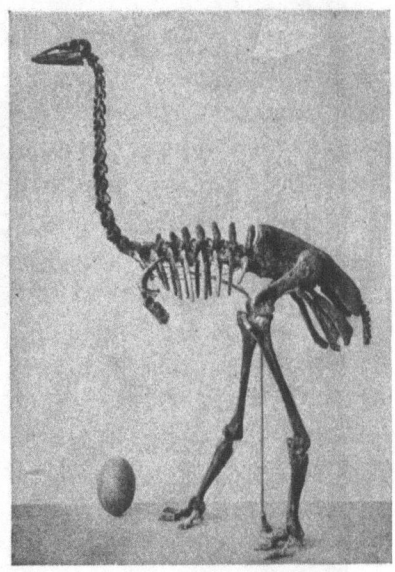

Fig. 92. At left, a restoration of *Ichthyornis*. About one-third natural size. From O. C. Marsh. At right, a skeleton of *Aepyornis*. Original about ten feet high. From L. Monnier.

feet in height — opinions differ. Its proportions were those of an ostrich, but since it was much taller, was larger in the body, and had more massive bones in the legs, it must have been a far heavier creature. The power of the legs must have been tremendous. We may, with Dr. F. A. Lucas, put the question, "If a blow from an irate ostrich is sufficient to fell a man, what would be the kicking power of an able-bodied Moa?"

Madagascar, as well as New Zealand, was inhabited in Pleistocene and more recent times by huge birds. All those now known belong to the genus *Aepyornis*, long famous for the size of its eggs. Eggs are notoriously fragile; yet in one way or another some escape destruction. Birds' eggs are known from rocks as old as the Cretaceous, and collectors come upon them from time to time in various Tertiary strata. *Aepyornis* (Fig. 92, at right) eggs have been found chiefly in the muck of swamps, although the most nearly perfect ones are those which have been transported by streams from the place of their original burial. One of the best specimens is said to have been taken at sea. The largest known has as its greatest diameters nine by thirteen inches. The egg of an ostrich measures four and a half by six inches; hence the shell of the egg of *Aepyornis* would hold the contents of six ostrich eggs. If the comparison be carried further, it appears that the shell would hold the bulk of more than a gross of ordinary hens' eggs and of some thirty thousand hummingbirds' eggs. In plain terms, the capacity is about two gallons.

Judged by the size of the egg, *Aepyornis* should have been about six times as large as an ostrich. But, as Lucas has pointed out, there is no definite relationship between the stature of a bird and the size of its egg. There still lives in New Zealand a flightless, four-toed bird, somewhat distantly related to the moas, famous for its eggs. The little *Apteryx*, smaller than our common barnyard fowl, produces an egg which measures three by five inches, and has about one-third the weight of its producer. Judged by the egg alone, *Apteryx* should be four feet high. But skeletons (one is on exhibition in the Museum of Comparative Zoölogy at Harvard) convince us that *Aepyornis* was not as much larger than an ostrich as the egg would lead us to believe. The largest *Aepyornis* was about as tall as the largest moa.

Archaeopteryx and *Archaeornis* lived at a time not far removed from that at which their stock diverged from some reptilian ancestor. But what was the ancestor, and how was the separation accomplished? No positive answers to these questions can as yet be given. We know so little that the most that can be done is to formulate theories and assemble facts to support or contradict them. Since the origin of flight is intimately connected with the origin of birds, that topic may be discussed first. Two theories purporting to account for the use of the anterior limbs in flying are current. Neither, it must be confessed, is really plausible or particularly well supported by evidence drawn from the anatomical characteristics of birds.

The first is the "cursorial theory," championed by the late Baron Nopcsa, an eminent Hungarian paleontologist. He pointed out that the hind limbs of birds are not so modified as to assist in flying but are essentially similar to those of bipedal dinosaurs. He thought that the first step toward the acquisition of the power of flight was the assumption of a semierect posture, in which attitude the hind legs were the sole organs of locomotion. On this hypothesis the ancestor was a slender, hollow-boned, bipedal creature which ran as readily as it walked. Everyone has seen athletes wave their arms as they run, seemingly trying, by beating the air, to pull themselves along a bit faster. Nopcsa maintained that the ancestral reptile-bird did the same and gained thereby a certain increment of speed, for the scales on the arms provided a surface sufficient to be of some use when pressed against the air. According to this theory, continued use of the arms as propellers brought constant increase in the length and breadth of the scales on them. As the scales lengthened, the pressure against the air caused their edges to become frayed, and so scales gradually changed to feathers. The broader the arm became, the greater its contribution to the speed of the animal. From generation to generation the wings increased in size, till finally, on a happy day, becoming at last more important than the legs, they lifted the body off the ground, and flight had been attained!

This theory seems a trifle fanciful, and verges upon the absurd in its assumption that feathers evolved first on the wings and then spread all over the body. The rather naïve idea that feathers are frayed-out scales should not be charged to the discredit of Baron Nopcsa. The same opinion has been held by many zoölogists and most paleontologists, and the statement that feathers are modified scales appears in most textbooks. It is only recently that it has been shown that they are fundamentally different structures, arising from different layers of cells in the skin. Feathers are as absolutely confined to the birds as hair is to the mammals. The cursorial theory of the origin of flight cannot be taken seriously enough to require detailed refutation. It should, however, be noted that there is nothing about the structure of *Archaeopteryx* or *Archaeornis* to suggest that they were rapid runners. The small pelvis and relatively feeble hind limbs are strikingly different from those of the cursorial birds of the Tertiary and Pleistocene, or even the bipedal dinosaurs.

The second suggestion is that of William Beebe and W. K. Gregory, who proposed what is known as the tetrapteryx theory. They suppose that the ancestors of the birds were active arboreal reptiles which were in the habit of jumping from branch to branch and so from tree to tree. Those individuals with the largest scales on arms and legs were able to make the longest swooping jumps; so from generation to generation they became better and better gliders. Beating the air with fore and hind limbs extended the distance coverable in such leaps, developing wings on all legs, the scales being transformed into feathers by the fraying of the edges. After acquiring considerable skill in flight by the use of four wings (the tetrapteryx stage), the birds discovered that they could do better by using the anterior pair only. Since the posterior wings became inactive, they gradually lost their large feathers and reached their present condition. The tetrapteryx stage served merely to tide the animals over that difficult period during which they were learning to fly.

The fundamental idea in this theory — that is, that the ancestors of the birds were scansorial, arboreal reptiles — is good, but there seems to be no evidence that birds ever had four wings. It is true that the hind limbs of *Archaeornis* appear to have borne a series of long quill feathers, but that they formed a wing is not so certain. It has also been asserted that the hind legs of the embryos of certain modern birds show lines of bristles which are vestiges of the wing feathers of the ancestor, but Heilmann has disproved this statement.

The tetrapteryx theory, with the posterior wings left out, a sort of Hamlet without Hamlet, is supported by Heilmann, the Danish ornithologist who has made the most important contributions to the solution of the problem of the origin of birds. His "gliding theory" includes the fundamental ideas of Beebe and Gregory with additions of his own, as will be indicated in later paragraphs.

As for the question of the ancestry of birds, three theories have been advanced. The first and most natural of these is that flying birds evolved from the flying reptiles, the pterosaurs. The similarities between the two are numerous. Both, in the early part of their history, have conical teeth, which in later ages are lost and replaced by a bill. In both the skull is held nearly at right angles to the backbone, and the brains are much alike. Both have large orbits for the eyes, a ring of sclerotic plates, hollow, air-filled bones with dense walls, and a large sternum, an element rarely ossified in reptiles. Yet despite the numerous similarities it is impossible that pterosaurs should have given rise to birds, for their chief specialization was the enormous development of the fourth finger, a digit of which no bird, even the most ancient, retains the least vestige.

A second theory, with less basis but, until recently, widely accepted, is that birds descended from bipedal dinosaurs. The chief features common to the two groups are hollow bones in theropodous dinosaurs and birds; posterior branches of the pubes in ornithischian dinosaurs, comparable in position to the pubes of birds; and ankle joints between the two rows of tarsals. The hind feet of many theropods have only four clawed toes, the fifth absent, and the first rotated behind the metatarsals, as in birds. Finally, the pelvis of some dinosaurs is large and long, with as many as eight vertebrae in the sacrum.

Although such a thought seems not to have occurred to many of the geologists and paleontologists who accepted this theory while it was popular, it would be impossible for birds to have arisen from the large heavy-boned ornithischians, in which group alone is to be found the so-called birdlike pelvis. Moreover, since birds have no prepubis, the homology between the ornithischian and avian pelvis is far from being exact. The only possible ancestors are the slender, hollow-boned theropodous coelurosaurs, which as Heilmann has shown in his book, *The Origin of Birds,* have many birdlike characteristics: "Hollow bones of very light structure, exceedingly long hind limbs with strong, elongate metatarsals and a 'hind toe,' a long narrow hand, a long tail and a long neck, large orbits and ventral ribs — these are bird-features immediately conspicuous."

As will be remembered, the coelurosaur lineage persisted from late Triassic to late Cretaceous times, which makes it possible to observe some of the evolutionary trends in the group even if individual lineages cannot be followed. These trends are such as to cause the coelurosaurs to become more and more birdlike with the passage of time. Triassic representatives of the group have conical teeth in sockets; Jurassic *Compsognathus* and *Ornitholestes* have similar but smaller teeth and fewer of them; Cretaceous *Struthiomimus* has lost all teeth and acquired a bill. Similarly, the bones of the skull became more slender, and the preorbital openings larger and larger, more like those of *Archaeornis*. The pelvis increased in length, the number of sacrals changing from three to five. Perhaps the most important feature of the evolution of the group is the fact that in the course of time the fore limbs, short in the Triassic *Podokesaurus* and *Procompsognathus*, became longer and longer; Heilmann was the first to point out this change, which is opposite to that characteristic of the other theropodous dinosaurs. As the arms became longer, the outer fingers were lost, until only three remain. The late coelurosaurs were, except for the absence of feathers, exceedingly birdlike. The obvious difference in skeletal structure is that in them the pubic bones extend downward and forward, instead of backward. But this is no insuperable obstacle to the derivation of the birds from them, for the pelvis of embryos of birds is of the typical reptilian triradiate type. The distal end of the pubis, it is true, is recurved, but the proximal portion projects downward and somewhat forward (compare Fig. 93). This is interpreted as indicating that the ancestor of the birds had a pelvis more like that of theropod than like that of ornithischian dinosaurs.

If one could stop here, it would seem logical to conclude that birds were derived from coelurosaurs. But there is a fly in the amber, a tiny flaw in the argument. Coelurosaurs had no clavicles. Pterosaurs managed to fly without the aid of these bones, but apparently birds cannot. A wishbone, or at least the elements thereof, is present in all birds, even the oldest. It is evidently a *sine qua non* of bird organization. No reptile which had lost it could have been ancestral to the group. The present opinion is that most of the birdlike features of the dinosaurs are due rather to the semierect position in which the animals walk than to the descent of one group from the other. Nevertheless, it is generally held that birds and dinosaurs had common ancestors, a position ably supported by Heilmann. After a detailed study of the osteology, anatomy, and embryology of birds, he has shown many reasons for believing that the pseudosuchian reptiles of the Triassic of South Africa, now generally accepted

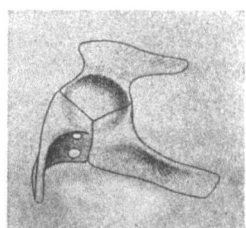

FIG. 93. One side of the pelvis of an ostrich (A), as compared with that of *Archaeopteryx* (B). Ilium above, ischium beneath, with backward projecting pubis below it. Redrawn after G. Heilmann. At right, the pelvis of *Euparkeria*, to show the downward curvature of the pubic bones. One-half natural size. From R. Broom.

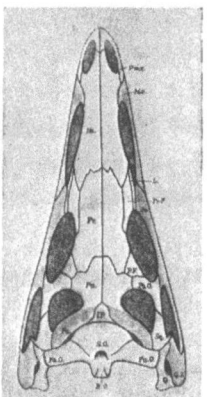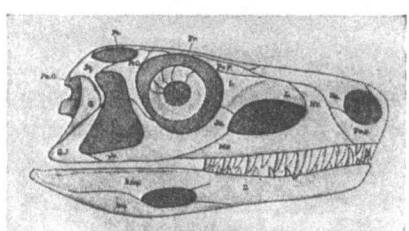

FIG. 94. Dorsal and lateral views of the skull of *Euparkeria*, a representative of the group (Pseudosuchia) from which both the dinosaurs and the birds are supposed to have been derived. Lettering as in Fig. 64. One-half natural size. From R. Broom.

FIG. 94A. A "quarrel scene" from the Upper Jurassic, with *Rhamphorhyncus* taking the part of Mercutio, and *Archaeornis* that of Tybalt. From a lithograph by Charles R. Knight, with his permission.

as the ancestors of the theropodous dinosaurs, were also those of birds. Unfortunately, his argument depends largely upon the interpretation of the elements of the skull of *Archaeornis*, a portion of the skeleton so badly crushed that more than one reconstruction is possible.

The movable quadrate and the lateral bar of the skull are so characteristic of birds that one might expect them to be present in the ancestor. Among reptiles, the free quadrate is found in the Squamata only, but one seeks in vain among the lizards, the most primitive of the Squamata, for any representative suggestive of a bird. They have no lower temporal arcade and hence no structure comparable to the lateral bar of the birds. Consequently, we are forced to agree with Heilmann that the movable quadrate is not a feature of prime importance but was formed independently in squamates and birds. Poorly preserved as the skull of *Archaeornis* is, it does seem to show that the quadrate was immovably articulated with the squamosal. Furthermore, the marginal bones were not of avian structure, for there was unquestionably an ascending process of the jugal behind the orbit, and the quadratojugal must have been short. Thus interpreted, the skull of *Archaeornis* is similar to that of the pseudosuchian, *Euparkeria* (Fig. 94), as described by Broom from specimens collected from the Triassic of South Africa. The chief difference is that *Euparkeria*, being a diapsidan, has a dorsal temporal opening, which is absent from *Archaeornis* and all later birds. The parietals, postorbitals, and squamosals of the latter appear to have grown together so as to obliterate the upper fenestrae. So far as the skull is concerned, all that can now be said is that the pseudosuchian presents a possible model by which the skull of *Archaeornis* may be reconstructed.

The pseudosuchians were bipedal reptiles with five toes on both front and hind feet. The arms were shorter than the legs but well developed. The body was covered with scales, and in *Euparkeria*, Broom says, "all the best preserved scales were about twice as long as broad and have the long axis lying anteroposteriorly . . . from this axis there are distinctly traceable ribs running sideways, in form almost representing a feather; we need merely imagine the ribs continued beyond the border of the scale." The obvious suggestion is that the scale was in process of transformation into a feather. One should always be cautious, lest one prove too much.

Two long bones, a part of the hyoid arch of *Euparkeria*, have led Broom to the conclusion that this animal had a birdlike tongue. Other avian characteristics mentioned by Broom or Heilmann are: cervical and presacral vertebrae of approximately the same number; ribs double-headed, with small uncinate processes; ventral ribs present; clavicles long and slender, forming part of a shoulder girdle which has all the elements of that of a bird. The pelvis of the pseudosuchian is particularly important, for, although it is attached to only two sacral vertebrae, it is like that of the embryo of a modern bird, the pubis projecting forward in its inner portion, downward and slightly backward in the outer. According to Broom, the tarsus foreshadows that of the birds; the metatarsals are elongate, the third best developed, an avian rather than a reptilian characteristic.

Heilmann has given a picture of a hypothetical animal intermediate in characteristics between the pseudosuchians and the Upper Jurassic birds. His *Proavis*, "no longer a reptile and not yet a bird," is described as an inhabitant of the trees of the Triassic forest, where it hopped from branch to branch and gained distance by gliding with outstretched arms. The exigencies of arboreal life led to the perfection of a grasping type of foot and to the elongation of the arms. The long slender body and long tail are described as covered with long, broad scales, which were beginning to change into feathers, particularly on the arms, then in the initial stages of transformation into wings. The stimulus derived from the pressure of the air on the tips of the scales is supposed to have caused their further growth, particularly where the pressure was strongest, as on the posterior margins of the limbs and along the sides of the tail. Thus eventually an *Archaeopteryx* stage was reached. About the skull one can speak with less assurance, but it appears to Heilmann to have been large, with numerous conical teeth, its structure much more reptilian than avian.

Proavis may or may not have existed. It is one of the "missing links"; toward its recovery, exploration should be directed. If the pseudosuchians were really the ancestors of the birds, one would expect to find the remains of the annectent group in the older Jurassic strata of Africa. For the present, however, it will be well to keep an open mind as to avian ancestry, for although Heilmann has presented cogent arguments for the acceptance of his theory, they are based largely upon the interpretation of the single badly preserved skull of *Archaeornis*.

FIG. 95. Restoration of *Archaeopteryx*, redrawn after W. P. Pycraft.

XIX
SQUIDS, DEVILFISH, AND CHAMBERED SHELLS

> He that useth many words for the explaining of any subject, doth, like the cuttle-fish, hide himself for the most part in his own ink.
>
> John Ray, *On Creation*

Some animals seem predestined to obscurity. Who has dramatized the life of a bryozoan or written odes to a brachiopod? Other creatures, because of size, color, repulsiveness, or beauty, gain a place in the popular imagination, and become immortalized in literature or art. The lowly scallop was carried to honor as the badge of the Pilgrims; the shell of the triton appeared as a decorative motif in Grecian times. No marine invertebrates, however, have been more fully adopted by artists and writers than the cephalopods. Where is he who has not shuddered with Hugo as Gilliatt fought the octopus, and whose adolescent ears have not been assailed by recitations of that quintessence of Victorianism with the ringing moral: "Build thee more stately mansions, O my soul"? The squid, octopus, cuttlefish, and pearly *Nautilus* have characteristics of one sort or another which have made them well known. Some are swift, dramatically ejecting ink to baffle pursuers; a few are giants, hasty glimpses of which probably inspire some of the sea serpent stories; others are of devilish repute because of their horrid appeal to the imagination. The pearly shell of the *Nautilus* has for centuries made it valuable, a thing to be encased in gold, an ornament to the table of the highborn and wealthy.

Although they are so well known, the cephalopods are at present a relatively small group. They fall readily into two subclasses. The more abundant, the squids, cuttlefish, and octopi, lack an external shell, so that they might be called the naked cephalopods, although, because they have only two gills, their scientific name is Dibranchiata. The others, represented at present only by three species of *Nautilus* in the Indian Ocean and the parts of the western Pacific adjacent to southern Asia and the Philippine Islands, are sometimes denominated the shelled cephalopods, or, technically, the Tetrabranchiata, since they have four gills.

Examination of a squid, cuttlefish, or octopus will show that all have more or less elongate or bulbous saclike bodies, set off from the head by a constricted region resembling a neck. The head of the squid has ten, that of the octopus eight, more or less elongated processes, the arms or tentacles. These are supposed to represent the foot; hence the name Cephalopoda, "head-footed." The tentacles are provided with suckers or hooks, effective organs for grasping and holding. Many dibranchiates have two especially elongated ones, modified to serve in transferring the male sperm to the females; these arms, in the largest of living squids, animals with bodies five feet in circumference and fifteen feet long, reach a length of some thirty-five feet. Sixty years ago tales of such creatures were classed as myths and sailors' yarns, and depicted in various paintings and engravings showing the attack and overthrow of skiffs and small sailing vessels by the "octopus." Now it is known that the tall stories have a basis in fact.

The eyes of the naked cephalopods are marvels in the invertebrate realm. Provided with a lens, cornea, and lid folds, they approach in perfection those of the vertebrates. It is doubtless the possession of this type of eye which has gained for the octopus the reputation of being the most intelligent of invertebrates. Lying concealed among the shadows of the reefs which form a congenial habitat, this keen-sighted beast has been observed to dart out only when there is obvious opportunity to capture its favorite prey.

Naked though such cephalopods are, most are not entirely shell-less. If a squid's back be cut open, a thin, flexible pen is discovered within the folds of the mantle which envelops the body (Fig. 96, at left). Too weak to be of any use as a support, it is obviously a vestigial structure. The octopus has no internal shell, but within the back of the cuttlefish is found the broad, porous, calcareous plate familiar to bird fanciers as the cuttlebone. Another modern dibranchiate has still more of a skeleton. Thousands of small, loosely coiled, chambered shells appear at times on the shores of certain islands in the tropical and subtropical oceans. They somewhat resemble *Nautilus*, except for their small size and the fact that the whorls are not in contact. Although long known by the name of *Spirula*, the living animal has been found only recently. This shell is not an external one, but is held within the mantle, near the posterior end of the body.

Without fossils it would be impossible to understand the significance of the curious hidden shells of the modern naked cephalopods, but instead of trying to trace their ancestry in detail it is easier to step back at once to the earlier deposits of the Mesozoic, there to examine some relatively ancient remains.

Residents of England, France, and Germany have for centuries been familiar with cigar-shaped stones commonly found in clay pits. Popularly known as "thunderbolts," they were long supposed to be of extraterrestrial origin, so it is not surprising that even as late as 1750 they formed a part of the current pharmacopoeia, being administered in powdered form as a medicine where a deficiency in calcium was indicated. The oldest "thunderbolts," or belemnites (Fig. 96), are found in strata of Triassic age, but they are much more common in Jurassic and Cretaceous deposits. Some are small; others are as much as eighteen inches long and more than an inch in diameter. The ordinary specimen is cylindrico-conical, pointed at the posterior end, deeply pierced by a conical cavity at the other. Except for this the body is solid.

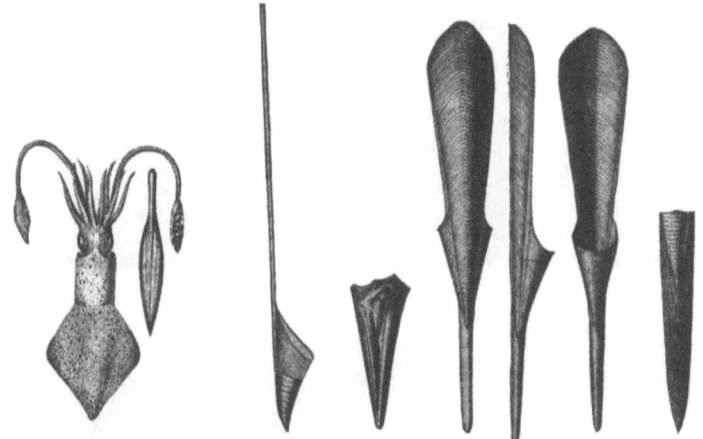

Fig. 96. At left, a modern squid with its pen. At right, various belemnites. The tall figures are restorations, showing the three parts of the shell; the short ones actual specimens. All from F. J. Pictet, *Traité de paléontologie*.

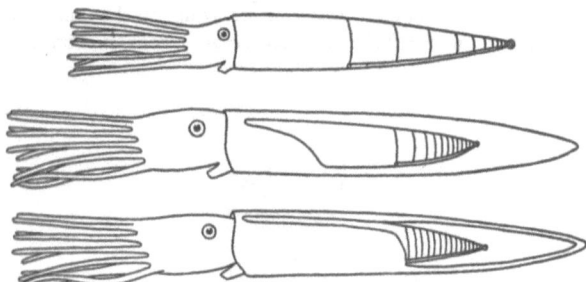

Fig. 97. Three theoretical stages in the evolution of the belemnites. Upper figure, animal within shell. Middle figure, animal outgrows and engulfs shell; part of the living chamber is resorbed. Lowest figure, all the living chamber except the dorsal proöstracum is lost, and a guard covers the phragmocone. Original drawings by Bradford Willard.

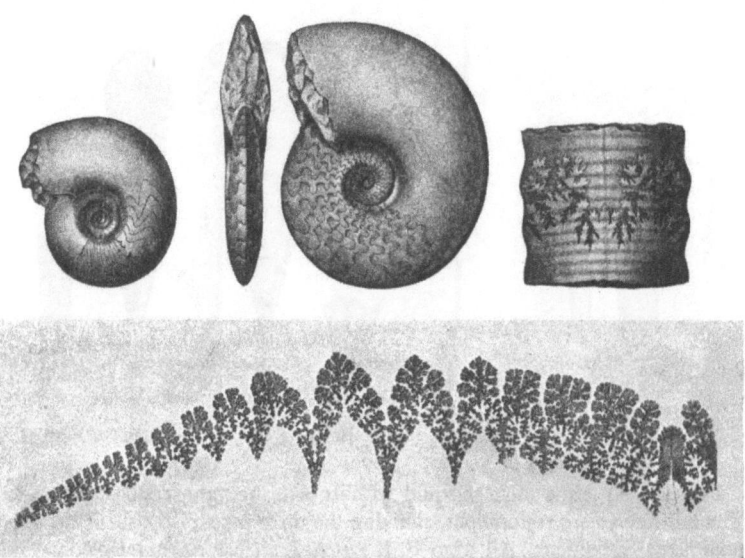

Fig. 98. Various types of ammonoid sutures. Above, at left, a goniatite with simple angular inflections; center, two views of an ammonite with the ceratitan suture, with denticulate lobes and smooth saddles; and at right, a fragment of an ammonite with a highly complicated suture. All from F. J. Pictet, *Traité de paléontologie*. Below, a drawing of the most complicated suture known, that of the Triassic *Pinacoceras*. From F. von Hauer.

Fig. 99. A complete ammonite shell, retaining the living chamber and lateral lappets. Note the position of the hyponome below the lappets and the notches for the eyes above them. The position of the shell is, of course, unnatural. From F. J. Pictet, *Traité de paléontologie*.

Broken transversely, it is seen to be made up of radially arranged prisms of calcite; when cut longitudinally it may be noted that the prisms are crossed by lines corresponding with the outline of the exterior. In other words, the belemnite was built up by the addition of successive external layers. Occasionally specimens are found with a conical shell in the cavity. This shell is subdivided by a set of curved partitions just as *Nautilus* or *Spirula* is. Each partition has a small circular opening on the ventral side, a characteristic which proves that the belemnite is really a cephalopod. At the apex is a minute globular chamber, the significance of which will be mentioned later. Still more rarely, even more complete specimens are found. In addition to the "thunderbolt" and the conical chambered shell imbedded in it, these have a broad thin extension of shelly matter which is an anterior, dorsal extension of the upper part of the latter. An entire belemnite consists, therefore, of three parts: a solid portion or guard, a chambered shell known as the phragmocone, and a thin flattened dorsal extension of the latter, the proöstracum.

It has already been pointed out that the guard was built up by the addition of successive external layers. This shows that it was formed later in life than the phragmocone, being deposited upon its surface. It also indicates that the shell of the belemnite was internal. This evidence is corroborated by the fact that many guards show on their surfaces ramifying canals which are impressions of vessels carrying the circulatory fluids. The belemnites are, therefore, skeletons of naked cephalopods. Fortunately, a few specimens retain impressions of tentacles. None is absolutely complete; hence there is some difference of opinion about a reconstruction. Many investigators have held that there were ten arms, as in modern squids; a careful student, Professor Abel of Vienna, finds evidence of only six. They appear not to have been provided with suckers but with hooks, equally useful but less specialized organs of prehension.

Since the belemnites are the oldest naked cephalopods, it is natural that paleontologists should think of them as the ancestors of the modern ones. Apparently the internal shell has been progressively reduced (Fig. 97). The modern cuttlefish seems to retain only the modified proöstracum, the incurved tip of which may represent a vestige of the phragmocone. The pen of the squid is regarded as but a vestige of the proöstracum; the octopus has lost even that remnant of shell. The Upper Jurassic strata at Solenhofen have afforded specimens of typical pens of squids, and Jurassic strata in England and Germany have produced excellent cuttlebones, some of them with sacs of fossil ink, a substance so like the modern sepia that, properly ground, it has been used by draftsmen in depicting the animal which furnished the pigment.

So much regarding the naked cephalopods. As far as they are now known, they appeared first, with heavy internal shells of threefold nature, in the Triassic, were extremely abundant in this expression throughout the Mesozoic, but lost a great part of the skeleton during the Tertiary. At the present time they have reached their culmination, in size at least, in the gigantic squids and octopi of Atlantic waters.

If the modern squid and octopus may be compared with those originally of good family, somewhat decayed, but capable of rising again to prominence by taking advantage of recent happenings, *Nautilus* is the peer without land or income, with a wonderful ancestry but no present position. So poor, indeed, is *Nautilus* that although we know his lineage for hundreds of millions of years we do not know the full life history of a single individual.

For the moment, however, we are interested chiefly in the adult, which is well known. The shell, like that of a snail, is coiled, but in one plane, after the fashion of a watch spring, rather than in a screwlike spiral. Unlike that of a watch spring, however, the last complete turn or volution envelops the others, almost completely hiding them. When the shell is cut along the median plane, the section reveals several volutions. About halfway back in the outermost there is a concave partition or septum which forms the posterior wall of the living chamber, the habitation of the animal. Numerous septa, approximately equally spaced, subdivide the cavity behind the living chamber into a series of empty compartments. Each septum is perforated below the center by a small opening, around the margin of which the shell is prolonged a short distance backward. The tube extending through these openings to the apical chamber is the siphon, and the shelly sheaths are the siphonal collars or funnels. When *Nautilus* emerged from the egg it formed a thimble-shaped shell about itself. As it grew, it added calcareous material at the anterior margin. When it reached a certain size, the animal for some unknown reason ceased for a time to increase the length of its shell, pulled the posterior end of its sac-shaped body forward, and secreted a calcareous partition behind it. The contour of this septum corresponds to the shape of the posterior end of the body. Growth was then renewed, and the process repeated until adult size was reached and all the septa were formed. Unfortunately this activity is known only by inference from the fact, recently observed by Dr. Rudolf Ruedemann, that there is a coincidence between the number of septa and the number of resting stages which can be identified by a study of the lines of growth on the shell. It is probable that the so-called resting stages represent periods when food was scarce and that the pulling forward of the posterior end of the body was due to shrinkage during a time of underfeeding. One of the most interesting experiments which could be performed would be to raise young *Nautili*, but although repeated attempts have been made no one has yet been able to obtain fertilized eggs or larval specimens. Professor Arthur Willey, who devoted a year and a half of his life to this problem, finally succeeded in keeping the animals alive in natural aquaria in the Philippines and actually got eggs, but no fertile ones. *Nautilus* presents a profitable subject of research for a zoölogist with plenty of time, money, and ingenuity.

Since the septa are deposited later than that part of the shell adjacent to which they lie, it is obvious that they are wholly internal to it. There is nothing whatever on the exterior which indicates the location of one. Each meets the interior of the shell along a line which corresponds with the shape of the partition. If the septum were simply concave, like a watch crystal, and the shell conical, then the line along which it met the shell would be a circle. As a matter of fact, the septum in modern *Nautilus* does not have this simple shape; hence the line, which is called the suture, turns forward and backward in a definite way. If the interior of a *Nautilus* were completely filled with plaster of Paris and the shell ground off, the edges of the septa, the sutures, would become visible, and it would be seen that they make sweeping curves backward and forward. Where such a curve is convex backward toward the apex, the suture is said to have a lobe; where it is convex forward toward the aperture, there is a saddle. Lobes and saddles alternate with one another all around the edge of a septum. When cephalopods are studied, it is necessary to have some specimens from which the shell has been naturally or artificially removed.

Naked cephalopods have had a relatively short history, but shelled forms, more or less like *Nautilus*, have existed since the Upper Cambrian — perhaps since the Lower Cambrian From early Ordovician days till the end of the Mesozoic they were abundant. At the end of the Cretaceous, however, some great calamity befell them, and only a few managed to survive through the Tertiary to the present. Little or nothing is known concerning the soft parts of these ancient animals, but the shells are so similar to those of living examples that all are grouped together as the great subclass Tetrabranchiata. This should not be taken to mean that paleontologists assert that the extinct representatives of the group had four gills. There is reason to believe that they had more than two pairs.

It is readily seen that the shell of *Nautilus* is, in effect, a coiled, chambered tube. If one could be straightened out and restored to a really conical form, it would resemble the sort of cephalopod which was common throughout most of the Paleozoic. Many of the older representatives of the group secreted long slender conical shells with simple septa and sutures. Such are called orthoceratites, or orthoceracones, because of their resemblance to straight horns. They reached their maximum in size and variety during the Ordovician and the Silurian, but are found in strata as young as the Triassic. A few of the Ordovician specimens are fifteen feet long, and some Silurian individuals had a length of seven or eight feet. Although they did not reach the gigantic size attained by some modern squids, they were large for their times.

The oldest cephalopods obviously related to *Nautilus* are a few found in Upper Cambrian rocks. In the Lower Cambrian, however, there are two small forms, one of which, *Salterella*, is common and widely dispersed, whereas the other, *Volborthella*, is found chiefly in countries adjacent to the Baltic. Both are represented by small straight or somewhat curved tubes which consist of a series of funnels nested together. They are generally considered to be the oldest cephalopods, but show some peculiarities which suggest that they may have other relationships.

By no means all Paleozoic cephalopods which resemble *Nautilus* are orthoceracones. Some have curved shells, in which case they are called cyrtoceracones. A few have one complete volution, although the whorls are not in contact, and are known as gyroceracones, a term sometimes abbreviated to gyrocones. Still others are fully coiled, but in such a way that the whorls are barely in contact, so that all can be seen in lateral view. This is a planospiral coiling which produces the evolute nautilicone. As a modification of this basic pattern there are some which in the earlier whorls are planospiral but the later-formed portion of the shell is a straight tube. Such are known as lituiticones. Some late Paleozoic nautiloids are as tightly coiled as modern *Nautilus*, the last volution embracing and covering the earlier whorls, forming a fully involute nautilicone. Needless to say, there are all gradations between evolute and involute shells. Finally, there are a few ancient relatives of *Nautilus* which resemble gastropods in having an asymmetrical spiral shell, a torticone.

It should be borne in mind that this is merely a rough and unscientific grouping by shell forms, which indicates nothing about relationships. Two orthoceracones may be closely or distantly related. They may belong to the same species or to different suborders. There are families in which nearly all forms are represented.

All Cambrian, Ordovician, and almost all Silurian cephalopods have simple septa and sutures. Many Devonian and later specimens, however, have more complicated ones. Some have angular, pointed lobes, or even angular saddles. Others have more elaborate sutures, the lobes and saddles being subdivided by secondary inflections. Such forms became increasingly common during the late Paleozoic and were the dominant Mesozoic shelled cephalopods. This obvious difference from *Nautilus* led to the investigation of other characteristics and to the consequent subdivision of the Tetrabranchiata into two great orders, the Nautiloidea and the Ammonoidea.

Most ammonoids (Fig. 98) can readily be distinguished from nautiloids by the greater complexity of the sutures, but since there are exceptions to this rule other criteria have been sought. One of the differences lies in the position of the siphonal tube. The funnels of *Nautilus* are short, extending only a millimeter or two back of each septum. Most fossil nautiloids and ammonoids, however, have a calcareous tube which is continuous through all the deserted chambers. In some it consists of elongated siphonal collars which extend from one septum to that behind; in others the collar is short, but a calcareous tube of another origin bridges the gap to the next partition. Such a continuous tube, however formed, is termed a siphuncle. In all ammonoids the siphuncle is in contact with the inside of the shell, and in every family but one it is ventral in position; that is, it lies just within the outer side of the whorl, if the shell be a coiled one. In the case of the one exception, it is on the dorsal side. In most nautiloids, on the other hand, the siphuncle is not in contact

SQUIDS, DEVILFISH, AND CHAMBERED SHELLS 195

with the shell; it may be central, or above or below the center. In cases where it is actually ventral it is large, whereas in all ammonoids it is small.

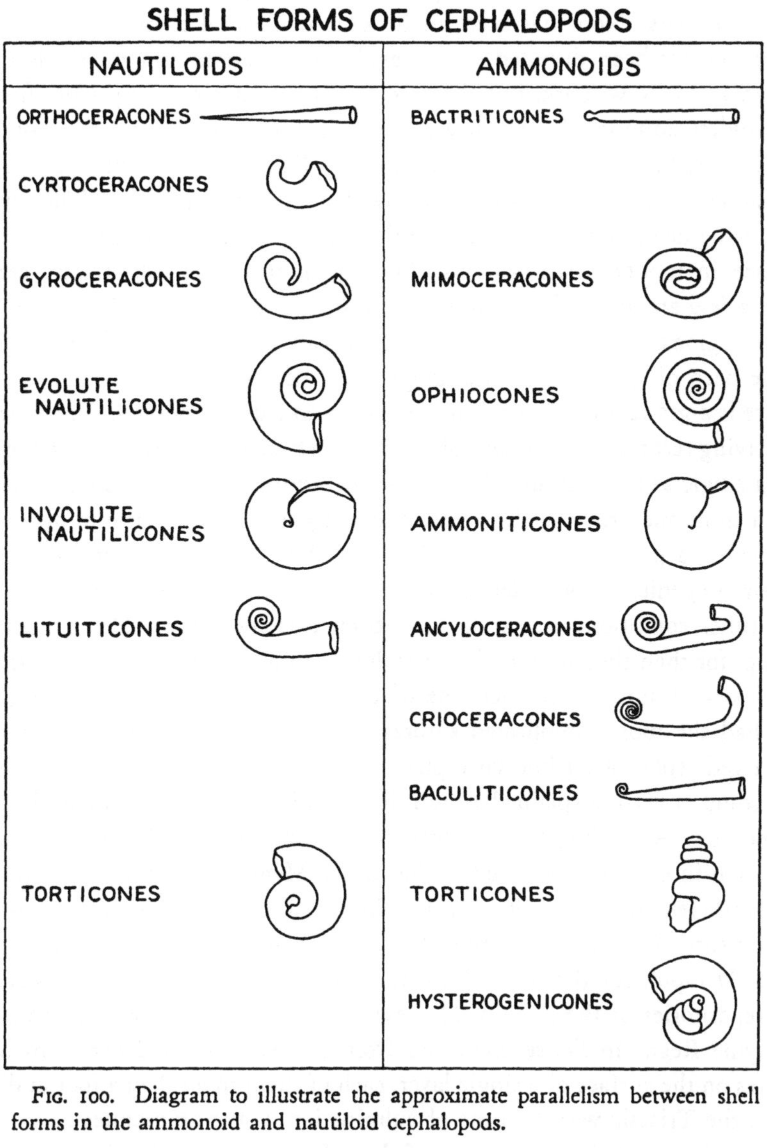

SHELL FORMS OF CEPHALOPODS

NAUTILOIDS	AMMONOIDS
ORTHOCERACONES	BACTRITICONES
CYRTOCERACONES	
GYROCERACONES	MIMOCERACONES
EVOLUTE NAUTILICONES	OPHIOCONES
INVOLUTE NAUTILICONES	AMMONITICONES
LITUITICONES	ANCYLOCERACONES
	CRIOCERACONES
	BACULITICONES
TORTICONES	TORTICONES
	HYSTEROGENICONES

FIG. 100. Diagram to illustrate the approximate parallelism between shell forms in the ammonoid and nautiloid cephalopods.

Another and more important difference is one that can be determined only if the apex of the specimen be visible. The initial shell of all ammonoids is a minute spherical test which remains throughout life attached to the apex of the cone. In most coiled shells it is completely buried by the later whorls but can be found by breaking back to the center. Nautiloids lack such an initial chamber. At the apex of a *Nautilus* is a scar which has been interpreted as indicating that the initial shell was membranous or chitinous, not capable of preservation. Similar scars are present on fossil nautiloids. None has been found with a spherical calcareous nucleus. This last statement has been disputed on the evidence of certain straight shells found in the Devonian of New York, but they were probably not correctly identified.

The oldest ammonoids, which are found in the Silurian, and most of the late Paleozoic ones, have relatively simple sutures. The majority have a few rounded or angular lobes and saddles, and a narrow V-shaped ventral lobe. The angularity of some of the lobes or saddles suggested the common name, goniatites, but the word is not used in modern classifications. Most goniatites (Fig. 98, at left) are small, although a few are as much as ten or twelve inches in diameter. Since they abound in certain regions, particularly in Upper Devonian and Upper Carboniferous strata, they have been of great use in stratigraphy and correlation.

There is a gradual transition from goniatites to true ammonites. The latter differ only in having secondary inflections of the suture (Fig. 98). Consequently, the appearance of a single saddle within a lobe, or a lobe within a saddle, would signalize the transition from one group to the other. Various goniatites gave rise to diverse stocks of ammonites, a progress accomplished during the Carboniferous and the Permian. Like their progenitors, the Paleozoic ammonites were all of small size, but during Triassic times conditions appear to have been extremely favorable to members of this group, for then they increased tremendously in abundance, size, and particularly in complexity of sutures. In fact, one of the Triassic forms, *Pinacoceras* (Fig. 98, below), had the most complicated sutures, which means, of course, the most complexly curved septa, of all known cephalopods.

For some as yet unexplained reason the end of the Triassic was a critical era for the ammonites. According to specialists, representatives of all but one of the many families then in existence perished at that time. Members of this one family, fortunately, weathered their adversities. Like Noah's children they increased, multiplied, and replenished the earth, or, rather, the sea. Their evolution must have been extraordinarily rapid, for the oldest Jurassic rocks, the Liassic, are the greatest of the world's repositories of the remains of these creatures. I have stood on strata of this age at Lyme Regis, in Dorsetshire, and from one spot counted over two hundred ammonites on the surface of a single layer, each of them more than a foot in diameter. Although the Triassic was the time of culmination of the ammonites in variety, the Upper Jurassic saw their giants, some of them five, a few ten feet in diameter. The group remained a dominant one in the seas till the end of the Cretaceous. Then, suddenly, they were all gone. Why? No one really knows.

Although much has been written about the overspecialization and degeneracy of the Cretaceous ammonites, their peculiar forms have lately been interpreted as due to

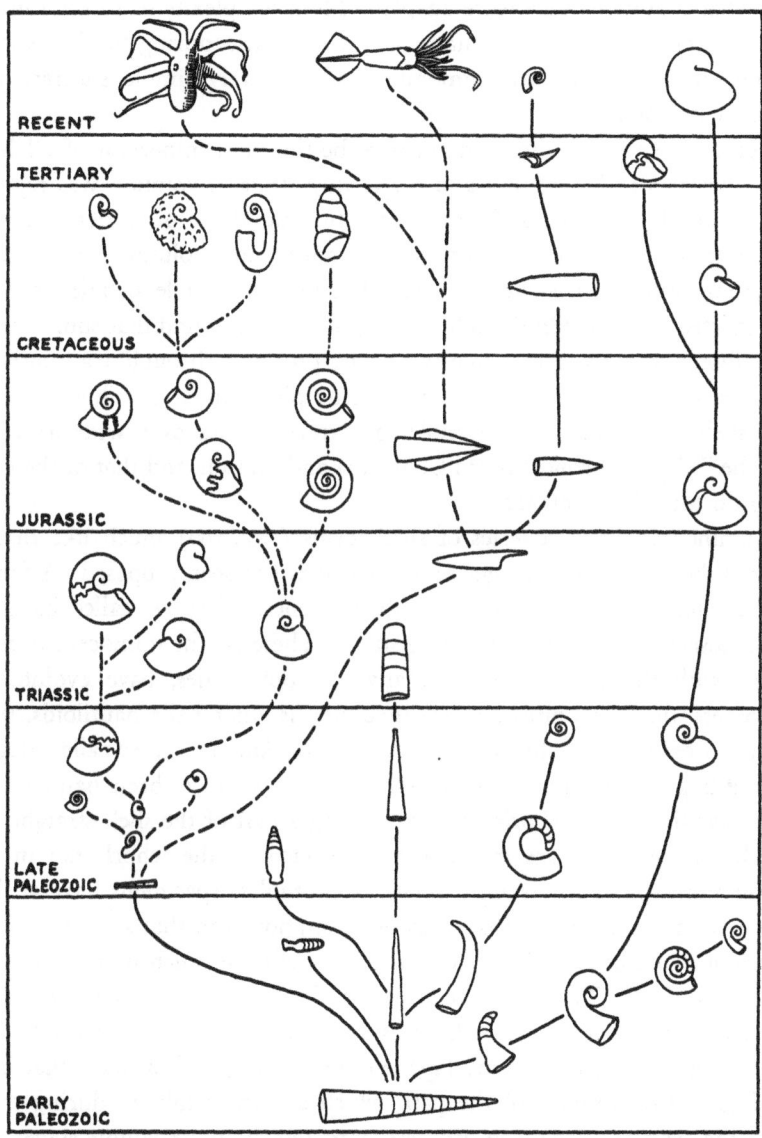

Fig. 101. Chart to show one interpretation of the history of the cephalopods. Solid lines connect members of the Nautiloidea, dashed lines the Dibranchiata, and dots and dashes the Ammonoidea. All figures conventionalized.

their habits. As to overspecialization, we have nothing but surmises, for we do not know to what conditions they were adapted. Mere complexity of suture cannot have been an important factor, for that had reached its climax during the Triassic. Moreover, many of the Cretaceous ammonites had relatively simple sutures; yet they succumbed with the rest.

Geologists have found the ammonites to be the most important of all Mesozoic fossils for purposes of identifying and correlating strata. Evolution was rapid in the group, particularly during the Triassic and Jurassic. Careful students have noted changes in ammonite faunas and progress in ammonite evolution from bed to bed in the vertical succession, a circumstance that makes possible detailed correlations within restricted regions. On the other hand, it has been found that some ammonites were distributed rapidly over extensive areas, probably through transportation by oceanic currents, either as members of the true plankton or possibly as tenantless, gas-filled shells. This makes it possible to correlate strata over wide areas, as, for instance, the Triassic of southeastern California and Nevada and that of the countries bordering on the Mediterranean.

The ammonoids have a series of shell forms (Fig. 100) much like that of the nautiloids, although an even longer series of names has been proposed. A few of the earlier goniatites are as straight as the orthoceracones. They are called bactriticones. A few are loosely coiled, at least in young stages. These are the mimoceracones, which correspond with the gyroceracones. Many, the ophiocones, have evolute coiling, whereas others, the ammoniticones, are involute; as among the nautiloids, there are all gradations between the former and the latter. Among the so-called degenerate forms there is great variety, although only a few types have been named. Ancyloceracones have the earlier whorls evolute, the larger part of the shell straight, curved, or hooked. Crioceracones are much like gyroceracones, the whorls not in contact. Baculiticones are apparently straight shells, but complete specimens have at the apex a minute involute coil. The coil is so seldom seen, however, that such are commonly called "straight ammonites." Turreted snaillike shells are much more common than among the nautiloids. They occur in strata as old as the Triassic and are known in general as torticones, more particularly as turriliticones, if coiled with the whorls in contact. Some shells, such as the Japanese *Nipponites*, with a shape that has been compared to that of a model of the path which a particle follows during an earthquake, or the western American *Emperoceras*, which in its young stages was an ammoniticone, later became an ancyloceracone, and in the adult was a torticone, practically defy classification. They may be called hysterogenicones.

It is of some interest to note that the Nautiloidea are the ancient and central stock of the Cephalopoda, that they reached their climax in the Mid-Paleozoic, and that they persist to the present practically unchanged from the state which they had reached in the late Paleozoic. From the Nautiloidea sprang the Ammonoidea, probably during the Silurian time. This group had a long period of luxuriance in the Mesozoic but disappeared abruptly. There is difference of opinion about the origin of the Dibranchiata, but the spherical initial chamber of the phragmocone seems to indicate that they are descended from the ammonoids. Late in their arrival, they flourished throughout the Mesozoic, became greatly modified during the Tertiary, and seem to be enjoying a second period of vitality today.

XX

INSECTS: THE FIRST AVIATORS

The bee is enclosed, and shines preserved, in a tear of the sisters of Phaëton, so that it seems enshrined in its own nectar.
Martial, c. A.D. 90

Whence we see spiders, flies, or ants entombed and preserved forever in amber, a more than royal tomb.
Francis Bacon, 1623

I saw a flie within a beade
Of amber cleanly buried.
Robert Herrick, c. 1648

At the meeting of the Royal Society were exhibited some pieces of amber sent by the Duke of Brandenburg, in one of which was a spider, in another a gnat, both very entire.
Evelyn, Diary, March 24, 1682

Ever since Neolithic times, four thousand or more years ago, Baltic amber has been an article of commerce. For at least twenty centuries the animals trapped therein have been a focus of interest, since by their singularly perfect preservation they were easily recognized. The Hispano-Roman epigrammatist Martial, the politician-philosopher Bacon, and the clergyman-poet Herrick, whose special claim to fame is that he taught his favorite pig to drink from a tankard, had no more difficulty in identifying these fossils than did the members of the Royal Society. Fossil insects are, therefore, no novelty. They were known hundreds of years before there was a science of paleontology. Nevertheless, it was only during the early decades of the twentieth century that this branch of the science was put on a satisfactory basis, primarily as the result of the researches of Anton Handlirsch, R. J. Tillyard, A. V. Martynov, and F. M. Carpenter. The last, a colleague and erstwhile student of mine, furnished material for the present chapter and wrote several pages of it.

It is impossible in a book of this nature to do justice to the insects. Although they are only a subordinate branch of the phylum Arthropoda, and although only one-third or one-fourth of them have yet been described, the known species outnumber all other animals by at least two to one. Several thousand (10,400 in 1930) species of extinct insects are now known, but recent discoveries indicate that we are just beginning to become acquainted with the fossils. As will be shown in the sequel, whole chapters of their evolution are still unknown.

Insects are commonly thought of as those invertebrates which have the power of flight. As a matter of fact, not all members of the class have wings. A better definition is suggested by the old name of the group, the Hexapoda. The class may be defined as that group of arthropods which is characterized by the possession of three pairs of walking legs. The body is divided into three regions, head, thorax, and abdomen, separated by more or less pronounced constrictions. The appendages of the head include a pair of antennae and the mouth-parts, consisting of a labrum or upper lip, a pair of mandibles, one pair of maxillae, and a median labium, or lower lip. In the more generalized insects the mouth-parts are mandibulate, that is, adapted for chewing; but in many others they are modified for piercing and sucking. Most adult individuals possess two sets of eyes, one pair of compound ones and three simple eyes or ocelli. The thorax, which bears the organs of locomotion, consists of three segments, each with a pair of legs. Wings are borne on the second and third thoracic segments of most living members of the subclass Pterygota, although in the true flies (Diptera) the hind pair are absent. The segmentation of the abdomen of living insects is distinct in most cases. Primitive forms, and the young of others, show eleven segments in addition to a terminal process. The abdomens of most lack appendages, but some have jointed cerci, which appear to belong to the eleventh segment. The males, which have the opening of the seminal ducts on the ninth segment, in some cases have claspers in that region. The oviducts of the females open between the seventh and eighth segments, where some have ovipositors, organs that guide the eggs into localities favorable to the hatching and feeding of the embryos.

The wings are of much importance to the paleoentomologist, and a detailed knowledge of their structure is essential, since about half of all the fossils are known from these organs only. Fortunately, the "veins" of the wings have been shown to be of the greatest value in the identification of the various groups. Although only a few primary trunk veins are present, the possible arrangements of longitudinal and cross-veins are almost unlimited. Experience with the modern forms, which constitute more than 95 per cent of all described species, indicates that the wing venation can be relied upon for the separation of genera and species of most orders. After a comparative study of the wings of all modern and some fossil insects, Professors J. H. Comstock and J. G. Needham concluded that the principal longitudinal veins of the wings could be homologized, and that in general the homologies could be determined by the structure of the immature wings. They were thus able to construct a hypothetical venational pattern representing their conception of the original wing structure. The insects which possess wings most closely resembling this hypothetical type are regarded as the most primitive. It is interesting to note that, although the Comstock-Needham interpretation has in general proved satisfactory to the student of recent insects and has been widely adopted, paleoentomologists have found it necessary to make modifications of the scheme. This is merely one example of the necessity of studying the ancient representatives of a group before making generalizations. Professor A. Lameere, who has studied fossil insects as well as living forms and has a knowledge of the results obtained by Tillyard and others during the first two decades of the twentieth century, has formulated a modification of the Comstock-Needham interpretation which bids fair to replace it.

The oldest insects now known are found in strata of the Alleghany series, which are, next to the Pottsville, the oldest rocks of the Upper Carboniferous (Pennsylvanian) system. Unfortunately, they are full-fledged Pterygota, with no indication of the particular group of arthropods from which they sprang. Many of them belong to an extinct order, the Palaeodictyoptera, creatures which, although they do not reveal their ancestry, are in some respects more simple than any later forms (Fig. 102, at left). True insects, with two pairs of well-developed wings (vestiges of a third pair are present on the first thoracic segment) and three pairs of legs, these animals nevertheless have primitive characteristics, such as the conspicuous segmentation of the abdomen, caudal cerci, and a simple venation, with seven principal longitudinal veins crossed by numerous transverse ones. Moreover, the wings were broadly joined to the thorax, and hence strictly lateral, incapable of being folded back over the body. The Carboniferous members of the order had chewing mouth-parts. The group is best represented by fossils in Pennsylvanian strata, but a few individuals indicate its continuance in Permian times. Among the latter is one species which had the mouth-parts modified for sucking purposes. This specialization has led some students to place this species in another order, the Protohemiptera.

The Palaeodictyoptera are not only the oldest insects now known but the probable ancestors of all other winged ones. It is not likely, however, that they will continue to hold this proud position, for simpler fossils of the class must inevitably be found in Mississippian and Devonian rocks. Even now there is evidence that a wingless form is present in the Rhynie chert (Mid-Devonian) of Scotland; antennae similar to those of modern springtails have been described, but the heads to which they are attached are so poorly preserved that many paleontologists have doubted the identification. It is understood, however, that bodies have been found which show that the animal really belongs in the order Collembola. Unfortunately, detailed descriptions have not yet been published. If this claim is substantiated, it will confirm the inference which entomologists have already drawn, that wingless insects (Apterygota) preceded those having the power of flight.

Although the Palaeodictyoptera were the simplest hexapods existing during Upper Carboniferous times, they were not the most abundant. This honor belongs to the Blattaria or cockroaches, an order which reached its culmination during that period. Particularly well suited for longevity by their omnivorous food-habits and their relatively simple structure, they have persisted to the present day. Their decline, nevertheless, was a rapid one, for Carpenter has shown that, although they made up about 60 per cent of the Upper Carboniferous insectan fauna, they were reduced to approximately 35 per cent in the Permian, 7 per cent in the Mesozoic, and only 1 per cent in the Tertiary and Recent. This refers to species, not individuals; the "Croton bugs" are still with us.

Cockroaches and Palaeodictyoptera were the most common Pennsylvanian insects, but considerable differentiation had been achieved at that time, for thirteen orders are represented, all but one (Blattaria) now extinct. Most striking of all were the Protodonata, creatures which resembled dragonflies. Most of them were larger than existing dragonflies and one of them, *Meganeura*, an inhabitant of western Europe, had wings a foot or more in length. These huge forms were not abnormal but, rather, oversized members of a group which, like the dinosaurs a million years later, tended to become giants. They spread widely over the northern hemisphere

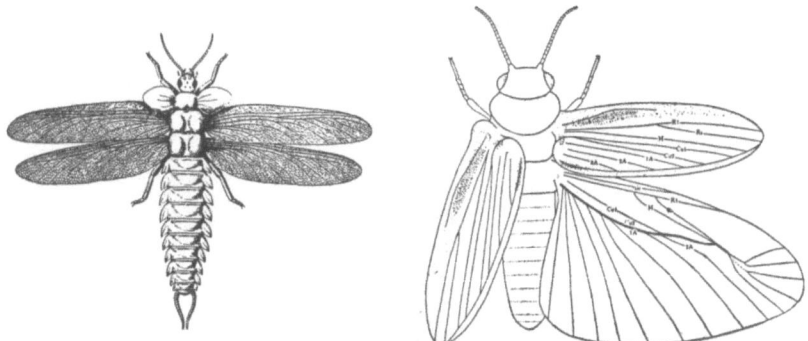

FIG. 102. At left, *Stenodictya*, a palaeodictyopteron from the Upper Carboniferous of France, possessing winglike lobes on the prothorax and a simple venation. After Anton Handlirsch. At right, *Protelytron*, a member of the extinct order Protelytroptera, from the Lower Permian of Kansas. The fore wings resemble those of beetles. Original drawing by F. M. Carpenter.

and were abundant during the early Permian. Recently Carpenter found in the Kansan Permian a species which was apparently somewhat larger than the Carboniferous one. All the Protodonata were predaceous, feeding on other animals, probably mostly on insects, and the large creatures must have consumed an enormous number of individuals. It is not improbable that they were in some measure responsible for the reduction in the number of roaches in the Permian. Another Upper Carboniferous order which deserves special mention is that of the Megasecoptera, a group represented in the Permian by a specialized branch termed the Protohymenoptera. Most of the other Carboniferous insects belong to orders with characteristics more or less like those of the grasshoppers, making up what has been called the protorthopteroid complex. Among them may be the ancestors of several modern orders which have not as yet been traced back to the Permian.

The insectan fauna of the Carboniferous is decidedly archaic; that of the succeeding period is much more modern, for in Permian strata at one locality or another are found representatives of seventeen orders, eleven of which still exist. Only five of them originated in Carboniferous times; the others are newcomers. For some reason, insectan evolution was rapid at this time. Since hexapods were not then specialized for the pollination of plants or for feeding on their nectar (there was none), it may be inferred that whatever stimulus led to the change was probably geographic rather than organic. Only two probable causes occur to one. The first is the general refrigeration which brought on the Permian glaciation; the second, the complicated series of changes which must have accompanied the draining of the Coal Measure swamps. These two factors may have acted together. Although their effect upon the vegetation over most of the northern hemisphere was neither sudden nor particularly striking, changes did occur, and they must have had some effect upon the evolution of the first aviators.

FIG. 103. *Protentomobrya*, a springtail (Collembola) contained in Cretaceous amber from Manitoba. Original drawing by J. W. Folsom.

Perhaps a few words about some of the more important orders which appeared in the Permian may be useful. The geological ranges of insects, as they appear in even the most recent textbooks, are so inaccurate that a summary, for which Carpenter is responsible, appears to be needed.

The existing orders which appear for the first time in the Lower Permian are: Mecoptera, Neuroptera, Odonata, Psocoptera, Plectoptera, Embiaria, and Homoptera; in the Upper Permian, Perlaria, Thysanoptera, and Coleoptera.

The Mecoptera (scorpion flies) of the Permian are minute, some with a wing expanse of only 10 mm. Relatively more numerous than now, they made up about 9 per cent of the species, whereas their present status is less than .04 per cent. The early Neuroptera were in an advanced state of evolution; both the Planipennia (lacewings) and the Megaloptera (alder flies) had been differentiated. The earliest true dragonflies (Odonata), unlike the Protodonata, were small creatures, scarcely more than an inch and a half across the wings. More vital than their gigantic cousins, the real dragonflies have persisted to the present, reaching their culmination in the Upper Jurassic, whereas the Protodonata died out in Triassic times.

The occurrence of true Homoptera in the Lower Permian is of much significance, since they appear to have been the most highly specialized members of their class at that time. Not only did they possess mouth-parts adapted for sucking juices from plants, but the wing venation was as greatly reduced as that of some existing members of the order. The Psocoptera or bark lice were abundant during the Lower Permian; over five hundred specimens have been found in the Kansan Permian alone. Unlike the modern species of the group, in which the hind wings are much reduced, the Permian bark lice had two pairs of similar wings with a venation much like that of the Homoptera. The Plectoptera, mayflies, are not uncommon in Lower Permian strata. Their nymphs were aquatic, like those of the present. The adult mayflies of the time, like the psocids, had two pairs of similar wings, whereas the modern ones have greatly reduced hind wings. One of the most surprising facts about these oldest known mayflies is that they had the peculiar adult molt, or ecdysis, so characteristic of living ones.

In addition to these modern orders of insects and a few extinct ones which persisted into the Permian from the Carboniferous (Palaeodictyoptera, Protodonata, Protorthoptera, Megasecoptera), the Lower Permian fauna also included two interesting orders which so far as we know did not survive beyond the Paleozoic. One was the Protelytroptera (Fig. 102, at right), small, beetlelike creatures, with the fore wings modified to form covers for the thorax and abdomen. The other order was the Protoperlaria, hexapods which closely resemble the existing stone flies and were probably ancestral to them (Fig. 104). The nymphs (Fig. 105, at right) were aquatic and breathed by means of nine pairs of lateral abdominal gills, the vestiges of which were present in the adults. Almost all of the recent stone flies lack abdominal gills, but in the more generalized species (family Eustheniidae) the nymphs have them on the first five or six abdominal segments and vestiges are visible in the adults. The adult Protoperlaria were so similar to some of the Protorthoptera that it may be said truly that the former are Protorthoptera which have become aquatic in the nymphal stages.

Three other recent orders make their appearance in the later Permian, one of these being the true beetles (Coleoptera), the second the thrips (Thysanoptera), and the third the true stone flies (Perlaria).

There are two important problems toward whose solution knowledge of the Paleozoic insects has contributed. One of these is the origin of the wings, for, unlike those of birds and other flying animals, they are not modified fore limbs but entirely new structures. The second and third segments of the thorax of living insects bear the wings, the first segment or prothorax being reduced or modified for some special purpose. But in the Carboniferous insects, generally speaking, and in some of the Permian ones (Protoperlaria, Protorthoptera), the three thoracic segments are nearly alike in form and structure, and, what is more significant, the first supports a pair of lateral outgrowths resembling small wings. The constant appearance of these wing-flaps in so many unrelated Paleozoic groups indicates that such structures were probably present in the progenitors of the Pterygota. It seems likely that the functional

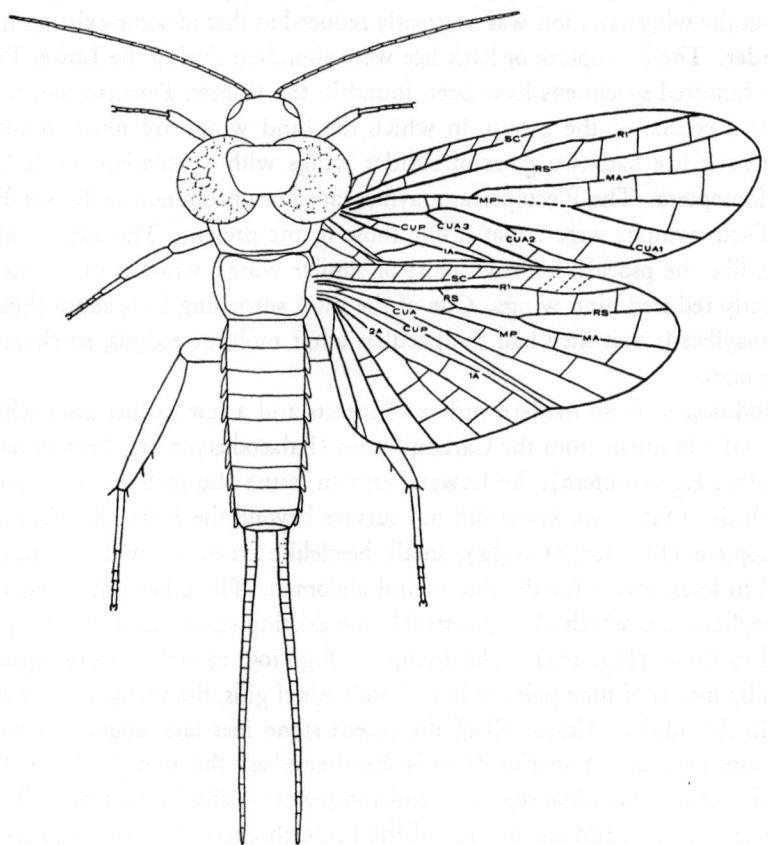

Fig. 104. Reconstruction of *Lemmatophora*, a protoperlarian from the Lower Permian of Kansas, with membranous prothoracic lobes, and vestiges of gills along the sides of the abdomen. The venation is like that of the Recent stone flies. Sc, subcosta; R1, radius; Rs, radial sector; MA, anterior media; MP, posterior media; CuA, anterior cubitus; CuP, posterior cubitus; 1A, 2A, first and second anals. From original drawing by F. M. Carpenter.

wings arose as such lobes and that at one stage in the history of the flying insects each segment of the thorax possessed similar lateral outgrowths. For some reason the pair next the head did not develop into useful wings and have disappeared entirely from Mesozoic and later forms.

The other important question upon which Paleozoic history sheds some light is

 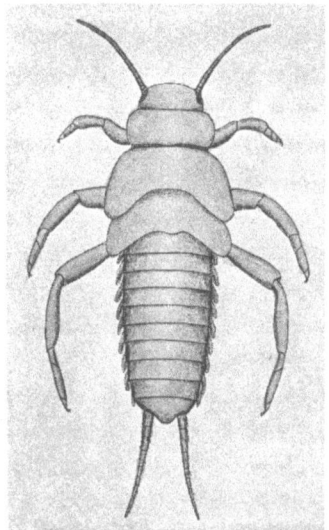

FIG. 105. At left, *Clatrotitan*, a protohemipteron from the Triassic of New South Wales, possessing an extraordinarily large sound-producing organ on the fore wing. The wing shown in the photograph is about six inches long. From C. Anderson. At right, restoration of a Protoperlarian nymph, from the Lower Permian of Kansas. These nymphs were aquatic and possessed lateral abdominal gills. Original drawing by F. M. Carpenter.

FIG. 106. An unusually well-preserved dragonfly of the genus *Protolindenia* from the Upper Jurassic at Solenhofen, Bavaria. From a photograph of a specimen in the Carnegie Museum, Pittsburgh, Pa.

Fig. 106A. A neuropteran, *Lithosmylus columbianus* (Cockerell), distantly related to the modern lace-wings. The specimen is from the Miocene of Colorado, and now in the Museum of Comparative Zoölogy. Much enlarged. Courtesy of Frank M. Carpenter.

the origin of a complicated life history or metamorphosis. Some hexapods — for example, the grasshoppers — have much the same general structure from the time they issue from the egg until they reach maturity; their wings appear at an early stage in their ontogeny and gradually increase in size. Such a development is termed an incomplete metamorphosis (hemimetabolism). Others, such as the moths, differ in the early stages from the adult; their wings develop inside the body and appear only during a quiescent stage, known as the pupa. This is complete metamorphosis, or holometabolism. So far as is known, all the insects of Carboniferous developed by incomplete metamorphosis. In the Lower Permian there are two orders (Mecoptera and Neuroptera) which are now holometabolous, and the presence of true larvae in those strata shows that complete metamorphosis had been acquired. Since that time such forms have gradually become more abundant at the expense of those with incomplete metamorphosis, until 88 per cent of the existing species have holometabolous development. Although no Carboniferous insects are known to possess complete metamorphosis, the fact that in the Lower Permian two orders are in that category suggests that this form of development originated during the earlier period and may have been due to late Paleozoic reduction in temperature.

Comparatively little is known about Mesozoic insects, especially those of the Triassic and Cretaceous. This is unfortunate because it is difficult to connect the modern fauna with that of the Permian. Enough early Mesozoic species have been found, however, to show that the fauna of that time was decidedly more advanced that that of the late Paleozoic. One insect, recently found in the Mid-Triassic of Australia, had a sound-producing organ on its wing much larger than that in any living form (Fig. 105). The sound produced by such individuals could have been heard for a great distance. It is interesting to reflect that they probably made more noise than any other animals of the time, for neither birds nor mammals are known to have existed then. Whether contemporary amphibians croaked or reptiles grunted is unknown.

Five modern orders appeared for the first time during the Mesozoic; one, the Heteroptera (true bugs), is found in the Triassic; the other four, Trichoptera (caddis flies), Dermaptera (earwigs), Diptera (true flies), and Hymenoptera, in the Jurassic. Since the oldest known Hymenoptera include some parasitic forms, it is obvious that the group is much older than the Jurassic. Curiously, the more highly organized members of this group did not appear until the Tertiary. Beetles were probably the most abundant insects of the Mesozoic; Orthoptera were common; and dragonflies (Fig. 106) were undoubtedly much more numerous than they are at present. The recent discovery of insect-bearing Cretaceous amber in northern Canada bids fair to extend our knowledge of late Mesozoic forms greatly.

Tertiary insects are abundant, especially those of the Oligocene and Miocene. The Lepidoptera (butterflies and moths) first appear in the Eocene, although the order is probably much older than that. The oldest termites are in the Baltic amber (Oligocene), and this is also true of certain obnoxious forms, such as the fleas. The modernization of the insectan fauna in the Tertiary was completed by the arrival on the stage at this time of the so-called social insects, bees, wasps, and ants. The extraordinarily rich faunas of the Baltic amber and of the Miocene shales of Florissant have enabled W. M. Wheeler, C. T. Brues, F. M. Carpenter, T. D. A. Cockerell, and others to describe the habits, as well as the morphology, of many of these ancient Hymenoptera. The results are summarized in a masterly way in Professor Wheeler's *Social Insects*.

Ants, among the most perfect social forms, are fortunately about the most abundant Tertiary fossils of this class. They are not common in the Eocene, but Wheeler was able to study 11,711 specimens from the Oligocene, and Carpenter almost as many from the Miocene. According to Wheeler the "amber" ants are as highly specialized as those of the present day. Six of the eight modern subfamilies are represented in the fauna. That social life was then highly organized is shown by the fact that the castes were as fully differentiated as at present, for in some genera there were workers of various forms. Even in those days the ants had their "cows," aphids or plant lice; blocks of amber have been found enclosing both the domestic animals and their warders. There is also some evidence that the ants welcomed certain beetles as guests in their nests. Only 44 per cent of the genera of amber ants seem to be extinct; eight of the species are practically indistinguishable from forms now living. So far as can now be seen, this group had reached its objective as early as the Oligocene and has made no progress since. Various sorts which lived in the north temperate zone in Mid-Tertiary times later retired to the tropics, the present stronghold of the Formicidae. Inferences might be drawn from these two facts. Will *Homo sapiens* eventually reach a culmination of physical evolution, cease to struggle with wintry weather, and lie down beneath a palm, each queen to be served by thousands of sterile workers? If so, will the race maintain itself as the ants have done, or shall we become as rare as tapirs?

XXI
ARCHAIC MAMMALS

<blockquote>Be commonplace and creeping, and you attain all things.
Beaumarchais, *Barbier de Seville*</blockquote>

A history of Mesozoic times related by some mighty dinosaur to an admiring group of smaller reptiles would probably have been in the main a recitation of how, from an obscure nativity, their race had made themselves masters of land, sea, and air. He might have mentioned the amphibians to substantiate the claim of progress, recalling the lowly stock from which the overlords had sprung. Even the birds may have won recognition as close, though unimportant relatives, but it is doubtful if the obscure little furry animals which lurked about in the underbrush would have been mentioned at all. Probably they had never been noticed. Although the history of the mammals began almost as early as that of the dinosaurs, it was not a startling or colorful story so long as the reptiles retained dominion. Apparently the Mesozoic world was exactly fitted for reptilian life. Warmth of blood and the protection of hair were no particular advantage in times when nearly featureless continents permitted the extension of subtropical conditions much further toward the poles than at the present day. When food was plenty, neither activity nor intelligence aided animals greatly, for there was little competition. Life was relatively easy; only with struggle is there progress.

Creeping from their cradle in southern Africa during the late Triassic, the mammals spread first into Europe and later, in the Jurassic, reached Mongolia and America. The Mesozoic fur-bearers, although small and unimportant in their day, are of great interest from the evolutionary standpoint.

Mammals are warm-blooded creatures which suckle their young. Most of them have a covering of hair or, in the case of the naked ones, such as elephants and whales, some trace of hair, at least on the young. Although some have a scaly covering, a hair is not a modified scale but an entirely different structure, confined to this group. The presence of a muscular diaphragm separating the thoracic and abdominal parts of the body cavity, a larynx, an external ear, a four-chambered heart, and other features peculiar to the class, are important specializations but not of much assistance to the paleontologist. The skeleton, however, shows many notable characteristics which allow ready identification. Only the more conspicuous need be enumerated here.

The skull differs from that of other vertebrates in being more compact and having fewer bones. Each ramus of the lower jaw consists of a single bone, in contrast to the numerous elements of the reptilian jaw. It articulates with a process of the squamosal, for the quadrate, the articulatory bone of the lower vertebrates, has been drawn into the ear, where it forms the incus. There is a single temporal arcade, but instead of being composed of the quadratojugal and jugal, as in the reptiles, the posterior part is a projection from the squamosal, whereas the anterior portion may be the jugal or a process of the maxillary. At the base of the skull are two occipital condyles laterally placed, as in the Amphibia. Finally, the teeth of most mammals are heterodont, subdivisible into groups known as incisors, canines, premolars, and molars. The incisors are in the premaxillae, the others in the maxillae of the cranium. The incisors and, in most, the canines are single-rooted, the premolars double-rooted, the molars generally double or, in the upper jaw, triple-rooted. With the single exception of some of the ceratopsian dinosaurs, all vertebrates other than the mammals have single-rooted teeth.

Nearly all mammals have seven cervical vertebrae, the principal exceptions being certain sloths with six, eight, or nine. The first of the series, a large, ringlike bone which receives the occipital condyles of the skull, is known as the atlas; behind it is the axis, with a peglike anterior process on which the atlas rotates. The trunk vertebrae are readily separated into two regions, an anterior series of dorsals, with ribs, and the posterior ribless lumbars. The faces of the centra of the individual vertebrae in these regions are flat, those of the neck opisthocoelous. The limbs are straighter and carried more beneath the body than is commonly the case with Amphibia or Reptilia. The coracoid element of the pectoral girdle is present as a distinct bone only in the monotremes, being vestigial and fused to the scapula in others. All of the elements of the pelvic girdles of most are so fused together in the so-called "innominate" bone that it is not easy to make out the exact boundaries of ilium, ischium, and pubis. And, lastly, the typical phalangeal formula is 2, 3, 3, 3, 3, although one or more fingers and toes have been lost by many mammals.

Present-day mammals are generally subdivided into three groups. The first is a small one, the monotremes, represented by the duckbill and the spiny anteater of Australia. These animals betray their primitive position by their reptilian habit of laying eggs, and the equally reptilian position of the limbs and structure of the shoulder and pelvic girdles. The pectoral girdle, particularly, retains the numerous bones of the lower animals. Adult monotremes are toothless, but the young have incompletely formed vestigial teeth.

The young of the marsupials, a second and more important group, are feeble at birth and are carried about for some time in the pouch of the mother. An extra pair of bones on the pubes, known as epipubic or marsupial bones, present in no other mammals except the monotremes, differentiates the pelvis from that of the placentals. Other unusual features are shown in the teeth, which are peculiar in that all execpt one

ARCHAIC MAMMALS

in each jaw are permanent; the only replaced teeth are the last premolars. Curiously, marsupials have three premolars and four molars, whereas the higher mammals have four premolars and three molars in each jaw. The modern marsupials are chiefly Australian, the exceptions being the opossums of North and particularly of South America.

The third and largest group of mammals includes those which while within the body of the mother are nourished through a spongy mass known as the placenta, whence they are called the Placentalia. With rare exceptions, these have forty-four or fewer teeth. If the typical number is present, there are in each jaw three incisors, one canine, four premolars, and three molars. For convenience, the dentition is written as a formula in which the teeth of one upper and one lower jaw are included. Thus, for a primitive mammal, one would write: $i\frac{3}{3}$, $c\frac{1}{1}$, $p\frac{4}{4}$, $m\frac{3}{3}$; but for ourselves: $i\frac{2}{2}$, $c\frac{1}{1}$, $p\frac{2}{2}$, $m\frac{3}{3}$, since we have on each side only two incisors, one canine, two premolars, and three molars. The formula must be multiplied by two to show the total number of teeth. Another feature of the dentition of the placentals is that there is a milk series consisting of the incisors, canines, and premolars, all of which are replaced by permanent teeth.

Because of the presence of two occipital condyles, it was at one time supposed that the mammals arose from the Amphibia, but with a fuller knowledge of the theraspid reptiles of the Triassic of South Africa it has become obvious that the latter is the group from which the class was derived.

The Theraspida, descendants of the pelycosaurs, include several groups, the most important of which is that of the carnivorous theriodonts, probably ancestral to the mammals. They were comparatively small animals which walked and ran rather than crawled; most of them had sharp conical teeth. One of the most obvious mammal-like characteristics is the division of the dental series into incisors, canines, and molars. This is especially well shown by *Cynognathus* or *Thrinaxodon* (Fig. 107), which has tricuspate posterior teeth. As in mammals the canines are the anterior teeth of the maxillaries, the incisors are in the premaxillaries, and the cheek teeth are the largest. The formula is variable, but some have $i\frac{4}{3}$, $c\frac{1}{1}$, $p+m\frac{9}{9} \times 2 = 54$. One of them, *Sesamodon*, said to be more like the mamals than any other member of the group, has the formula $i\frac{4}{1}$, $c\frac{1}{1}$, $p + m\frac{7}{1}$, or only two more than the placental mammal. Other mamalian features of the dentition are the bite of the mandibles inside the upper teeth, with the lower canine crossing in front of the upper one, and the method of replacement of teeth. Broom has shown that in some species at least the later teeth come up in the old sockets directly beneath the milk teeth instead of between the teeth, as in other reptiles, and there is some evidence that the posterior cheek teeth are not replaced. Other mammalian characteristics shown by one or another of the theriodonts are: the presence of a partially divided occipital condyle, or of three condyles, the loss of the median of which would produce a mammalian condition; a mammal-like palate which pushes the posterior nares far back; a zygomatic arch formed by squamosal and jugal; a lower jaw formed mainly by the dentary; seven cervical vertebrae; fused pelvic bones; a mammal-like carpus and tarsus; and the phalangeal formula 2, 3, 3, 3, 3.

Although so similar to mammals, the theriodonts still retained many reptilian characteristics. The brain was small, with a pineal opening present in most; the lower jaw did not articulate with the squamosal but with a small quadrate; and a vestige of the quadratojugal still remained. In the most primitive ones a large postorbital bone extended backward from the eye along the margin of the parietal, thus bordering the temporal opening as in the typical Synapsida. Prefrontals were also present. Although the dentary made up the greater part of the lower jaw, other elements were present on the inside and at the posterior end.

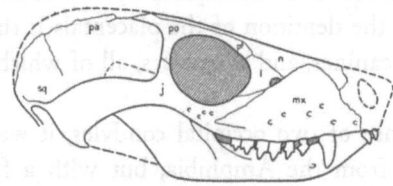

FIG. 107. *Thrinaxodon*, a small Triassic mammal-like reptile: *pa*, parietal; *po*, postorbital; *pf*, prefrontal; *n*, nasal; *l*, lacrimal; *mx*, maxilla; *j*, jugal; *sq*, squamosal. About natural size. Redrawn after W. K. Gregory.

Recently Broom has described the remains of an important reptile from the Upper Triassic of South Africa. The specimens are incomplete, but they appear to furnish a needed connecting link. The best-known genus is *Ictidosaurus*, a small creature about the size of a rat. The skull is almost entirely mammalian, the prefrontals, postorbitals, and pineal foramen being absent, and two occipital condyles present. The single reptilian feature is the presence of a small quadrate bone with which the lower jaw articulates. The lower jaw likewise retains one reptilian bone at the posterior end where it articulates with the cranium. Except for this articulation of the jaw, *Ictidosaurus* is, so far as is now known, a mammal. It is not at all likely, however, that it was a direct ancestor of any later mammal, for its small conical teeth are less mammal-like than those of the cynodonts. It is hoped that the unearthing of this new type will be followed by the recovery of the bones of many of its relatives, for among them we expect to find creatures more nearly in the direct line of ancestry.

The mammals found in the Mesozoic strata are known chiefly from minute separate teeth and from jaws less than an inch long. Such fossils are rare in the Triassic but fairly common in certain places in the Upper Jurassic. Many species have been described, although not much can be learned about any one of them. No whole skeleton has yet been found in Mesozoic rocks; recently a few skulls have been collected from the Cretaceous of Mongolia.

The oldest remains of mammals are found in strata of Upper Triassic age in South Africa and Germany, but the specimens are so few in number that knowledge of these earliest representatives of the group is meager. A single incomplete skull of *Tritylodon longaeus* (Fig. 108, at left) was long ago described by Sir Richard Owen from the Karroo formation (Triassic) of Basutoland. At first supposed to be a mammal, it was later considered one of the mammal-like reptiles; now opinion seems to have returned to a belief in the original determination. From the Keuper of Germany small isolated teeth have been obtained. All these specimens represent

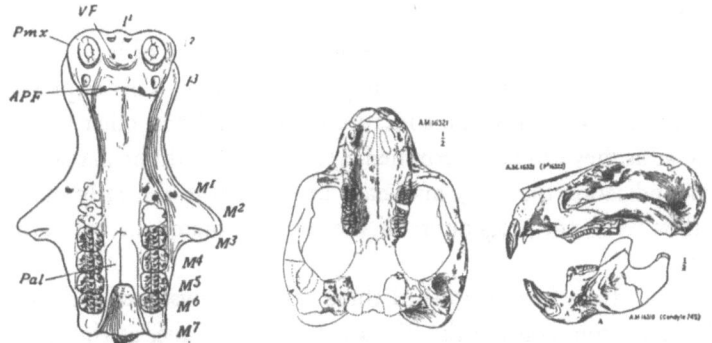

FIG. 108. At left, the Triassic allothere, *Trytylodon*. I^1, I^2, I^3, incisors; M^1 to M^7, molars; *Pmx*, premaxilla; *Pal*, palatines. One-half natural size. From G. G. Simpson, *Mesozoic Mammals*, by permission of the Trustees of the British Museum (Natural History). At right, palatal and lateral views of the skull of *Taeniolabis*, a Paleocene allothere from New Mexico. One-sixth natural size. From Granger and Simpson.

an ancient group which have peculiar cheek teeth with many tubercles on the crowns, a characteristic which suggested the name Multituberculata, by which they are commonly known, but the present scientific usage is to call them the Allotheria. Although they represent a line of mammalian evolution which cannot at the present time be connected with that of the higher representatives of the group, they are entitled to some consideration, since they spread through Europe, central Asia, and North America. They became increasingly abundant in Jurassic and Cretaceous times and reached their maximum in size in the oldest period of the Tertiary, the Paleocene, just before their complete extinction. No entire skeleton of a member of this group has yet been discovered, most species having been founded on teeth. The greater part of a skull has been described from the Upper Cretaceous of Mongolia, and good ones, with fragments of the skeleton, from beds of Paleocene age in central Montana and New Mexico (Fig. 108).

The most characteristic feature of this group is the presence of large, elongated cheek teeth, with numerous tubercles on the crowns. The lower ones have two, the upper, three longitudinal rows of low blunt cones, obviously better adapted for crushing than for cutting or tearing food. Canines are lacking, but the incisors are elongate, erect in some forms, in others procumbent, projecting far forward. Comparing the dentition of the best-known species with that of modern animals, it is inferred that many of them were rodentlike in their habits, probably using the erect incisors for gnawing or pulling the bark from cycads or other plants. Those with procumbent incisors may have fed on fruits, nuts, eggs, insects, and such small animals as they were able to catch. Like some modern rodents, they were more or less omnivorous. All were small, the largest the size of a modern beaver, but most of them comparable to mice.

It is obvious from the lack of canines and the presence of large, peculiar molars that all were specialized, in spite of their great antiquity. This impression is borne out by what little is known of the structure of the skull and other bones. The relationships, so far as they can be determined, appear to be with the marsupials, although there is some indication that they may have been allied to the early monotremes. In any event, they were not progenitors of the modern mammals but form a subclass of their own.

The Jurassic strata have furnished three sorts of jaws and teeth of a primitive nature. The more common are the Triconodonta (Fig. 109, at right), a name suggested by the fact that each cheek tooth has three fangs in a line, the median high, the anterior and posterior ones subordinate to it. Triconodonts have been found in England and North America, but the number of species is not great. They were small, conjectured to have been about the size of mice. Little is known of the skeleton, since the best material so far found retains no more than lower or upper jaws and parts of the skull associated with the palate. Most of them had numerous teeth, the lower jaw commonly carrying a canine, four premolars, and from three to six molars. Not many specimens retain the incisors, but such as have them exhibit three or four small, slender, conical teeth on each side of the jaw. All the cheek teeth, and in most the canines, are double-rooted, although the roots of the premolars are not in all cases fully divided.

The general appearance of this dentition is very much like that of some of the cynodont reptiles, especially such a one as *Cynognathus*. The triconodonts were, however, true mammals, as is shown by the double-rooted cheek teeth and the fact that the dentary makes up the whole of the lower jaw. Intermediate structurally between the cynodonts and the triconodonts are the mammal-like reptiles from the Upper Triassic of North Carolina. Two little jaws, each less than an inch in length, were found long ago by Ebenezer Emmons and described by him as *Dromatherium sylvestre* (Fig. 109, at left). One of them was later removed to another species and genus. Both resemble triconodonts, and until recently they were supposed to be the most ancient American mammalian remains. W. K. Gregory, and lately G. G. Simpson, have shown, however, that there is reason to doubt this identification, for the jaws display some reptilian and no positively mammalian characteristics. The teeth are similar to those of the triconodonts, *Dromatherium* having three incisors, a canine, and ten cheek teeth, the last seven of which are definitely three-cusped. Detailed study, however, shows that the cheek teeth are not actually double-rooted, the root being merely laterally compressed, with a median depression which does not fully divide it. They are therefore intermediate in structure between the single-rooted ones of the cynodonts and the double-rooted ones of the triconodonts. Moreover, although the jaw appears to consist of only one element, the dentary, there is reason to believe that a part is missing from the posterior end and that another small bone

FIG. 109. At left, *Dromatherium*, a mammal-like reptile from North Carolina. One and two-thirds natural size. From G. G. Simpson. At right, a reconstruction of *Trioracodon*, a Jurassic triconodont. One and six-tenths natural size. From G. G. Simpson, *Mesozoic Mammals*, by permission of the Trustees of the British Museum (Natural History).

was present at the articulatory angle. The triconodonts seem to have been derived directly from cynodonts similar to *Thrinaxodon*, but they perished at the end of the Jurassic without giving rise to any other group.

Modern marsupials are of two kinds. Some, the kangaroos and their allies, have two prominent incisors, whence they are called the Diprotodonta; others, more primitive, with numerous small incisors, are the Polyprotodonta. A well-known example of the latter group is the American opossum. Opossums like modern ones existed in late Cretaceous times, and animals that seem to be closely related are found in earlier Cretaceous and Jurassic strata. They are characterized by the possession of numerous teeth, more or less like those of later marsupials and placentals. The cheek teeth of members of this group are approximately triangular in form, each with three principal cones, one at each corner of the triangle. The same pattern is found in the early opossums and insectivores, the latter the most ancient of the placental mammals. Fossils recently found in the Cretaceous of Mongolia have been definitely proved to be the remains of placentals, and, in fact, of true Insectivora. Because of the likeness of the dentition, the insectivores and the polyprodonts and, inferentially, all the marsupials and placentals are now supposed by paleontologists to have had a common ancestry in the oldest known mammals possessing triangular teeth. These ancestral animals, the Trituberculata or Pantotheria, made their first appearance in England in Mid-Jurassic times. Their oldest representative, *Amphitherium*, had numerous teeth, one species having a formula which, by reduction, could be modified to produce either the marsupial or the placental dentition. This scheme of derivation of the marsupials and placentals from a common ancestor, rather than one from the other, gets over one or two manifest difficulties inherent in the older idea that the placentals, more specialized by virtue of their longer period of gestation, must have been derived from the marsupials. Although obviously in many respects more primitive than the placentals, marsupials are more specialized in that the milk dentition is so reduced that only one tooth on either side is shed, whereas in the placentals all but the true molars are replaced. This theory explains why marsupials, although not descended from the placentals, show in one modern group a sort of rudimentary placenta.

The trituberculates are, therefore, highly important, for they are probably ancestors of most if not all living mammals. Although their study has not yet given a clue to their own ancestry, they seem to have been derived from some sort of theriodont reptile. During later Jurassic times this group spread to North America and East Africa, where their remains occur in strata associated with those which have furnished the bones of the great sauropod dinosaurs. All were small long-jawed creatures outwardly like mice or rats. Their incisors were conical, their canines sharp, double-rooted cones, and the premolars small, sharp, piercing or bladelike cutting teeth. The molars were characteristic, mostly of triangular form, with four or more cusps on the crowns of the upper ones, three on the lower. Primitively there seem to have been four premolars and eight molars, but some show considerable reduction from this large number of cheek teeth. It is unfortunate that so little is known of the skeleton of these animals. So far as can be judged from the jaws, they were particularly well equipped for catching and devouring insects and other soft-bodied invertebrates, or any vertebrates smaller than themselves. They probably lived chiefly upon flesh, although they may have eked out their diet by eating such fruits or seeds as were available. They vanished with the Jurassic, but the presence of their triangular molars in the marsupials and insectivores of the Cretaceous is evidence that they were transformed into higher types of mammals.

The triconodonts and the trituberculates are the most abundant of the three groups of mammals which made their first (and last) appearance in the Jurassic. The third group, known as the symmetrodonts (Fig. 110), is characterized by a simple triangular molar that was at one time considered to be intermediate between those of the two sorts already described. These teeth are primarily simple cones, but each bears two secondary cusps, which appear on the outer side of the upper teeth and on the inside of the lower ones. The teeth are much like those of the triconodonts, except that their form is triangular. It was natural that paleontologists a couple of generations ago should believe that the triangular tooth of the trituberculate arose by a rotation of the subordinate cusps of the triconodont, outward in the upper jaw and inward in the lower. But since no intermediate forms have been found, this idea has been discarded by those who have recently studied the evolution of teeth. Apparently the symmetrodonts were never numerous, and so little is known of them that it is impossible to say whether or not they were closely related to other Jurassic mammals.

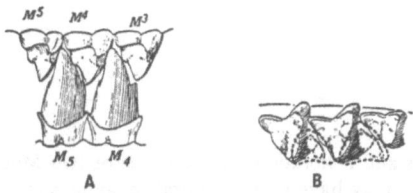

Fig. 110. Diagrams to show, in lateral and crown views, the possible occlusal relationship of symmetrodont teeth. From G. G. Simpson, *Mesozoic Mammals*, by permission of the Trustees of the British Museum (Natural History).

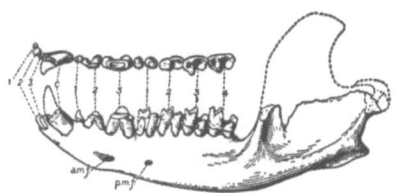

Fig. 111. The jaw of the Cretaceous opossum, *Eodelphis*. After W. D. Matthew.

The oldest true marsupial so far found has been named *Eodelphis* (Fig. 111). It was described from jaws collected in the Belly River formation of the Upper Cretaceous. Although an opossum, it cannot have been a direct ancestor of the modern ones, for it has one less tooth in each jaw. It is, however, sufficiently like the modern representatives of its group to prove that, so far as lineage is concerned, the opossums carry the true blue blood, no other modern mammalian family being traceable so far back as the Cretaceous.

The expeditions conducted by Dr. Roy Chapman Andrews in central Asia have evoked tremendous public interest and achieved lasting fame by the discovery of dinosaur eggs. But nothing they have produced can vie in scientific importance with a small part of the collection made in 1925, a few fossils in which the public has evinced no interest whatsoever. These are seven small, incomplete skulls, the oldest known remains of placental mammals. They have been described by Dr. George G. Simpson, who has contributed greatly to our knowledge of Mesozoic mammals by redescribing all the known species of the world. He had the courage to undertake the study of a group that had previously defied analysis, and developed a technique which enabled him to bring order out of chaos. Four genera and five

FIG. 112. Skulls and restorations of the heads of two Mongolian placentals. The one at the left may have been ancestral to some sorts of modern insectivores. The other has a more carnivorous type of dentition, suggesting relationship to the later creodonts. One-half natural size. From W. K. Gregory, in the *Scientific Monthly*, by permission of the Science Press.

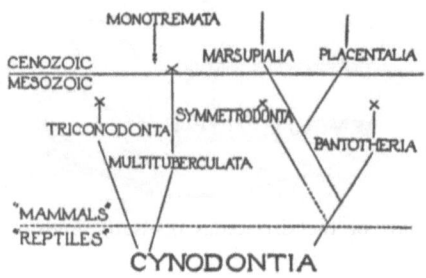

FIG. 113. Diagram to show the probable relationships of the various groups of Mesozoic mammals. From G. G. Simpson.

species are represented by these seven specimens. All appear to be related to the modern Insectivora, a group generally accepted as the most primitive in its subclass. Two families, showing divergent evolutionary trends, are described. The teeth of one are in a continuous series, the canines enlarged and single-rooted, the premolars trenchant, and the triangular molars sharp. Although the members of this family are insectivores, their teeth suggest carnivorous habits, and it is believed by Simpson and Gregory that the oldest Tertiary carnivores, the creodonts, as well as many of the later Insectivora, may have sprung from this stock. The other family represents a somewhat specialized group of insectivores, with the lateral upper and median lower incisors elongated. There is a somewhat long toothless portion of the jaws behind the incisors, followed by triangular cheek teeth. Although not closely related, the animals of this group may have been allied to the ancestral line of the rodents.

These few skulls, obtained by a party working under difficult conditions, hindered by lack of time and by inhospitable surroundings, are only a sample of what the Mongolian Cretaceous may have in store for paleontologists. They encourage us to expect wonderful additions to our knowledge as exploration proceeds. Fossils of this region may eventually make it possible to write a satisfactory introductory chapter to the history of mammals.

At the present time six groups of Mesozoic mammals are known. Most abundant are the multituberculates, which certainly arose from some theriodont reptile of the Triassic. They survived to the Eocene but did not give rise to any other group. Next are the triconodonts, probable offspring of the cynodonts, possibly of a form not unlike *Cynognathus*. They were little mouselike creatures, which, like the similar symmetrodonts, did not outlive the Jurassic and left no descendants. More important are the trituberculates, Pantotheria, known from the Jurassic only, but ancestors of the modern mammals. From them came the primitive marsupials, the opossums of the Cretaceous of America, and the primitive placentals, the insectivores of the Cretaceous of Mongolia. Asia, long traditionally held to be the birthplace of man, may have been that of the placental mammals.

XXII

THE MAMMALS INHERIT THE EARTH

> What! will the line stretch out to the crack of doom?
> *Macbeth*, Act IV, scene 1

The later part of the Mesozoic witnessed a transfiguration of the earth. Vast shallow seas disappeared from the interiors of the continents and receded from their borders; great mountain chains surged up into billowlike crests, and the courses of ocean currents were altered. This geographic revolution changed climatic conditions in various regions and brought about a new vegetation. As early as the Lower Cretaceous new trees appeared, trees which bore conspicuous flowers, shed their leaves annually, and yielded fleshy fruits of great food value. By the end of that period deciduous angiosperms had replaced, in many areas, the pines and other evergreens, the ginkgos and cycads, which had been the common trees of the Mesozoic. During the early Tertiary, as the climates of the continental interiors became drier, incapable of sustaining a population of larger plants, grasses appeared. These changes had a profound effect upon all the animals of the globe. Exactly how and why are still mysteries, but life on both land and sea showed considerable alteration. The great reptiles of the land and the humble mollusca of the sea found the times equally difficult. Some survived this period of stress and change; others vanished completely.

It was a new earth that the little mammals inherited in the Tertiary, an earth furnished with vegetable food in greater abundance and variety than had ever previously existed. Danger from reptiles other than crocodiles, snakes, and turtles had passed; the overladen table called for eaters of fruits, nuts, leaves, and grasses. The world was, as never before, fitted for vegetarians. It is not surprising, then, that the mammals began to thrive, to wax strong, and to differ one from another. As has been seen, the first placentals, the insectivores of the Mongolian Cretaceous, had tirangular cheek teeth with high cones on the crowns. When the jaws were closed, the upper and lower teeth sheared past one another. Such jaws would be useful in catching and cutting insects, worms, small mammals, fruits, seeds, and succulent plants. One infers, therefore, that the early insectivores were really omnivorous. They are called insectivores because they are most closely related to such present-day mammals as live chiefly on insects, not because of their own diet.

Although only a few Cretaceous placental mammals are now known, there must have been hundreds of species and myriads of individuals of the little creatures. Modern insectivores, which retain to a surprising degree the primitive dentition and

THE MAMMALS INHERIT THE EARTH

skeletal characteristics of their distant ancestors, occupy many habitats. Many, such as the hedgehogs, are terrestrial; some, like the moles, burrow; but others, the shrews, dwell in trees, and still others are semiaquatic, subsisting largely on fish. It is probable that the Cretaceous ancestors were equally versatile, seeking food in various situations. As time went on, the different clans became accustomed to particular types of food. Some sought it constantly in trees, their habits of climbing and jumping gradually bringing about changes in the skeleton, lengthening the arms and developing the hands and feet for grasping. The more ambitious of them learned to glide, and finally to fly. Others ran about on the ground eating leaves and twigs. In these, constant running developed the central toes, enabling them to forsake the flat-footed gait, lengthened the legs, and eventually produced hoofs. Still others fed on bark; gnawing strengthened the front teeth, whereas the molars became grinders. These mammals sought for roots and their feet gradually changed into implements for digging. But even with an abundance of vegetable nourishment the old Mesozoic blood-lust lingered. Flesh is, after all, the most concentrated of foods. Those who eat it have time for something more than feeding — time to sleep and clean their coats, to make themselves attractive and to purr, time to gain strength to fight and make themselves masters of the world. The sharp teeth of the early placentals indicate that all would eat flesh when they could get it. There was not enough for all, so the bolder, the larger, the cleverer, got what there was, and finally became solely beasts of prey.

These changes were not instantaneous but gradual. The fossils are sufficiently abundant to allow some lines to be traced; in other groups the records are so scanty that with our present limited knowledge it seems almost as though they had been specially created. Our experience of the discovery of one connecting link after another leads us to believe, however, that sooner or later the whole lineage will be revealed.

The oldest Tertiary mammals are found in Paleocene formations, which have been productive of fossils at only a comparatively few localities. The Paleocene has been to students of the Tertiary what the Cambrian is to geologists in general. It was the last epoch to be recognized, and only recently has its distinctive fauna come to be known. Within the past decade or two many new fossils have been described, but, as is the case with the Cambrian, most of them from incomplete, unsatisfactory material. The three classic localities have been the San Juan basin of northwestern New Mexico, Cernay, near Rheims, northeast of Paris, and the Fort Union strata in central Montana. The last of these has recently yielded numerous fossils, and the productive area has been traced through southern Montana into northern Wyoming. Paleocene fossils are now known from various outcrops from Alberta to New Mexico, and one hundred and sixteen genera of mammals have been described from them. One of the most important of the newer discoveries was that of eleven genera in strata of this age in Mongolia. Interesting, but less significant, is the presence of twenty-one genera in central Patagonia.

These data are from a census published by Dr. G. G. Simpson in 1936, and are a revelation to one who has not kept abreast of recent developments. They show that sixteen of the twenty-eight orders of Tertiary and recent mammals were then in existence. Three, the Allotheria (Multituberculata), Marsupialia, and Insectivora, survived from the Cretaceous. The remainder are new arrivals on the scene. Except for the first two orders listed above, all are placentals, a fact that suggests an extremely rapid expansion in that group. Where did the differentiation take place? The multituberculates and marsupials had been in North America since the Upper Jurassic and the Upper Cretaceous respectively, but the placentals are all new to this continent. Since the oldest placentals so far found are Mongolian, paleontologists appear to have jumped to the conclusion that northern central Asia was their cradle. This may be true, but does the present evidence warrant such a belief? It may be that North America was the real center of origin, for at the present time ninety-five genera of placentals are known here, as compared with nine in France, nine in Mongolia, and fifteen in Patagonia. Ten orders are represented in North America, six in Mongolia, and five in France. Combining France and Mongolia to represent Eurasia, there are seven. Only three are known from Patagonia.

It is true that in the present state of exploration statistics do not have much significance. Those just quoted may be interpreted as meaning (a) that the North American strata have been searched much more thoroughly than those of Mongolia, (b) that the first great differentiation of the placentals may have taken place in North America, even though they originated in Asia, or (c) that the placentals actually did originate on the continent of North America early in Cretaceous times.

That the last is possible seems to be supported by the following arguments, the first of which has already been cited and to some extent discounted.

1. The Paleocene mammals were much more numerous and much more highly differentiated on this than on other continents.

2. The most primitive of placentals, the insectivores, are much better represented here than in Eurasia. Twenty-three genera are known from North America, two each from Mongolia and France. Furthermore, the primates, which stand next in the scale of organization, are represented by twelve genera here and only one in Eurasia.

3. Such evolution on this continent was possible. Everyone agrees that the placentals descended from the pantotheres, which were just as common in North America during late Jurassic times as they were in Europe. They have not so far been found in Asia, although there is reason to believe that they once existed in that region. They disappeared from Europe and North America at the same time, the end of the Jurassic. It is as probable that their further evolution was in North America as in Eurasia —

more probable, in fact, for the greatest abundance of trituberculates is in western Europe, away from, rather than toward, a possible center of distribution in Eurasia.

4. It seems to be generally agreed that both the placentals and the marsupials were derived from the pantotheres. The oldset known marsupials are found in the North American Upper Cretaceous. They are not as yet known from the Mongolian Cretaceous or Paleocene, although they were in France by Eocene times. Perhaps this merely balances the argument from the presence of Cretaceous insectivores in Mongolia.

For the sake of the record it is necessary to insert here a list of the orders of mammals known to have been in existence in Paleocene times. They are the Allotheria, Marsupialia, Insectivora, Tillodontia, ?Dermoptera, Chiroptera, Primates, Taeniodonta, Edentata, Rodentia, Carnivora, Condylarthra, Amblypoda, Litopterna, Notoungulata, and Astrapotheria. Only the most important can be discussed. The last three are South American herbivores. Only one of these orders, the Notoungulata (Toxodontia), is represented outside that continent. It has one genus in the Paleocene of Mongolia and one in the Lower Eocene of North America.

All of the Paleocene mammals are primitive; nearly all have five fingers and five toes; most have the typical forty-four teeth. The upper premolars are triangular, the molars triangular or quadrangular, with blunt or elongated piercing cones. The legs are short, the lower bones of nearly equal size, and the gait of all was digitigrade. The terminal phalanges are blunt, neither claws nor hoofs. The animals are small, the giants among them not much larger than sheep; this, however, is a considerable increase over any Mesozoic mammal. Although all are much alike, still there are recognizable differences.

Insectivores (Fig. 115) other than moles are not familiar to the average American, but the common mole, although somewhat stouter-limbed than most of its allies, gives a fair idea of the group to which it belongs. Most of the insectivores are small, with slender, somewhat elongate heads. Many have a complete dentition, although some have lost one or more incisors and premolars. The molars of most are triangular and retain many primitive characteristics. The Old World hedgehogs, not to be confused with that prickly American rodent, the porcupine, are relatively slightly modified descendants of some of the Mongolian Cretaceous placentals. They have not always been restricted to Eurasia, for Paleocene and Oligocene representatives of the group have been found in North America. The oldest of North American insectivores are still so imperfectly known that opinions about their relationships differ, but some of them are thought to be allied to the squirrel-like tree shrews of the Indo-Malayan region.

The rapidity of diversification among the mammals at the beginning of the Tertiary is made evident by the appearance of flying mammals — the bats — at that time. It is true that only a single imperfect upper jaw has yet been found in the Paleocene (this in Colorado), but more and better-preserved specimens, even those retaining wing bones, are known from the European Eocene. Although the pterosaurs are the most specialized of the reptiles and the birds the most specialized of all vertebrates, the bats do not represent the greatest known departure from the primitive structure of the mammals. That honor has been preëmpted by the baleen whales. But the bats occupy a position which is only slightly subordinate. Had not most of them retained their primitive insectivorous habits, they would probably have stood (morphologically) at the head of the list. Bats, and perhaps bats alone, can be accused of having failed to live up to their full opportunities. It is true that some of them have gone beyond most mammals in adopting the insectan habit of blood-sucking, but apparently they did not become vampires early in their history. At any rate, no bat has as yet succeeded in losing its teeth, so the group can hardly be put on the plane of the pteranodonts, the post-Cretaceous birds, or the whalebone whales. But bats and men are on their way. Another million years may see toothless man at the top, for he can easily dispose of bats and whales, should they threaten to take a technical supremacy.

As a matter of fact, a flying mammal is more to be expected than a flying reptile. As the late Dr. W. D. Matthew pointed out, there are many reasons for believing that the primitive trituberculate mammals, both placental and marsupial, were arboreal. They had two reasons for living in trees: the first, the great size of their reptilian neighbors, and the second, the abundance of food available to scansorial creatures. The dinosaurs were not competitors of such animals as could climb.

If, as seems probable, the late Cretaceous insectivores were arboreal, it is not at all surprising that some of them should have learned to fly. That the ability to fly came about through the growth of skin between all the fingers except the first and second is strange, in view of what happened to the pterosaurs and birds. It might have been expected that one or the other of the schemes adopted by those animals would have been followed. But the various creatures which learned to fly were differently constructed at the time when the new habits were initiated. The ancestral pterosaur had already lost the little finger; the reptile which gave rise to the birds had lost the fourth as well. The bats started from scratch, with all five. In the sequel those apparently most handicapped fared best, probably because of feathers.

The lemuroids were the first and most primitive of the primates, that division of mammals with hands and feet equipped with nails rather than hoofs or claws. The Paleocene representatives of the group leave no doubt of the close relationship between them and the insectivores, for they are so much alike that specimens are distinguished only with difficulty. The lemuroids resembled the modern lemurs in having long skulls but were without some of the latter's specialization. Though primates were more or less common in North America during the Paleocene and Eocene, their later history can be traced only on other continents. Those found in the Paleocene were all tiny creatures about the size of squirrels or mice, with slender limbs, long fingers, and the thumb (pollex) and great toe (hallux) partially opposable to the other fingers and toes.

Other early offshoots of the insectivores were the Tillodontia, large rodentlike creatures which did not survive the Eocene, and a few small creatures which are tentatively placed in the Dermoptera. The only living representatives of this order are the so-called flying lemurs of the East Indies, somewhat batlike animals which glide from branch to branch like flying squirrels. They have an expansion of the skin extending from the neck to hands and feet, and ending on the tail. Although their hands are enlarged, there is no indication that they represent an intermediate stage between insectivores and bats. It is curious that there is no record of this group between the Eocene and the present time.

The creodonts are the first carnivores. Appearing in the Paleocene, only slightly modified from insectivores, they became more and more abundant during the successive ages of the Eocene but died out in America early in the Oligocene, in Europe in the Miocene. They are discussed further in the chapter on beasts of prey. They were the sole predators of the Paleocene and Eocene, the most peaceful ages in the history of mammals.

The condylarths (Fig. 114) are the most primitive hoofed mammals, interesting because they differ so little from the contemporaneous creodonts. As with the latter, the limbs were short and massive, the gait semidigitigrade, the tail long and heavy. The Paleocene condylarths have trituberclar teeth much like those of the creodonts and even similar tusk-like canines, but the later Eocene representatives had quadrangular grinding molars of the bunodont type. The principal differences of the early condylarths from the creodonts appear to have been the narrower skulls, weaker jaws, and somewhat blunter toes of the former. The two groups are so much alike, however, that unless material is fairly complete it is almost impossible to distinguish them. All this indicates that the herbivorous hoofed mammals (ungulates) arose from a group of insectivores closely allied to that which produced the carnivorous creodonts. The condylarths did not survive the Eocene; in fact, they appear to have died out before more than half of Eocene time had elapsed. It cannot be proved that they left descendants.

The Amblypoda are another group of unprogressive hoofed mammals, important in Eocene times but extinct since the end of that period. The earliest representatives of the race were of relatively small size, about that of pigs. They have stout limbs and the clumsy, blunt-toed, elephantlike feet which suggested the name for the group. Curiously enough, the early amblypods show many of the characteristics of the creodonts, the body and tail being long and catlike, the skull bearing similar tusklike canines. The head was narrow, however, and the molars had crowns of the grinding type. The teeth seem to be more fully modified for an herbivorous diet than those of any other Paleocene mammals. The ancestor is to be sought in some Cretaceous condylarth-like creature.

The amblypods (Fig. 116) must have been abundant during the Eocene, for one museum has accumulated about two hundred more or less complete skeletons of them. This unusually large amount of material has permitted the tracing of two or three lines of evolution. The one best known culminated, just before the extinction of the race, in the largest Eocene animals, some of them with the bulk of an elephant. These great uintatheres had remarkable heads, with three pairs of knoblike horns and extremely long, saberlike upper canines. One pair of horns was on the nose, another in front of the eyes, and the last in front of the ears; it is not known whether their bony cores were covered with horn or with a callous skin, although probably with the latter. The long tusks were in all likelihood used chiefly to pull down branches of trees and to strip the leaves from them, although it has been suggested that they could have been used for fighting if the animal struck with the mouth open, after the fashion of a snake.

The Amblypoda are sometimes called the dinosaurs of the Eocene because the brain was smaller in proportion to the total bulk than in any other living or extinct mammal. Moreover, its surface was smooth, a large portion of it occupied by the lobes devoted to the sense of smell. These creatures seem to have reached, for mammals, the climax of brute strength as contrasted with brain power. Their extinction was probably due to the smallness of the brain and the great size of the body. They had too little intelligence to protect their young from the crafty carnivores and too much bulk to be provided with food when hard times came upon them. Thus as early as the Eocene the dominance of Mesozoic brute force ended, and evolution along the line of increasing brain began.

The edentates, chiefly South American animals, are characterized by loss and degeneration of teeth; such as remain are peglike and devoid of enamel. Even their oldest known ancestors, the Paleocene representatives of the group in North and South America, are surprisingly specialized in this respect. *Palaeanodon* did, it is true, have four fairly good cheek teeth with some enamel, but the Eocene forms have a much feebler dentition. Although they may possibly have been ancestral to the armadillos, there is no evidence that they had a bony armor. Taeniodonts, which are allied to the edentates, are more common than the latter in North America, but apparently their line led nowhere.

At one stage or another during the Eocene most of the other major groups of mammals made their appearance. Late in the epoch the modern types of carnivores emerged from the creodont stock. Perissodactyls were the predominant types of hoofed mammals, represented by primitive horses, tapirs, and rhinoceroses among groups which persist to the present day, as well as by the extinct titanotheres and amynodonts. The artiodactyls were almost as numerous as the odd-toed mammals,

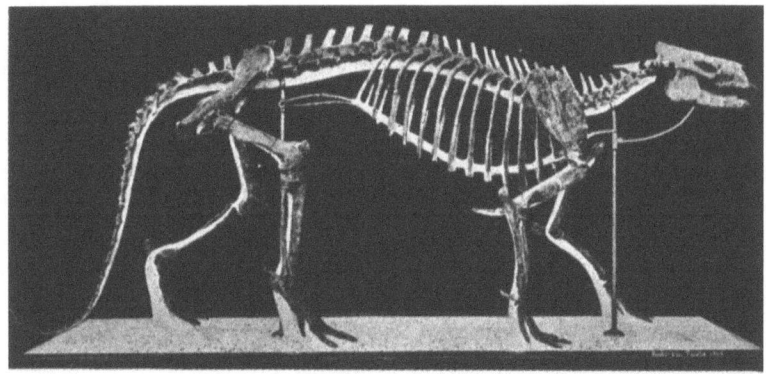

FIG. 114. The primitive condylarth, *Phenacodus*. Photograph by courtesy of the American Museum of Natural History, New York City.

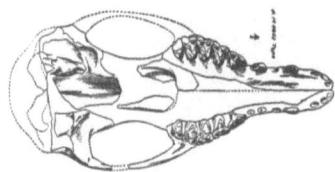

FIG. 115. *Diacodon*, an Eocene insectivore, figured to show the tritubercular teeth. One-third natural size. From Matthew and Granger.

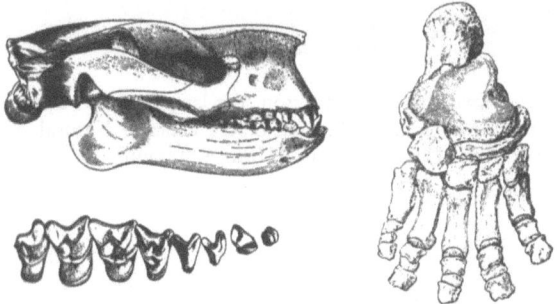

FIG. 116. Skull, teeth, and right hind foot of *Barylambda*, a large Paleocene amblypod. Note the triangular teeth. The skull is about twelve and a half inches long. From Bryan Patterson.

Fig. 116A. *Barylambda faberi* Patterson, the largest of the Upper Paleocene amblypods, restored by John C. Hansen of the Chicago Natural History Museum. Courtesy of the Chicago Natural History Museum.

including the first of the camels, of the deer, and of the peccaries, among animals still existing, as well as several extinct groups, most outstanding of which is that of the merycoidodonts. In the Eocene of Africa are found the oldest remains of elephants and whales, representatives of the latter reaching America in late Eocene times.

By the end of the Eocene the major differentiation of mammals had taken place. During the remainder of the Tertiary, the Oligocene, the Miocene, and the Pliocene, mammals became more and more like those now existing. Most of the surviving groups belong to families which were fully differentiated during the Oligocene or, at latest, the Miocene. Nothing really new, with the exception of man, has appeared since the latter epoch. Several groups prominent in Oligocene and Miocene times have long been extinct. Others died out during the cold periods of the Pleistocene, and still others have been or are rapidly being exterminated by man. The heyday of mammals, other than man, has passed. Two-thirds of their history, from the Triassic to the end of the Mesozoic, was passed under the suppressive rule of the reptiles. They made little progress during that long period, but when in the Paleocene they inherited the earth they rapidly made full use of it. Finally one of their own number has decided that he alone shall dominate. In taking possession he is exterminating such of his brethren as cannot be brought into profitable bondage. The history of the mammals has been one of constant increase in use and size of brain; with man, brain has finally superseded all other factors in survival value.

XXIII

BEASTS OF PREY

Nought treads so silent as the foot of Time;
Hence we mistake our autumn for our prime.
Edward Young, *Love of Fame*, Satire v

Claws and conical teeth are the equipment of the beast of prey, be it amphibian, reptile, or mammal. The carnivorous reptiles inherited their dentition directly from the Amphibia, but since the Jurassic ancestors of Tertiary mammals had triangular cheek teeth, a conical or a bladelike mammalian tooth behind the canines is a specialized, not a primitive, instrument. Few mammals have become so specialized as to have teeth like those of primitive flesh-eating amphibians or reptiles. The nearest approach is, perhaps, in the single-rooted, conical ones of certain porpoises. A few truly carnivorous mammals have simple peglike teeth in the anterior parts of the premolar or molar series, but such are commonly vestigial, practically without function. The primitive trituberculeat dentition was about as well adapted for chewing animal as vegetable food; hence it is only the abundance of the latter which can account for the predominance of herbivores over carnivores during the Tertiary.

The principal trend among the carnivorous mammals has been toward the evolution of progressively more efficient cheek teeth. It will be remembered that the triangular teeth of the lower jaws of the pantotheres slid into the reëntrants between the upper ones, producing a shearing action. This shear, because of the shape of the teeth, was irregularly zigzag, and hence inefficient. The earliest carnivores, the creodonts, inherited the triangular tooth, but throughout the Tertiary the flesh-eaters, evolving along various lines, came to have more and more bladelike cheek teeth. Thus the direction of cutting became less and less irregular, until in the more highly specialized groups the action of the jaws is like that of ordinary household shears. If such a comparison be made, however, it should be, perhaps, with an old pair, badly nicked toward the tips but retaining good cutting edges near the rivet which holds the blades together. Not all of the cheek teeth became modified for cutting, but only one or two pairs in each jaw. Anyone who has seen the family cat at work upon its plate of liver must have noted that most of the chewing is done within a limited region on each side of the mouth. The teeth which do most of the work have become enlarged, highly specialized, the most important of the whole series. These shearing organs are called the "carnassials" or "sectorials"; to the tax-

Fig. 117. *Sinopa grangeri*, a doglike Eocene creodont from Wyoming. Note the arched back, long tail, and heavy limbs with almost plantigrade feet. Photograph by courtesy of the American Museum of Natural History, New York City.

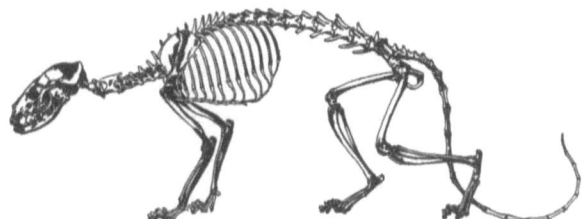

Fig. 118. *Pseudocynodictis*, a slender Oligocene dog of a type about which there has been much discussion. Some consider it to be in the main line of descent of modern dogs; others think that it is the last member of an extinct group. From W. D. Matthew.

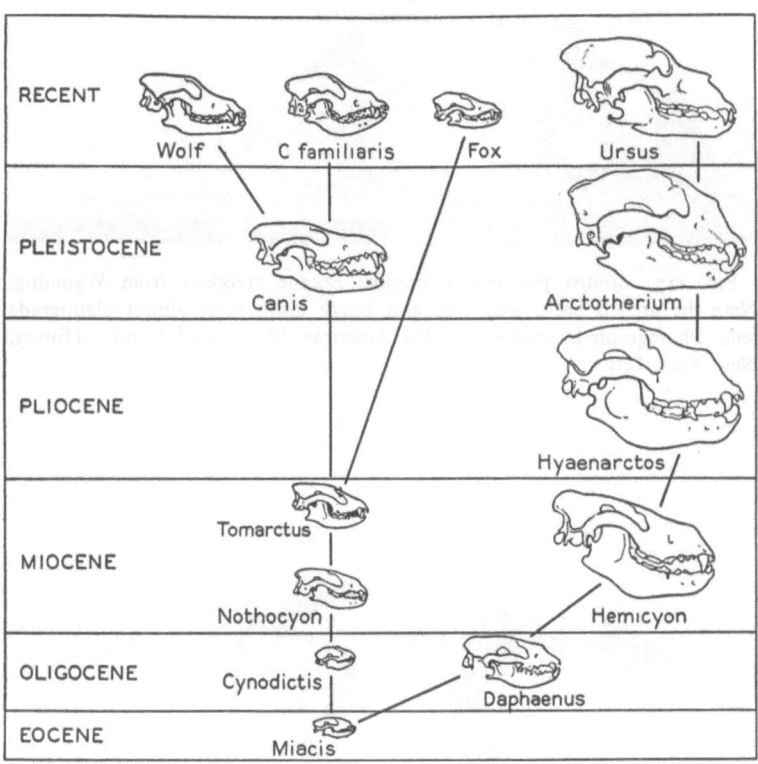

Fig. 119. The family trees of the wolf, domestic dog, fox, and bear, as understood by W. D. Matthew. Such family trees are subject to constant revision, as "new" fossils are found. Redrawn after Matthew, with modifications after F. B. Loomis. This *Cynodictis* is now called *Pseudocynodictis*.

onomist they are of fundamental importance in the division of the beasts of prey into subordinate groups.

The order Carnivora is commonly treated as composed of three lesser sections or suborders. The most specialized are the Pinnipedia, the seals and walruses, adapted for life in the sea; the members of the other two are, in the main, terrestrial. One, the Fissipedia, includes all the modern land carnivores, of which there are seven families: the dogs; the raccoons; the bears; the mustelids, including martens, weasels, and skunks; the cats; the hyenas; and the civets. The last two never reached North America, and the bears arrived only in the Pliocene; but the others have a more or less full record in the Tertiary of this continent. All of these animals have but a single carnassial in each side of each jaw. The fourth premolar of the upper series is blade-like, this tooth being opposed by the first molar of the lower jaw. All Fissipedia have less than forty-four teeth. The third and most primitive suborder is that of the creodonts, early Tertiary carnivores with, for the most part, the typical forty-four teeth of the placental mammalian dentition. The carnassials are variable in number or not fully developed as shearing teeth, for, as is the case with the Fissipedia, some creodonts were omnivorous rather than strictly carnivorous. The few known Paleocene members of the group were small; the Eocene and Oligocene were the periods of greatest differentiation and also of greatest size, some creodonts of the latter time being as large as modern wolves or bears — one, *Andrewsarchus*, much larger. Only one family survived till the early Miocene.

Although diversified, all creodonts (Fig. 117) share certain characteristics. The canines are large, the premolars simple, laterally compressed, the molars fundamentally tritubercular but variously modified in the several families. The head, which is large, shows a typical characteristic of the carnivore in its width at the orbits, bulging zygomatic arches giving it unusual breadth. The brain case was small, and the sagittal and occipital crests consequently high, to afford room for the attachment of muscles. The body was long, and the tail long and heavy, a characteristic inherited by many modern carnivores. "Lauk! what a monstrous tail our cat has got!" The limbs were short and stout; the feet, except in specialized members of one family, pentadactyl and digitigrade, or semiplantigrade.

As has already been pointed out, some of the Mongolian Cretaceous insectivores have characteristics which suggest the creodonts: in the Paleocene are found scanty remains of animals belonging to the most primitive genera of the group. The individuals are small, and their sharp-cusped tritubercular molars recall those of insectivores. This family, the Oxyclaenidae, may be shown, when better known, to have been ancestral to all other carnivores.

One family of the creodonts, the Miacidae, has the same arrangement of carnassials as the modern carnivores; hence Professor W. B. Scott's suggestion that it should be linked with the Fissipedia rather than with the Creodonta. The oldest members are found in Paleocene and Lower Eocene formations, the youngest in the Upper Eocene. They are the only creodonts with $p\frac{4}{4}$ and $m\frac{1}{1}$ functioning as carnassials. The brain case was larger than that of their relatives, and the feet had five toes, arranged in spreading fashion, each armed with a small, sharp, partially retractile claw. The dogs are the only modern carnivores as yet definitely connected with this group, but all Fissipedia must trace to it even though some of the connecting links are still unknown.

Turning now to the modern terrestrial carnivores (Fissipedia), no attempt will be made to follow the evolution of all seven families, but three of the best known, the dogs, bears, and cats, will be discussed.

Second only to the higher primates in development of brain and intelligence are man's chosen companions, the dog and the cat; of the two, the useful dog ranks somewhat higher than the ornamental cat. There can be little doubt that the domesticated dog is a close ally of the modern wolf. In fact, the only obvious difference is that the pupil of the eye of the dog is circular, that of the wolf oblique.

As compared with cats, modern wolves and dogs show relatively little departure from the structure of their Eocene ancestors. Almost the only carnivores which capture prey by continuous pursuit, speed and stamina are their prime requisites. Speed has been improved through the ages by the lengthening of the legs, but this has not been accompanied by any great reduction in the number of toes. The small size of the ungual phalanges of the clawed animals makes it almost impossible to evolve a one- or two-toed foot (though to all intents and purposes this has been accomplished in the kangaroo). Moreover, although the carnivores pursue the hoofed mammals of the plains, their own habitations are commonly on the softer soils in the forests or along creek bottoms, where all toes come in contact with the irregular surface of the ground. Nevertheless, there is some reduction, for the thumb of the modern wolf or dog is vestigial, and there is no great toe. Functionally, therefore, the feet are four-toed. An artiodactyl-like feature is that the toes are in pairs of equal size, the inner somewhat larger than the outer ones; in this respect the feet suggest those of pigs. All dogs are digitigrade, with rather elongate feet, the weight borne on the pads under the distal phalanges. The claws are blunt, not retractile, of little use in the capture or dismembering of the prey but excellent for scratching on doors. The bones of the forearm show another adaptation to cursorial habits in that they have no power of rotation, the radius and ulna lying side by side. The skull is long, the brain large, with well-convoluted cerebrum. The dentition is primitive in many respects. Forty-two teeth are retained, the formula being $i\frac{3}{3}$, $c\frac{1}{1}$, $p\frac{4}{4}$, $m\frac{2}{3}$. The last upper molar is absent, and the third lower one is small, on the verge of disappearance. The upper molars retain to a remarkable extent the primitive triangular shape and tritubercular arrangement of cusps.

Such are, in general, the characteristics of *Canis familiaris*, the dog, and *Canis lupus*, his uncivilized brother, the wolf. Dogs or wolves, henceforth in this section called dogs, were common during Pleistocene times. No representative of the direct ancestors is yet known from the Pliocene, but according to F. B. Loomis, whose outline is followed here, *Tomarctus*, a small Mid-Miocene animal, differed but slightly from a true wolf. It is supposed to have been ancestral not only to *Canis* but also to *Vulpes*, the fox. Ancestral to *Tomarctus* is *Nothocyon* of the Lower Miocene, which is intermediate in size between a modern red fox and a coyote. Its dental formula is that of the dog, but the teeth are smaller and less widely separated. The thumb is much less reduced than that of modern canids, although distinctly shorter than the fingers. The foot has five digits of different lengths, for the fingers and toes are not in pairs as in the later forms. The claws are thin, sharp, somewhat retractile, and hence more catlike than those of dogs. It probably was not a particularly rapid runner. Fortunately, brain casts are known, and these show a smaller size and fewer convolutions than those of modern wolves.

In Oligocene strata are found skeletons of the somewhat smaller, foxlike *Cynodictis* (Fig. 118), with long body and tail, short, weak limbs, and five-toed feet, armed with sharp claws. The spreading wrist and ankle bones connote the large, loosely articulated feet of an animal not swift in pursuit but probably a stalker, awaiting the opportunity to make its kill. Incidentally, the big head and feet of your awkward puppy are souvenirs of his creodont ancestors. The dental formula is that of *Canis*, but the short face enforced a compact dentition, without diastemata.

Although *Cynodictis* (now *Pseudocynodictis*) may be the first dog, its progenitor is found in the creodont family Miacidae (Fig. 119). There seems to be no doubt that the Eocene *Miacis*, the genus which has given its name to the family, included the ancestors of the dogs. Small in size, more like weasels than wolves, animals of this genus retained the typical forty-four teeth and had five functional toes on the short, spreading feet, each toe provided with a sharp, somewhat retractile claw. The brain was a little larger than that of other creodonts but only slightly convoluted. These creatures can hardly be called dogs, for they appear to have been ancestral not only to that family but to various other groups, some now extinct, others, such as the bears, still with us.

Modern bears differ in striking fashion from dogs both in appearance and habits; yet they have the same ancestry. The skull is, it is true, rather doglike, although the width at the arches, reminiscent of the creodont ancestor, makes the face broad. *Ursus*, the bear, has the same number of teeth as *Canis*, but the anterior premolars, instead of being strong, shearing teeth, are small, three of them single-rooted, and of so little use that many individuals lose them early in life. Premolar $\frac{4}{}$ and molar $\frac{}{1}$ are not shearing but crushing teeth, entirely unlike true carnassials. The molars are longer than wide, with numerous tubercles on the crown, suggesting the pig rather than the carnivore. Like the pig and man, the bear is omnivorous. Those species which live where they can get it seem to prefer vegetable food, though they are not averse to such tidbits of flesh as come their way and are particularly keen for the nectar which bees have stored for their own use. Polar bears, on the other hand, are carnivorous perforce, their diet largely piscine. Other characteristics of the bear are distinctly un-doglike. The body is short and heavy with a short tail, the limbs short and massive, the feet large, plantigrade, with five functional toes bearing long sharp claws. Unlike the dogs most of them climb trees readily, although, fortunately for man, the least amiable of their species, the grizzly, lacks this power.

At the present day bears, although chiefly holarctic, are found on all continents except Australia. Apparently they were widespread during the Pleistocene also, some of the species then in existence belonging to the modern genus *Ursus*. During that period, however, the short-faced bear, *Arctodus*, or *Arctotherium* as some call it, ranged throughout North America from Pennsylvania, South Carolina, and California through Mexico, and penetrated far into South America. It was the only bear to reach the southern continent. It appears to have been ancestral to *Ursus* (Fig. 119). In Pliocene times bears were rare in North America, although *Hyaenarctos*, which was probably ancestral to *Arctodus*, did reach this continent. This genus appears to have been at home in southern Europe and Asia, where its remains have been found in France, Greece, and the Siwalik hills of India. It was more primitive than the modern bear in that only the first premolar was small and that the upper molars were quadrangular, rather than longer than wide.

The ancestor of *Hyaenarctos* was French, not American, but this ancestor is classed by taxonomists not as a bear but as a dog, *Hemicyon*. Its remains are found in the Upper Miocene deposits at Sansan. It was digitigrade rather than plantigrade. The first upper molar was rounded-triangular rather than quadrate, and the second, oval and somewhat diagonal in position, as would be expected of a tooth modified from one of originally triangular form.

According to W. D. Matthew, the Oligocene *Daphoenus* of America was the ancestor of *Hemicyon*. Several species are known, and paleontologists differ as to the interpretation of the fossils. They are classed as dogs, although they show many catlike characteristics. In spite of these features, they seem to have given rise to two lines, in one of which the teeth remained sharp and doglike, whereas in the other the bunodont condition of the bears was attained. This doglike line did not lead to modern dogs but produced huge, long-headed animals which died out during Pliocene times. Not so perfectly adapted to their environment as the true dogs, they were unable to compete with them and so disappeared, whereas the omnivorous bearlike creatures, which competed with neither dogs nor cats, have persisted to the present. *Daphoenus* was small, not so large as a coyote. The teeth were small and closely set, the full number, forty-four, being present. Carnassials were present, not the sharp shearing blades of true sectorial teeth but forerunners of the blunt crushers of the bears, although they retained some of the shearing power of those of the Miacidae. The skull had a short face and relatively small brain case; the feet were spreading and digitigrade but did not show as great a departure from the plantigrade type as in the contemporaneous ancestor of the true dogs (*Pseudocynodictis*). The claws were partially retractile. The fact that the lumbar vertebrae were large suggests powers of leaping, a catlike feature. All in all, this probable ancestor of the bears was little like the modern representatives of the group. The ancestor of *Daphoenus* was undoubtedly some member of the creodont family Miacidae, probably *Miacis* itself, ancestor, as we have seen, of the dogs.

Why the antipathy between domestic cats and dogs? Is it because they have a common ancestry? Or is it, as owners of pets often suspect, merely a jealousy which has been engendered in the comparatively recent days in which both have been the associates of man? The latter explanation is the more probable, since the habits of cats and dogs are so unlike that, although both are carnivores, they are actual competitors in only a few habitats. The cat hides, lurks, crouches, and springs. Dogs openly give chase, generally in packs, their baying progress markedly in contrast to the silent, individualistic activities of the felines. These diverse activities bring into use different muscles; consequently the structures of the animals of the two groups have become increasingly unlike in the progress of time. In most respects cats are more specialized than dogs. Only in the quality of the brain does the dog excel all other carnivores.

Cats have short, broad heads, which provide little room for the teeth. The dentition is correspondingly reduced, chiefly by the loss of most of the molars; typical cats have only one molar in each jaw, the upper one small, peglike, practically useless, the lower a large sectorial which shears past the last upper premolar. The most specialized depend largely upon the last upper premolar and the last lower premolar and first lower molar for slicing their meat into pieces small enough to be swallowed; others have one more functional premolar of the shearing type on each side. But the shortening of the jaw is not the only cause of loss of teeth. Practically all felids have large canines, with which their prey is seized or killed. It is commonly the case that when canines — or other teeth, for that matter — are much enlarged the neighboring ones become more or less functionless and tend to disappear. Thus, in cats, the first premolar is absent, the second is absent or vestigial, and the third, although generally present, at least in the upper jaw, may be small.

The Felidae are divided into two subfamilies, the true cats (Felinae) and the saber-toothed tigers (Machairodontinae). The dental formula of the former is $i\frac{3}{3}$, $c\frac{1}{1}$, $p\frac{2-3}{2}$, $m\frac{1}{1} \times 2 = 28 - 30$. The incisors are small, the canines large, oval in section, the lower nearly as large as the upper. Each jaw has two large premolars. The neck is short; the body and tail are long in most, although the tail of the lynx is short. The hand has five fingers, the thumb rather short; the foot has four toes, the first having been lost.

Unfortunately, little is known of the genealogy of the true cats. They appear first in Lower Miocene deposits in both North America and Europe; thenceforth they seem to have been fairly common on both continents, but no direct lineage is traceable. The Lower Miocene *Nimravus* differs from the modern cat chiefly in having an additional small premolar in the lower jaw and somewhat more primitive carnassials. Curiously, the upper canine was longer than the lower, laterally compressed, bladelike rather than oval in section, and hence comparable to the tusk of the saber-toothed tiger rather than to that of the cat. This suggests that the two

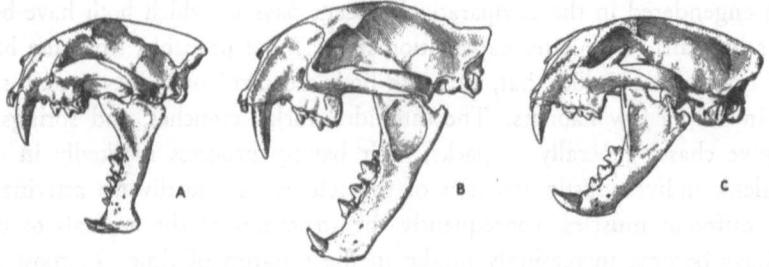

Fig. 120. Skulls of three cats, showing the small amount of change during their known evolution. A, *Dinictis*, Oligocene; B, *Nimravus*, Miocene; and C, *Felis*, Recent. Redrawn from W. D. Matthew. Recent investigations throw doubt on Matthew's conclusions, expressed in the text, that *Dinictis* and *Nimravus* are ancestral to the Felinae. The former is probably a "sabertooth," and the latter is considered the type of a third subfamily, intermediate between the two usually recognized.

subfamilies of cats had a common ancestor, probably already identified by the vertebrate paleontologists in *Dinictis*, a primitive cat of the Oligocene. Although the latter has the long canines of the saber-tooths, it has many of the characteristics of the true cats.

The saber-toothed tigers were the most spectacular and most highly specialized of the cats. As a subfamily they differed from the true cats in having bladelike canines, a flange on the lower jaw, five toes instead of four on the hind feet, a bony outgrowth on each ungual phalanx which served as a partial shield for the retractile claw, and a short tail. The history of these animals is as little known as that of the true cats. Like many other North American animals they achieved their greatest perfection and suffered extinction during the Pleistocene. Because of the abundance of its remains in the tar pits at Rancho La Brea, *Smilodon*, the last representative of this race, is commonly regarded as a Californian, perhaps a "native son." As a matter of fact, however, individuals belonging to various species of this genus ranged during Pleistocene times over the area from Pennsylvania and California in the north to the pampas of Argentina in the south. Larger and more terrible than any lion or tiger, imbued, seemingly, with a blood-lust — for many paleontologists believe that it killed for blood rather than for flesh — *Smilodon* must have been the scourge of its age. In strength and ferocity it outranked the contemporaneous American cat, *Felis atrox*, largest and strongest of all lions. Its terrible scimitar-like canines have gained for it the interest of all. These teeth, eight inches long in the largest individuals, are laterally compressed, with serrate posterior edges, perfect weapons for the severing of the jugular vein. The small size of the lower canines indicates that the saber-tooth killed by striking, after the fashion of a snake, rather than by

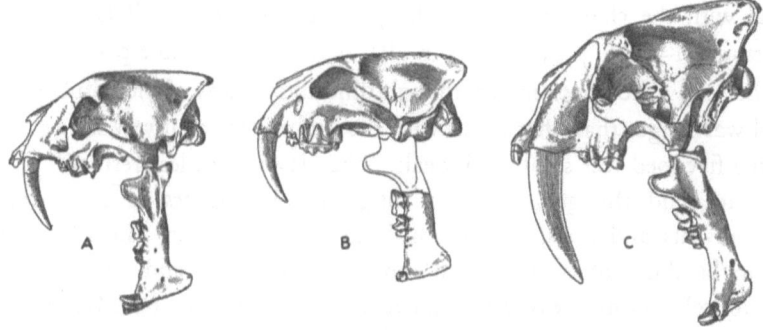

FIG. 121. Skulls of three saber-toothed tigers, showing the relatively small changes which have taken place during their evolution. A, *Hoplophoneus*, Oligocene; B, *Machairodus*, Pliocene; C, *Smilodon*, Pleistocene. Redrawn from W. D. Matthew.

seizing with the jaws and rending with the claws, as true cats do. It may well be that this extraordinary development of the canines was what led to the extinction of the group, just as the overgrowth of the incisors of squirrels and other gnawing rodents locks the jaws of certain individuals and results in starvation.

Little is known of American Pliocene or Miocene saber-tooths, although several species are known from our Oligocene. In Europe, however, several species of *Machairodus* are found in deposits of Pliocene to Pleistocene age. These differ from *Smilodon* in that the body is smaller, the canines shorter, the flanges on the lower jaws larger, the brain case smaller, and the dentition less reduced. *Smilodon* has a dentition more reduced than that of any true cat; that is, $i\frac{3}{3-2}$, $c\frac{1}{1}$, $p\frac{2}{2-1}$, $m\frac{1}{1}$. All species of *Machairodus* have two lower premolars. The upper molar, although vestigial in both general, is behind the carnassial in the French genus, not internal to it as in *Smilodon*.

Several species of saber-tooths are known from the Oligocene, some of them showing an admixture of truly feline characteristics. The Lower Oligocene *Hoplophoneus* appears to be ancestral to all later members of its race. The size was variable in the different species, the oldest being smaller than a modern wildcat. The simplest dental formula in the genus is i$\frac{3}{3}$, c$\frac{1}{1}$, p$\frac{3}{3}$, m$\frac{1}{1}$, a total of 32. The carnassials are less specialized than those of other cats, in some respects doglike. The canines are long and bladelike, serrated on both edges, but do not project beyond the flanges of the lower jaw, these being deeper than in later forms. According to J. C. Merriam's interpretation, *Hoplophoneus* probably had habits exactly the opposite of those of the modern cats. He believes that it held its prey with its strongly clawed, grasping feet, while it struck repeatedly with its knifelike canines.

A contemporary of *Hoplophoneus*, but somewhat more primitive, was *Dinictis*, an animal with the characteristics of the possible ancestor of all later Felidae. Its dental formula, i$\frac{3}{3}$, c$\frac{1}{1}$, p$\frac{3}{3}$, m$\frac{1}{2}$ (34 in all), was distinctly more primitive than that of any other cat. The upper molar, though diminutive, was of the tritubercular type. The skull was longer than that of the cats, the limbs were relatively longer and more slender, the five-toed feet small and weak. The claws were less retractile than those of other cats, and the gait almost plantigrade. It appears to be a connecting link between cats and dogs, and although its short tusks are laterally compressed it seems to be the prototype of both saber-toothed tigers and true Felinae. The views as to relationships expressed above are those of W. D. Matthew. Recent studies have thrown doubt upon some of his conclusions, but without setting up new lineages.

Mammals seem to have been only moderately successful as terrestrial carnivores. An attempt has been made in the preceding pages to trace the lineages of a few of the more conspicuous groups. There are many others, but the creodonts, dogs, cats, hyenas, and bears are the ones that really deserve the title of beasts of prey. As has been seen, their evolution was parallel to that of the other mammals. On the whole they have been more variable, though less prolific in the production of species, than the herbivores. This statement seems contradictory and needs some elucidation. The point is that the terrestrial carnivores show but little of that program evolution which is so conspicuous a feature of the history of the hoofed quadrupeds. Horses, camels, rhinoceroses, tapirs, and other groups differentiated in the Eocene or earlier, and each pursued its own path. There was wide variation within narrow limits. Among the beasts of prey, on the other hand, there is no clear-cut lineage extending back to the Eocene. The dogs furnish the nearest approach to one, but an Eocene dog was not a dog in the same sense that an Eocene horse was a horse. The creodonts held the stage during the first half of the Tertiary, variation producing among them doglike, catlike, hyenalike, bearlike, and other sorts of creatures. With their decay in the Oligocene, the Fissipedia emerged and through a similar series of variations produced the modern carnivores. But it was not till Miocene times that dogs, cats, and bears were fully differentiated. These animals reached their maximum in the Pliocene and Pleistocene. Their rapid downfall has been due to such a wholly unpredictable combination of circumstances that they can hardly be blamed for having mistaken their autumn for their prime.

FIG. 122. Restoration of the Pleistocene saber-toothed tiger, redrawn after a painting by Charles R. Knight.

XXIV

VEGETARIANS SEIZE THEIR OPPORTUNITY

No man can tell what the future may bring forth, and small opportunities are often the beginning of great enterprises.
Demosthenes, Ad Leptinem

The formation of great plains high above sea level east of the Cordilleran-Andean ranges of the Americas and north of the Himalayas in Asia was a result of tremendous earth movements at various times during the Tertiary. None too well watered, the vegetation of these plains gradually changed; forests became thinner and thinner, giving opportunity for the spread of grasses and shrubs, till then held in check by the shade of great trees. Finally a condition was reached in which trees and lush herbage were restricted to relatively small areas in river bottoms, whereas extensive regions were occupied by grasses. These were tender in the spring of the year, but as the season advanced they became tough and wiry. Dry grasses have high food value; even in winter they serve to sustain life, though animals may have to seek them beneath a cover of snow. Food of any sort is always at a premium; hence various mammals left the forests and invaded the ever-increasing areas of the plains. Some may have done so from choice, others by accident, but environmental pressure was probably the chief motivating cause.

As the restriction of the forests reduced the amount of food in them, their area became too small to support the growing population. It is the same sort of pressure which is troubling the world today, that pressure of overpopulation which causes wars, sometimes of physical combat, sometimes battles of wits called diplomacy. Just as human beings are forced by economic necessity to seek homes in unaccustomed areas where they must learn to eat new foods and protect themselves from new enemies, so were the Tertiary mammals compelled to adapt themselves to new circumstances. Fortunately for the success of mammalian evolution, they did not suffer the sudden transitions which have fallen to the lot of man. The Englishman or the German, transplanted to equatorial Africa or the islands of the South Seas, tries to take his environment with him, and is more or less successful. It is largely this which enables him to "carry on." Mammals with less brain could not do so, nor was it necessary, for climates change slowly. Young horse or deer or antelope never remarked, "Leaves are scarcer here than they were in great-grandfather's time. We'll simply have to eat grass." Mammals had to eat, and they ate what they could get. Times were hard, undoubtedly. There was a worldwide depression. Many a rich and noble family went to the wall. But those which learned to eat grass found a new source of income. "Natural resources" were not by any means exhausted. On the basis of grass new family fortunes were founded.

Carnivores could not flourish in this new environment, in which there were few opportunities for concealment. Decrease in number of enemies allowed greater expansion of the herbivores. So in the Mid-Tertiary a new dynasty arose, that of the hoofed animals, the Ungulata, whose reign was checked only in part by the glacial climates of the Pleistocene but has been brought almost to a close by the prowess of man, first in hunting and later in agriculture.

The first hoofed animals lived in the Eocene forests, feeding on leaves, coarse vegetation, and probably to some extent on such animal food as happened to be available — insects, worms, snails, perhaps occasionally a bird, a lizard, or a small mammal. Their teeth, as has already been pointed out, did not differ greatly from those of contemporary flesh-eaters. Primitive ungulates differed only slightly from primitive carnivores, and there are indications that the former group is a branch of the latter. The chief specializations of the hoofed quadrupeds have been the lengthening of the legs, the reduction in the number of toes as the animals became more and more adapted to running on the open prairies, and the evolution of grinding crowns on the molar teeth, a modification which enabled them to subsist on tough dried grasses. Some races of hoofed animals, however, continued throughout their history to feed upon coarse vegetation, which can be obtained on the plains as well as in the wooded areas; others remained in the forests. All such browsers have retained a rather primitive dentition, teeth better adapted for crushing than for grinding.

The Ungulata are readily subdivided into two great orders, the Perissodactyla and the Artiodactyla. Numerous species belonging to both orders still exist; even more diversified are the extinct forms. North America has been the principal home of the perissodactyls throughout their history; Eurasia and Africa the locus of the evolution of the artiodactyls. There are notable exceptions to this generalization, for the Siberian-Alaskan land-bridge repeatedly permitted the interchange of faunas during Tertiary times.

There are several marked differences between the two orders. The Perissodactyla are the odd-toed hoofed mammals, the horse and the rhinoceros being the most familiar modern examples. Although the modern horse has one toe and the modern rhinoceros three, the numbers obviously odd, the tapir has four on the front feet, and various extinct perissodactyls had an even number of toes. What is really meant when one speaks of an odd-toed animal is that the foot is mesaxonic; that is, the middle toe is the largest, the axial plane passing through it (Fig. 123 B). Hence this digit is bilaterally symmetrical. Another outstanding characteristic of the Perissodactyla is that in all existing animals of the order all the premolars except the first resemble the molars. This modification came about gradually, for few of their ancestors have teeth so constructed. One of the upper bones of the ankle, the astragalus, is also useful in distinguishing the two groups. That of a perissodactyl is deeply grooved above (Fig. 123 B, C), where it articulates with the principal lower bone (tibia) of the leg, but flat beneath; the astragalus of the artiodactyl (Fig. 123 D) is grooved on both upper and lower surfaces. There are, of course, many other characteristics of the skeleton which afford the initiates in vertebrate paleontology instant information but which have little significance to the novice.

Since the history of the perissodactyls is best documented in North America, it follows that they have been intensively studied by American paleontologists, who have described in detail the ancestry of the horses and the titanotheres and have made considerable progress in tracing the various families of rhinoceroses. The discussion

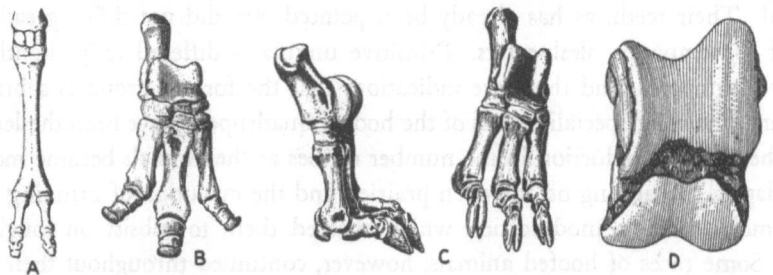

Fig. 123. A, fore foot of the camel to illustrate the typical artiodactyl structure. After W. H. Flower. B, hind foot of *Hyrachyus*, an Eocene rhinoceroid, as an example of a perissodactyl foot. After E. D. Cope. C, two views of the hind foot of *Moropus*, the "clawed" perissodactyl. After O. A. Peterson. D, the doubly grooved astragalus characteristic of the artiodactyls.

of two of these groups is undertaken in a later chapter entitled "Some Genealogies." This leaves only the tapirs and the strange, aberrant ancylopods to be described here.

Because of its many primitive characteristics, the tapir has repeatedly been called a "living fossil" (Fig. 125). Nevertheless, it cannot be considered an unchanged survivor of the Eocene fauna. No race of vertebrates, so far as is known, has survived any really long period of time without some change. All that is implied in the term quoted above is that the specializations are relatively inconspicuous.

Modern tapirs are more specialized than their ancestors in that they are larger, although none has reached elephantine or even rhinocerine proportions. They are primitive in that they retain the typical forty-four teeth. But the teeth themselves have been somewhat specialized. The third upper incisor has become so large as to take over the function of the canine, the true canine is small, and there is a long diastema between the canine and the first premolar; the premolars have much the same structure as the molars; the latter, however, are simple, for they are quadrangular, each bearing but one pair of cross crests, obviously no great modification of the primitive cusps of the tritubercular tooth. The nose and upper lip of the modern

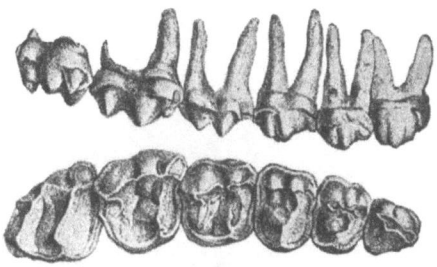

Fig. 124. Teeth of *Lophiodon*, a French Eocene tapiroid, as an example of brachydont teeth with lophodont pattern in the arrangement of the cusps. Natural size. From F. L. Pictet, *Traité de paléontologie*.

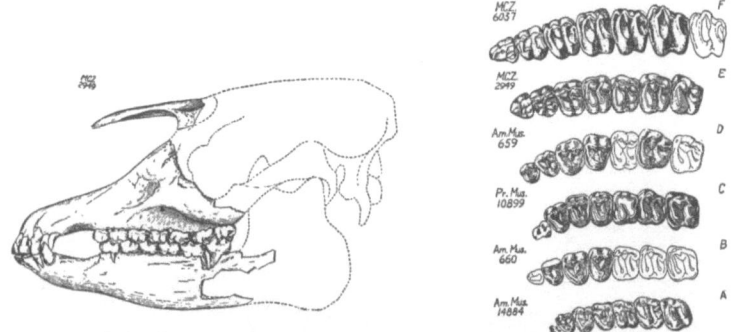

Fig. 125. At left, the skull of *Miotapirus*, a Lower Miocene tapir. At right, drawings of the left upper cheek teeth, to show the evolution from the Eocene to the present. A, *Heptodon*, Eocene; B, C, D, three species of *Protapirus* from successive zones in the Oligocene; E, *Miotapirus*, Lower Miocene; F, *Tapirus*. From Eric Schlaikjer.

Fig. 126. Teeth of a true pig, *Choeromorus*, from the Upper Eocene of France. Natural size. From F. L. Pictet, *Traité de paléontologie*.

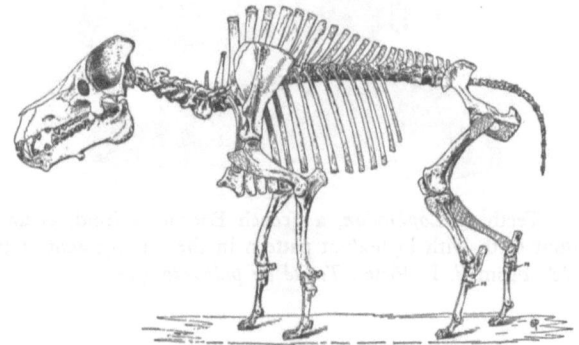

Fig. 127. The Lower Miocene *Dinohyus*, the largest of the entelodonts. From O. A. Peterson.

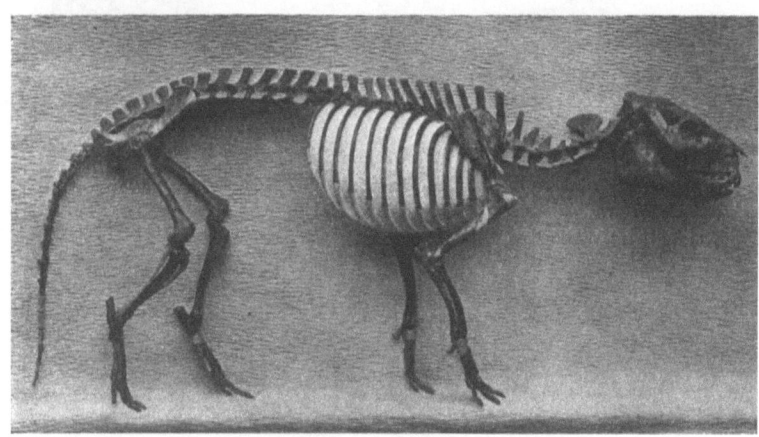

Fig. 128. *Merycoidodon*, from a specimen, seventeen inches high at the shoulder, in the Museum of Comparative Zoölogy, Harvard University.

tapir are combined in a short proboscis, reflected in the skull by short frontal bones and the separation of the nasals from the maxillaries and the premaxillaries; but this specialization is hardly comparable to that of the elephants. The legs are slender but not long. The front feet have four, the hind three short toes, in no way comparable to those of the horse. On the whole the primitive outweigh the specialized characteristics.

The geographical distribution of modern tapirs is that which is to be expected of a primitive group of mammals. As Matthew demonstrated, many archaic creatures have survived till the present day because, as climatic conditions became increasingly rigorous in the regions which they commonly inhabited, they retreated, somewhat ingloriously, to the tropics, where food is so abundant as to enable even the most backward to survive. Tapirs illustrate this admirably, for they are found now only in Central and South America, in southern Asia, and in the adjacent islands. Such a distribution is considered to indicate the great antiquity of the race, for there has been no possibility of direct migration of terrestrial animals from tropical America to tropical Asia within the period covered by the known history of mammals.

Because tapirs have always been forest animals, and because the fresh-water formations from which most Tertiary mammalian fossils are obtained were deposited in flood plains and lakes on the prairies rather than in the forests, little is known of the lineage of the family. During the Pleistocene, representatives of the modern genus *Tapirus* appear to have been widespread, particularly in the northern hemisphere, their remains having been found in the United States, Europe, and China. A few penetrated to South America at this time, but none reached Africa or Australia.

Little is known of Pliocene or Miocene tapirs (Fig. 125); just enough scraps and fragments of bone have come to light to show that they were present in the northern hemisphere during the later part of the Tertiary. The older Oligocene beds of North America and the Mid-Oligocene of Europe have, however, produced skulls and skeletal material pertaining to a genus which has been aptly named *Protapirus*, for it is obviously a primitive tapir. Only about half the size of modern members of the family, it nevertheless foreshadows those now existing, for its molars are of the same pattern and the position of the nasals shows that a short proboscis was present. If observed in greater detail, however, it will be noted that the true canines were functional, since the outer incisors are not enlarged. Furthermore, the premolars are not like the molars but have a distinctive pattern, retaining some resemblance to the primitive triangular shape. In these respects *Protapirus* is more primitive than *Tapirus*, but there is no such striking change in appearance and structure as there is between Oligocene and Pleistocene horses.

Although the history of the tapirs during the Eocene is not yet supported by full information, its general outline can be grasped. The earliest Eocene tapirs, like the contemporary horses, were inhabitants of the forests. Both were small animals; both fed on coarse vegetation which was crushed rather than chewed. As conditions changed, the more adventurous horses emigrated to the grass-bearing plains, where their teeth and legs were more and more modified in such ways as to adapt them to successful life in the new environment. The more timid tapirs, on the other hand, risking less, clung to the wooded areas. From an evolutionary standpoint it is interesting to note that even though they remained in their original environment so far as possible they could not avoid sharing in some measure the effects of the general trend of changes undergone by most mammals during the Cenozoic. In the course of time all members of the family increased in size; the brain became larger; there was some, although not great, reduction in the number of toes. Most striking of all changes, the premolars gradually became molariform. This last cannot be ascribed to change in diet, for it is a characteristic of odd-toed mammals as distinguished from those whose digits are paired.

Can it be inferred from this that there are two sets of factors which cause changes in animals, one external, the other internal? Increase in size and sagacity and decrease in number of toes can all be ascribed to adjustment to the habitat, although terrestrial environments are not everywhere the same. Throughout the Cenozoic there was constant change of temperature, of kind and amount of vegetation, and of degree of competition for food. Reaction to none of these changes, however, will explain why the premolars of perissodactyls of every kind — plains-dwelling horses, woods-loving tapirs, cursorial, semiaquatic, and giraffe-like rhinoceroses — exhibit the same tendency to become like the molars. That such a change appears in all lines suggests that it is a matter of heredity. Was there something in the "blood" (chromosomes) of the ancestor which came to light sooner or later in all the offspring?

The tapirs are the most primitive of the perissodactyls, the horses the most specialized, but the chalicotheres are the most curious. They are classified as hoofed mammals; yet they have claws and no hoofs. They are called perissodactyls; yet the second, not the third toe is the largest (Fig. 123 C). On the other hand, although they have claws, they are not carnivores, for the teeth are of the typical herbivorous type, much like those of the titanotheres to which some paleontologists believe they are related. Most of them have a foot in which the second toe is the largest, but there is evidence that their ancestors were mesaxonic, as the other perissodactyls are. The chalicotheres are commonly cited as the great exception to Cuvier's "law" of the correlation of parts. When the first fragmentary specimens were found, the skull was interpreted as that of a perissodactyl and the feet as those of a pangolin (scaly anteater), the parts being thus distributed in two unlike orders. The large, deeply cleft ungual phalanges are not, however, exactly duplicated among the known carnivores.

Remains are mostly fragmentary and decidedly scarce; yet chalicotheres are known to have had a long history in the northern hemisphere. The oldest representatives are found in the North American Eocene; they occupied this country till the Miocene; in Europe and Asia they existed from the Oligocene to the Pliocene; they may have reached Africa in the Pliocene, but the record is doubtful. The best-known American ancylopod is the early Miocene *Moropus,* nearly complete skeletons of which are to be seen in the museums in Pittsburgh, New York, and Cambridge. It is a large animal, somewhat giraffe-like in aspect, the fore legs longer than the hind, and the neck elongate. The skull is small, the body long, sloping from shoulder to rump. Each foot had but three functional toes; nothing remains of the first, but the fifth is represented by a small, vestigial metacarpal in the manus, although it is entirely absent from the pes. The fore feet, which are larger than the hind, are somewhat perissodactyl-like in that the middle toe is the longest, although the second is stronger, with a larger claw. The European *Chalicotherium,* of Mid-Miocene and Pliocene age, was larger than *Moropus* but had a shorter neck. The Oligocene and Eocene members of the group were of smaller size.

Although the chalicotheres were so giraffe-like in general appearance, there is no evidence of any close relationship to that animal. The chalicotheres, the ancestral giraffes, and the huge rhinoceros, *Baluchitherium,* appear to have had similar propensities for gathering their food from the branches of trees. Of the three, only the extremely long-necked giraffe succeeded in changing to a diet of grass. And an exceedingly awkward job he makes of feeding, if we are correctly informed by the motion pictures. Nevertheless he has survived, whereas the others have not.

The artiodactyls of the present day are much more numerous and much more diversified than the perissodactyls. Pigs, cattle, and deer are familiar, but all of the first and the domesticated members of the second of these groups were brought to this continent by man. North America has had but few of the cloven-hoofed beasts as compared with Asia and Africa, where they now exist in great numbers and variety. This is unfortunate, for occurrences of fossiliferous Tertiary deposits of any great extent are rare in the Old World. A few localities in France, Greece, India, Mongolia, and China are the only ones in the northern hemisphere which have yielded many specimens. The record for Africa is even less satisfactory, for aside from the remarkable Eocene and Oligocene deposits in the Fayum desert of Egypt, the continent has produced almost nothing older than the Pliocene. Because of this dearth of material it has not been possible to trace artiodactyl lineages in any great detail, except for that of the camel, which belongs to one of the two groups which are North America's chief contribution to the even-toed mammals.

The general trend in artiodactyl evolution is the same as in that of the perissodactyls. The teeth become more complicated as the change is made from a diet of tender herbage and leaves to one of dry grasses. The body size increases, the cerebrum becomes more convoluted, and the legs are longer. The lengthening of the legs is accompanied by loss of one or more toes. Since the weight is carried chiefly on the third and fourth toes, the foot becomes symmetrical with regard to a plane passing between them (paraxonic) (Fig. 123 A). The first toe is vestigial or is lost; the second and fifth are in most cases smaller than the third and fourth. In all the advanced groups, the really rapid runners, the lateral toes are reduced to mere vestiges, the radius and ulna, and tibia and fibula are united, and the two functional metapodials are fused, producing the cannon bone. It is this condition, in which two toes are articulated to one bone, which gave rise to the appellation "cloven hoof." This expression has too often been interpreted as meaning that a single hoof has been divided into two, whereas in reality the upper bones of two toes have coalesced. The other striking peculiarity of the artiodactyls, the doubly-grooved astragalus, has already been mentioned.

Modern artiodactyls fall readily into natural groups, but paleontologists have experienced great difficulty in formulating a classification which will include all the fossil forms. In general, two great series may be recognized, one composed of the relatively primitive creatures which have a bunodont dentition, and another of a more advanced type with selenodont molars. The bunodont tooth (Fig. 126), short-crowned, with blunt conical cusps, like that of the pig, is well adapted for crushing any sort of food, hard or soft, animal or vegetable. Although they are primarily vegetarians, artiodactyls with such teeth are really omnivorous; most of them retain some characteristics of the creodonts. The selenodont molar (Fig. 129), short-crowned (brachydont) in its earlier manifestations but high-crowned (hypsodont) when perfected, is the common possession of cattle, deer, giraffes, camels, and other artiodactyls. This tooth is remarkable in that the crests are longitudinal instead of transverse, as is the condition in most grinding teeth. Each molar has two pairs of lunate "lakes" of dentine, bordered by thin sharp rims of enamel. In some groups there are small modifications of this simple plan, but in general all selenodont molars are so much alike that they are of comparatively little use in the identification of animals. Fortunately, there are few instances among the artiodactyls in which the premolars become molariform, and F. B. Loomis has shown that the structure of the last premolar is an important guide to the recognition of the various subdivisions of the group.

The Old World pigs and the New World peccaries are the best examples of the bunodonts. Although superficially much alike, they are assigned to separate families which, when united with the Hippopotamidae, make up the superfamily of the Suioidea, piglike creatures.

The Old World swine are the most primitive, for the typical dentition of forty-four teeth is retained and each foot has four toes, although the lateral ones are functional only on marshy ground. It is well known that the pig will eat anything, the tubercle-studded molars crushing nuts and bones as readily as they do the refuse on which the domestic animal is commonly fed. The records of boar hunts sufficiently attest the ferocity and agility of the carnivore-like creatures, armed with triangular canines of the most dangerous sort. Once widespread throughout the forested regions of Europe, northern Africa, and central Asia, *Sus scrofa* was one of the great beasts of the chase during the Middle Ages. The pursuit of the boar was the sport of kings; hence, by reflected glory, the animal became royal, and one of the four used in heraldic devices. Richard the Third, "Crookback," murderer of the Princes, employed it on his shield. Remains of pigs found in strata of various ages from the Oligocene onward show that they have long been inhabitants of Eurasia and Africa, but they never migrated to North America. Since the living creatures are so primitive, it is not surprising that the fossils show no spectacular changes during the evolution of the group.

Peccaries, the New World swine, have much the same appearance as the true pigs, although the absence of a tail would probably be noted by a sharp-eyed small boy. In reality, however, the peccaries are somewhat specialized. Instead of forty-four teeth, there are only thirty-eight, the third upper incisors and the first premolars having been lost. The upper canines point directly downward, not outward and upward as in the pigs. The molars are quadrangular, with four principal cusps, to which small wartlike tubercles are added, whereas in the European swine the cheek teeth are longer than wide, the chief cusps irregularly arranged. The last lower premolar of the peccary is molariform, an unusual condition for an artiodactyl. The fore foot has as many toes as the pig, but the hind has only two, for the second is vestigial and the first and fifth are absent. The metapodials are fused at their upper ends, forming a rudimentary cannon bone. At the present day peccaries are chiefly South American, although one species is found throughout Mexico and as far north as Texas and New Mexico. They are, however, primarily North American animals, for they have been here since the Oligocene and did not enter the southern continent until the Pleistocene. The Eocene ancestors of true pigs and peccaries are still unidentified. It will probably be shown eventually that they arose from the same stock, either in Eurasia or in North America.

Another family of bunodonts which flourished in North America during the Oligocene and Miocene (remains are also known from the European and Asiatic Oligocene) is that known as the entelodonts (Fig. 127). Although they left no descendants, they are of interest because of the great similarity of their teeth to those of carnivores. The incisors are conical, the canines powerful tusks, and the premolars trenchant, reminding one of those of the dog, but the molars are small and piglike. Despite this savage dentition, specimens in the Princeton museum seem to show that the entelodonts fed principally upon roots, for the canines and incisors are scored with grooves which could have been produced only by such sand-covered food.

The last of this line was *Dinohyus*, the largest of bunodonts other than the modern hippopotamus. This piglike giant, six feet high at the shoulders, was discovered by Olaf A. Peterson in the Lower Miocene strata of northwestern Nebraska, and its skeleton is mounted in the Carnegie Museum. Its most striking feature is the enormous skull, three feet in length, armed with splendid teeth. Long dorsal spines, for the attachment of the muscles necessary to support so large a head, project above the dorsal vertebrae, forming the nucleus of a hump at the shoulders. The legs are rather slender, the ulna and fibula united with their corresponding bones, the toes reduced to two. Curiously, no cannon bone was formed. This "terrible hog" was certainly not the ancestor of modern pigs or peccaries, for its feet were more specialized than those of either group.

The entelodonts present a curious mixture of characteristics to those who like to speculate upon the probable habits of extinct animals. One would judge from the structure of the feet that they were accustomed to life on the plains; yet they disappeared just at the time when the prairies were in their period of greatest expansion. If they were root-eaters, one would suppose that a forest would have been their natural habitat, particularly a moist forest. Perhaps they were harvesters of wild turnips, carrots, and other tuberous roots, although these could not have been easily dug by a hoofed mammal.

Every collector of vertebrate fossils and every curator of a paleontological collection knows that the "oreodons" were the most numerous of all North American Tertiary mammals. *Merycoidodon* (Fig. 128), to give the genus the name considered correct at the present time, seems to have been a gregarious inhabitant of the Great Plains during the Oligocene. Not only do the merycoidodonts appear to have existed in great herds, but individuals seem to have been almost unduly prone to early death and rapid burial. Had quicksands a morbid fascination for them? It would be interesting to know how many skulls of *Merycoidodon culbertsoni* have been collected. Dealers have handled them for years; they evoke endless sniffs from curators as collectors unpack the summer's spoils. Yet it is doubtful if enough material has yet been amassed to allow a really thorough study of this one species. Museums hesitate to discard specimens, no matter how many duplicates they may accumulate. Scientific inquiry is so many-sided that no one can foretell the moment when old material may be studied from a new standpoint. Thus, when M. R. Thorpe, trying to ascertain the reason for their extinction, studied "several hundred specimens of the Great Plains *Merycoidodon* and the John Day *Eporeodon* for the purpose of determining the relative age of the individuals at death," he would probably have reached more satisfactory results if he had had several thousand.

The merycoidodonts are among the simplest of the selenodont (Fig. 129) mammals and, as a group, one of the most successful, for they appeared first in late Eocene times and survived until the Mid-Pliocene. Why they were extinguished at that particular time is a mystery, for it was not a critical period for mammals. Some thirty genera and more than a hundred species of this family are known.

No merycoidodont was large or highly specialized. The form of most was piglike, though rather longer in the neck; the legs were short, with four toes on each foot. There was no real relationship to the pigs, however, for the cheek teeth are typically selenodont. A feature which at first glance seems peculiar is that these animals appear to have four lower incisors on each side of the symphysis. This is due to the fact that the first lower premolar has the shape and function of a canine, whereas the true canine is small, like an incisor. The result of this change is that the lower "canine" closes behind the upper instead of in front of it as in most mammals. This is not, however, a condition restricted to the family, for all ruminants except the camels show the same arrangement. The dentition is not that of a grazing animal, all the premolars and molars being sharp cutting and crushing teeth. Thus, although their remains are found associated with those of true ruminants, it must be inferred that the merycoidodonts dwelt by the water courses, living largely on leaves, young

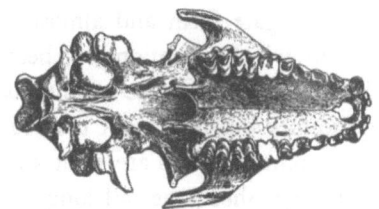

FIG. 129. The merycoidodont, *Merychyus*, to illustrate the selenodont pattern of the molar teeth. One-fourth natural size. From O. A. Peterson.

twigs, and succulent vegetation. They seem to have been ill adapted to withstand winters or life away from areas of permanent water. This may explain both the final extinction of the group and their relative abundance as fossils.

The Eocene merycoidodonts were small, but their dentition is so much like that of later ones as to give no clue to their origin. They are supposed to have been immigrants to North America during the late Eocene. There is, however, no suggestion as to the locality of the original home. It is not unlikely that ancestors will eventually be found in our older Eocene formations.

The group became considerably differentiated and widespread in North America during the Oligocene. The central stock seems to have been a cursorial type from which were derived some forest dwellers, some creatures of semiaquatic habits, and, most specialized of all, a few browsers with a tapir-like proboscis. The largest merycoidodont, *Promerycochoerus*, lived during late Oligocene and early Miocene times. Individuals were as large as modern hogs, though they can hardly have been so fat. One of the most remarkable groups of fossils ever collected consists of skeletons of a species of this group, obtained by Peterson in western Nebraska, and now in the Carnegie Museum in Pittsburgh. On a single slab, lying in their natural positions, are three individuals, remains of animals which had crept close to one another for mutual warmth or consolation at a time when some disaster overwhelmed them. They may have succumbed to the cold of a western blizzard, but more probably they were the victims of a sandstorm which suffocated and buried them. A similar group of merycoidodonts may be seen in the American Museum of Natural History in New York.

Although there are no definite connecting links, it may be confidently asserted that the hoofed mammals were derived from some as yet unknown carnivore of the creodont family. This relationship appears to find more obvious expression among the artiodactyls than among the perissodactyls. It has already been pointed out that the dentition of the entelodonts was strikingly carnivore-like. Thorpe has listed many characteristics of the merycoidodonts which attest their relationship to primitive flesh-eaters. As early as 1896 Joseph Leidy, the Philadelphia physician who brought this family into the scientific world, commented on the "decidedly wolfish aspect" of the skull, as indicated by the elongate form and almost continuous series of teeth. Among other characteristics, to which attention has been drawn by Thorpe, the most obvious are that the head is large, the face is long, as is the area in which the lower jaws are joined, the auditory bullae are not ossified, a common condition in the early creodonts, and the zygomatic arches are stout and heavy. So much for the skull. The body was long, the legs short, the tail long — all characteristics of creodonts rather than of ungulates. In addition Thorpe lists various characteristics of the bones of the fore and hind legs which are significant to the osteologist but of too technical a nature to be repeated here. The evidence is cumulative in support of the theory that clawed and hoofed mammals had a common ancestor.

All in all, the history of the merycoidodonts is comparable to that of successful and contented burghers rather than to that of noble families. They lived long, without much change; they produced no giants, but no pygmies; they were numerous without being important; we learn nothing in particular from their history; their coming marked no epoch; they were not missed at their departure. How like the average citizen!

Most important, most numerous, and most specialized of the artiodactyls are the ruminants, the cud-chewing grazers so important a source of food and clothing for man. Without the docile cattle, the stupid sheep, the hardy goats, and the rugged reindeer, man, primitive or civilized, could hardly exist. Members of this group, known technically as the Pecora, lack upper incisors, the tongue performing their functions. Probably most readers have seen the cow gather wisps of grass, twist them against the lower incisors, and pull. This process is different from the cropping bite of the horse, but fully as effective, as sheep raisers know. Few ruminants have upper canines; the corresponding lower teeth are small, incisor-like; all lack the first premolar. The molars are selenodont, less cuspate than those of the merycoidodonts and therefore better grinders.

The legs are long and slender, perfectly adapted for rapid motion. Some of the antelopes are, indeed, the fleetest of living animals. Although there are two toes, the leg is practically a chain of single bones, for the ulna is coössified with the radius, the fibula practically absent, and the two complete metapodials united as a cannon bone. It is noteworthy that the terminal processes, which articulate with the phalanges, are parallel; hence the toes are close together, not divergent, as are those of the somewhat splay-footed camel. "Dewclaws," a pair of vestigial fingers and toes (second and fifth), are present on many Pecora but fail to impede the action of the axial portion of the leg, for their short metapodials do not in most cases articulate with the bones of either wrist or ankle, nor are they attached to the cannon bone. Although apparently excess baggage, they are not large enough to be of any great hindrance in running and may be of some use in traversing marshy ground.

Males of most of the Pecora have horns, the few exceptions occurring amongst the deer. In the absence of any marked differences in the teeth, the horns have been used largely in classification. Although not at the moment in favor, the old terms Cervicornia (deerlike, solid-horned) and Cavicornia (hollow-horned) serve well to distinguish the two great groups of ruminants. The Cervicornia include the deer and giraffes, the former being much the more important.

The antlers of the deer are bony outgrowths formed and shed annually. During the period of growth they are said to be "in the velvet," for they are then covered with a layer of skin which is shed when the horns are complete. After the rutting season a part of the bone just below the "burr" near the proximal end of the antler is resorbed, and the greater part of the structure falls off, leaving only a short permanent projection of the frontals, known as the pedicle. Some deer have simple, undivided antlers; others carry magnificent but burdensome structures, many-branched, with the numerous tines coveted by sportsmen.

The oldest deer now known are small creatures without antlers, described from European Oligocene deposits. This simple type is apparently represented today by the Chinese water deer, which, like other deer without antlers, or with small ones, has a pair of long sharp recurved upper canines, effective weapons. In this line the primitive canine has been increasingly developed throughout the successive generations. Lower Miocene strata, again in Europe, have produced the earliest deer with antlers, forms referred to *Palaeomeryx*. Animals belonging to this genus have a pair of simple bony processes, probably covered with skin, over the orbits. From such ancestors, according to Loomis, who has studied the evidence in detail, were derived on the one hand the deer with deciduous antlers and on the other the giraffes. In this latter group the horns are not shed but are skin-covered bony outgrowths, comparable to the pedicles of the deer.

Evidence obtained from European fossils shows that the evolution of true deer with antlers was rapid after their first appearance in the early Miocene. A few straggled to America at this time but seem not to have prospered. Our oldest is the hornless *Blastomeryx*, whose remains have been found in strata of various ages from Lower Miocene to Pliocene. It was a small, lightly built creature, with saberlike canines and legs about as specialized as those of modern deer. It seems to have been ancestral to *Merycodus*, species of which existed in North America from Mid-Miocene to early Pleistocene. *Merycodus* had simple, two- or three-pronged antlers but was probably not ancestral to any modern deer, for its teeth were more fully hypsodont than are those of any living member of the family. The earliest *Merycodus* was only eighteen to twenty inches high at the shoulder; yet the legs were long and slender, without traces of the lateral toes. Thus both teeth and legs were more specialized than those of modern deer.

It was not until late Pliocene or early Pleistocene times that deer became really plentiful in North America. All that have left descendants appear to have come from Asia, although there is a possibility that some of the more southern forms may trace to some as yet unknown American Miocene animal. According to W. B. Scott, there were two immigrations, the first of which, in the Pliocene, brought in the ancestors of the Virginia deer and its allies. These animals spread not only over all North America but into the southern continent, where they multiplied and became considerably diversified. From an evolutionary standpoint it is interesting to note that, since they were the only South American ruminants, they found no great competition from animals with their own food habits when they arrived in Pliocene or early Pleistocene times. The second invasion of North America was in the Pleistocene and was marked by the immigration of deer which have remained northern in their distribution. Among them were the reindeer, the palmate-horned moose, and the many-pronged Wapiti, all animals which strayed into what is now the southern United States only during periods of glacial advance.

The history of the giraffes, which, as has been seen, appears to have begun in the Lower Miocene with *Palaeomeryx*, is obscure at the moment. The living members of the family, the giraffes and the rare okapi, are confined to the part of Africa south of the Sahara, but remains of extinct species have been found in deposits of early Pliocene age in Hungary, Greece, Persia, northern India, China, and East Africa. The modern giraffe is notable for its elongate neck and strong forelegs. The smaller okapi, with a much shorter neck and approximately equal fore and hind legs, appears to be much more like its Pliocene forebears than the giraffe is. This seems to indicate that the most striking specializations in this line have been acquired rather recently. European records older than the Pliocene are strangely silent as to the ancestry of the giraffes; hence it is probable that Loomis is right in asserting that the American Miocene *Dromomeryx* is the real connecting link between this group and the Lower Miocene ancestor of the antlered deer, *Palaeomeryx*. *Dromomeryx* (Fig. 130) is a small, Mid-Miocene artiodactyl with a pair of simple, skin-covered horns above the eyes. That it has been considered the oldest American antelope by one student of fossil mammals and the oldest giraffe by another is of interest to the layman chiefly as evidence of the affinity of the Cervicornia to the Cavicornia.

The latter are the most specialized of the ruminants, best fitted for life on the prairies. Their hypsodont teeth are adapted to a diet of grasses, fresh or dry, whereas the short-crowned teeth of most of the Cervicornia indicate that they were primarily browsers. The term "hollow-horned" refers only to the horny sheaths which cover the bony outgrowths from the frontal bones, present in females as well as males, except for the does of some species of antelopes. Neither bony core nor horny sheath is

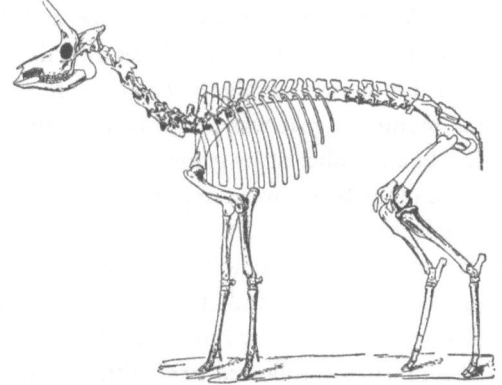

FIG. 130. A simple-horned Miocene deer, *Dromomeryx*. One twenty-fourth natural size. From E. Douglass.

shed, as a rule, nor are the outgrowths branched. Yet every rule has its exception, exemplified in the case of the Cavicornia by the prongbuck, which sheds its branched horny sheaths or antlers. The relationship of this "antelope" to other mammals is still, however, an unsolved problem.

So numerous and so varied, at least in external form, are the animals called antelopes that "authorities" differ as to the definition of the family. Employed even in its widest sense, it includes few denizens of continents other than Asia and Africa; it is most abundantly represented in the latter area, although Asia has its hordes of gazelles, central Europe and central Asia their chamois, and America the Rocky Mountain goat. So far as is now known, antelopes did not reach Africa till Pliocene or Pleistocene times, but since their arrival they have thriven wonderfully. Travelers, keepers of zoölogical gardens, sportsmen, and naturalists have made us familiar with the brilliant colors, the polished, straight, ringed, or twisted horns, and the great variety and range in size of these creatures. Some are but little larger than rabbits; others have the bulk of oxen; some are slender, graceful, cursorial dwellers in deserts; others, sluggish, phlegmatic, semiaquatic occupants of swamps. The names bongo, kudu, oryx, saiga, steinbok, and hartebeest have so caught the imagination of makers of books and stories as to have become almost household words. Like most of the modern fauna of that country, the antelopes are not of African but of more northern derivation. The oldest yet known are gazelle-like creatures found in the Miocene of Europe and India. The eland, kudu, and hartebeest appear to have been in India during a part, at least, of the Pliocene. At this same time antelopes of some sort seem to have invaded North America, but the fossils so far found are too fragmentary to be understood. The Pleistocene saw their migration as far south as Brazil and the Argentinian pampas, but they failed to get a foothold on the southern continent.

According to some zoölogists, the chamois of Europe and Asia and the Rocky Mountain goat are intermediate between the antelopes and the goats. Others maintain that our shy, surefooted mountain climber is no goat but a true antelope. Whatever its relationships, the mountain goat was probably a Pleistocene immigrant from Asia, for bones belonging to a member of its genus have been found in caves in California. True goats, on the other hand, were imported by man, never having risked the Bering passage themselves.

Some students of mammals divide the hollow-horned ruminants into three families, the first containing the prongbuck and its extinct allies (Antilocapridae), the second the antelopes (Antilopidae), and third the cattle, including bison, sheep, goats, et cetera (Bovidae). This practice is followed here, although no attempt is made to enter upon the long, involved discussions which would be necessary if one were to offer definitions of the second and third families. Others, with considerable justification, place the antelopes with the bovids in the family Bovidae. There is some indication that the Miocene antelopes may have been ancestral to all later Antilopidae and Bovidae, that the Pecora branched into the two characteristic groups in the Oligocene, and that the Pecora and Tylopoda (camels) had common ancestors in the Eocene. Lack of fossils prevents, at the present time, anything more than the general suggestion. True bovids appear to have been of Eurasian ancestry, none reaching North America till Pliocene or Pleistocene time. For that reason they will not be discussed here but will be referred to on a later page.

This old world of ours has been a long time in the making. Only recently was the time ripe for the multiplication of the vegetarians. When the opportunity came, they seized it. But man in a few thousand years has nullified the effects of millions of years of evolutionary change, and his efficiency as a destructive factor has been accelerated in recent decades. Most of the vegetarians are doomed, despite the efforts of a few farsighted naturalists to save them.

XXV

SOME GENEALOGIES

> 'Tis he, I ken the manner of his gait;
> He rises on the toe: that spirit of his
> In aspiration lifts him from the earth.
>
> *Troilus and Cressida*, Act IV, scene 5

Many question the value of human pedigrees and deride as ancestor-worshipers those who are interested in genealogical pursuits; yet these same people will pay a premium for a pedigreed dog or horse or bull. All are agreed, however, that one of the most laudable pursuits of the paleontologists is the tracing of lines of descent. This is, in fact, the only possible test of evolution. If the idea is sound, it should be possible to find in the rocks the ancestors of many, if not all, of the living animals and plants. Realizing as he does the manifold imperfections of the geological record, the student of fossils hardly expects perfect results. Yet much has been accomplished.

Invertebrate fossils are much more abundant than remains of vertebrates and have allowed the construction of numerous phylogenies. But these are not sufficiently spectacular to arouse popular interest. The description of the ancestors of familiar mammals has appealed to more people. As examples of this sort of work, it is most convenient to survey briefly the evolutionary history of the horse, the rhinoceros, and the camel.

Darwin put together in a coherent theory the rather nebulous ideas of evolution which were a part of the thought of his day. The genealogy of the horse, as deciphered by O. C. Marsh, was hailed as the first actual proof of the theory. Although much has been done in tracing other lines, this family remains at the present time the best known, far better understood now than when Marsh first pointed out the salient facts of its evolution. The story has been told so often that it is familiar to all, but a summary is inevitable. The present account follows that of the late W. D. Matthew, America's most learned vertebrate paleontologist.

The specializations of the skeleton of the modern horse are those of a large, swift-running, hoofed quadruped which depends for its food on the dry grasses of open plains. The skull and neck are long, as are the legs and feet; the gait is unguligrade — that is, the animal walks on the last or ungual phalanx of its toe. The dentition is somewhat reduced, canines being absent in females and the first premolars vestigial in both sexes. All the teeth are hypsodont, roots forming only with advancing age. The incisors are strong, powerful croppers, each with a deep cement-filled pit in the center. The premolars and molars are alike — almost rootless prismatic teeth that continue to grow out as they are worn down, the complex infolding of the enamel into areas of dentine and cement producing a grinder which cannot be worn smooth. The eyes are far back, a position necessitated by the depth to which the teeth are inserted. The legs are long slender pendula, actuated by masses of muscle in fore quarters and flanks. Grooved articulations between the bones allow fore and aft and prevent lateral movement. The mid portions of ulna and fibula are lost, the vestigial ends being coössified with radius and tibia respectively. Only the third digit of the foot is functional, the first and fifth toes having been entirely lost, whereas the second and fourth are represented by "splints," remnants of metapodials. The metapodial of the third toe is the long, large cannon bone. The ungual phalanx is broad and covered by a large horny hoof.

More than half a century and hundreds of thousands of dollars have been spent in finding the links between the modern horses and their little four-toed ancestor, *Eohippus* (Fig. 140). Several species of this Lower Eocene genus are now known, animals varying in size from that of a cat to that of a small fox; that is, eighteen inches to two feet long. All had the complete dentition, the premolars being small cutting teeth, simpler than the molars. The latter had low crowns, each with two or three pairs of blunt, conical cusps, the whole entirely unlike the complicated pattern of the teeth of the modern horse. The gait of *Eohippus* was digitigrade; probably pads were present under the toes, as in the carnivores. The bones of the lower arms and legs were separate from one another. Four toes, all functional, were present on the front feet, whereas on the hind there were three useful and two vestigial ones, the latter represented only by tiny nodules. The orbits were at about mid-length of the skull.

Eight genera show characteristics which illustrate the stages intermediate between *Eohippus* and the modern genus *Equus*. Many other kinds of horses are known, but these suffice to demonstrate the chain of significant changes.

Orohippus of the Mid-Eocene shows the first change from the ancestor, for the vestigial toes have disappeared from the hind feet, the third toe is slightly larger on all four feet, and the fourth premolar is molar-like. The third stage is represented by the Upper Eocene *Epihippus*, which is of larger size and has two molar-like premolars. All of the Eocene horses are small; all have four functional toes on the front feet and three on the hind. Moreover, all walked on the fingers and toes, not on the tips of them; that is, they were digitigrade.

Mesohippus (Fig. 132) of the Lower and Middle Oligocene shows a considerable advance, since the outer (fifth) toe of the front foot is reduced to a splint. The third digit of all feet is elongate and strong, being much the most important, although the two lateral ones still remain functional. Three premolars are molariform, and the ulna and fibula are attached to radius and tibia. The animals were about the size of sheep, up to about forty inches in length. *Miohippus* of the Upper Oligocene was somewhat larger than *Mesohippus*, with feet a little longer and stronger. Oligocene horses, then, had short-crowned teeth with a rather more complicated pattern than those of the Eocene, three-toed, digitigrade feet, and coössified lower limb bones.

The Lower Miocene *Parahippus* (Fig. 133) is distinguished as the first among the horses to have cement on the crowns of the teeth, which show increase in height. The second and fourth toes were vestigial, the third elongate. Horses now began to walk on the ends of their middle toes. *Merychippus* of the late Miocene is repre-

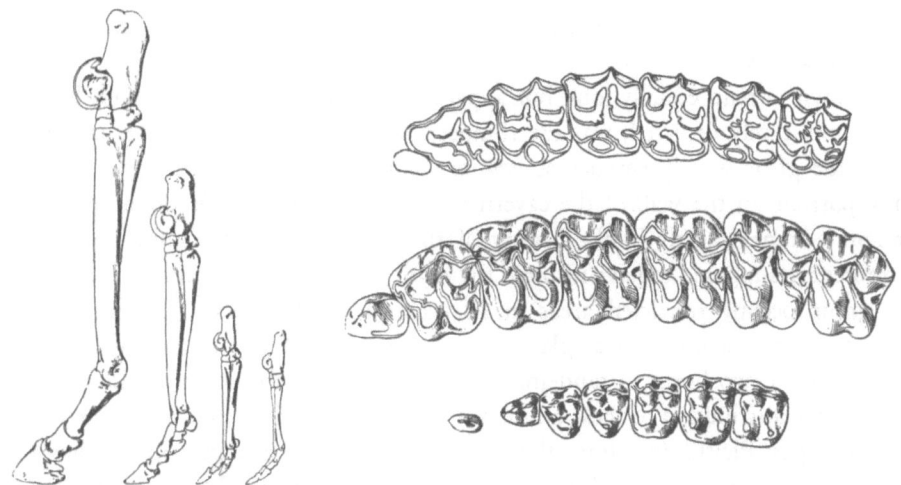

FIG. 131. At left, sketches showing the principal stages in the evolution of the hind foot of the horse. The tallest is *Equus*, then *Merychippus*, Miocene; next *Mesohippus*, Oligocene; and last, *Eohippus*, Eocene. At right, three stages in the evolution of the cheek teeth. Below, *Eohippus*; in middle, *Mesohippus*; at top, *Merychippus*. From Matthew and Chubb, *Evolution of the Horse*. By permission of the American Museum of Natural History, New York City.

sented by several species which in successive zones show higher and higher crowns on the teeth, the cement forming a heavy coating. Miocene horses have three toes, but the lateral pair are so much reduced that the animals walked on the tips of the middle toes only, as modern horses do. The teeth are high-crowned, the eye is back of the middle of the head, and increase in size has continued. The Upper Miocene and Lower Pliocene *Pliohippus* has cheek teeth three times as high as wide, and the side toes are actually reduced to long splints in some species. These animals were larger than Miocene horses.

Plesippus of the Upper Pliocene forms the link between *Pliohippus* and the true *Equus* of Pleistocene and Recent times. Nearly as large as a modern horse, it differed from it only in small details of the structure of teeth and feet. Pliocene horses were practically one-toed animals, with most of the characteristics of modern ones, differing chiefly in the longer splints of the lateral toes, the somewhat shorter teeth, and the smaller size.

All of this evolution took place in North America, although the ancestral *Eohippus* may have been an immigrant from the Old World, for it seems to be identical with the *Hyracotherium*, long since described by Sir Richard Owen from bones found in the Eocene London clay.

The horse is familiar to all, because he has been enslaved by man. Numerous pictures on the walls of caverns show that he was an object of human interest to the first known race of our species, *Homo sapiens*, some twenty thousand years ago, but that interest seems to have been evoked by the tastiness of his flesh rather than his quality as a beast of burden.

Known to man equally long but apparently less closely associated with him, for it was only rarely depicted, was the rhinoceros. Never domesticated, perhaps rarely killed, the great woolly rhinoceros was probably looked upon with awe. His red ochre portrait on the wall of the cavern of Font-de-Gaume in Dordogne may commemorate some tragedy in the history of the local tribe. Later single-horned rhinoceroses are supposed to have inspired the myth of the unicorn. Existing members of this group occupy restricted areas in Africa and southern Asia. Those living in India and Java have only a single horn; in Africa and Sumatra are found similar creatures with two horns in tandem, one above the nose, the other on the forehead. The present distribution of these animals is a relatively modern one, for in Pleistocene and early post-Pleistocene times they were abundant in Europe. Still earlier they were even more widespread, for from the Oligocene onward they roamed throughout Europe, Asia, and North America. In the Eocene of the last are found the oldest of them.

The modern rhinoceros bears little resemblance to the horse. It is a much less graceful creature, with relatively short head, neck, and legs, long heavy body, and thick, hairless hide. The feet have three toes each, the median one the largest. Most rhinoceroses dwell in marshes and forests, feeding on leaves and shrubs; that is, they are browsers. The dentition is reduced in all; the full series of cheek teeth is retained, but canines are absent and the number of incisors is less than that of primitive mammals. The cheek teeth have a characteristic pattern, that known as the lophodont, more easily illustrated than described but far simpler than that of the horse. Premolars and molars have the same crests. The horns, although long and conspicuous, have no bony cores, their presence being indicated solely by the thickening and roughening of the nasal bones.

The American rhinoceroses, represented by fossils from Eocene, Oligocene, Miocene, and Pliocene deposits, appear to fall readily into three major groups. First and oldest are the light-boned, cursorial forms, the hyracodonts of the Eocene and Oligocene. The oldest of these belong to the Mid-Eocene genus *Hyrachyus*, some

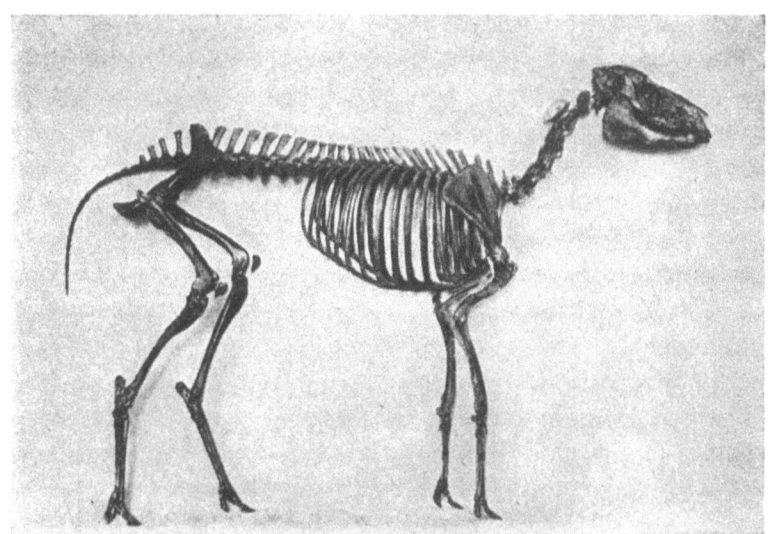

FIG. 132. *Mesohippus barbouri*, a three-toed Oligocene horse, twenty-two inches high at the shoulders. Skeleton in the Museum of Comparative Zoölogy, Harvard University.

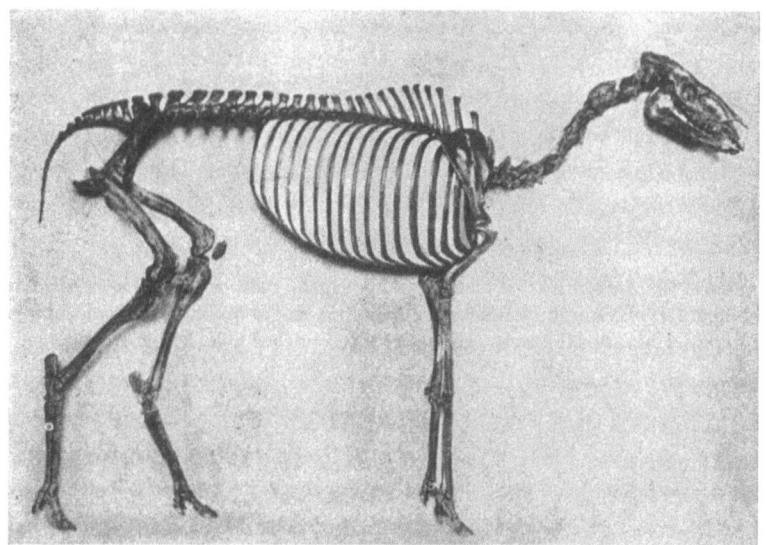

FIG. 133. *Parahippus wyomingensis*, a Lower Miocene horse, thirty-eight inches high at the shoulders. Note reduction of lateral toes. Skeleton in Museum of Comparative Zoölogy, Harvard University.

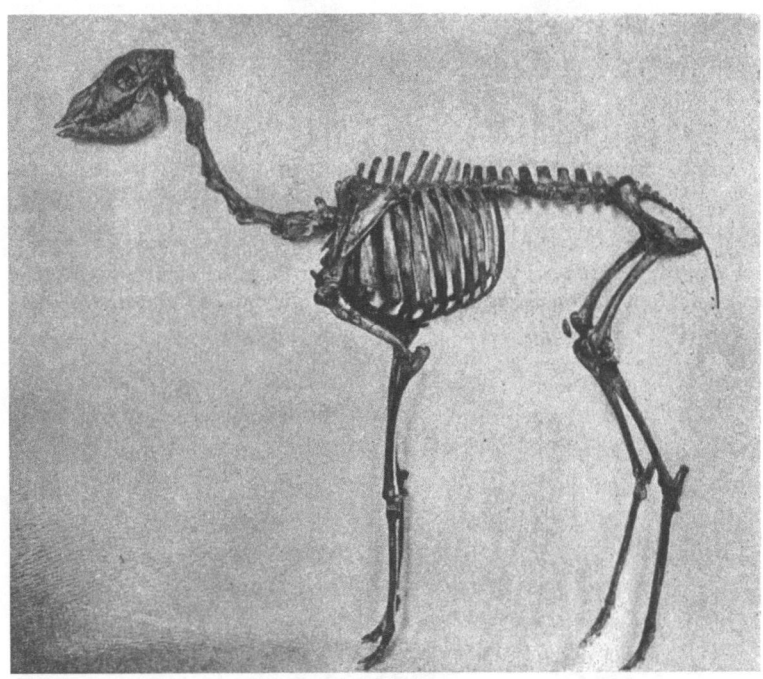

FIG. 134. *Stenomylus*, a slender, long-legged, swift-footed Miocene camel. Photograph by George Nelson of the specimen in the Museum of Comparative Zoölogy, Harvard University. It is twenty-six inches high at the shoulder.

species of which are as small as a fox, others as large as a tapir. It is probable that all later rhinoceroses were descended from members of this group, although the genealogical lines have not yet been proven in detail. *Hyrachyus* itself may have been descended from *Homogalax*, the reputed Lower Eocene ancestor of the tapirs. All species of *Hyrachyus* had long bodies, short necks, four toes on the front feet, and three on the hind. The skull must have been hornless, for the nasals are thin. All of the primitive forty-four teeth were present, the premolars smaller than the molars. Only the upper true molars were lophodont, of the true rhinocerine type, the lower ones resembling those of tapirs.

Somewhat larger cursorial forms are found in the Upper Eocene, animals that had one less toe on the front foot. These serve to connect with the Oligocene *Hyracodon*, the last of this line. *Hyracodon* was somewhat taller and heavier than a sheep, with a large head and stocky limbs. It was, however, proportionally much lighter in build than any modern rhinoceros, with a longer neck and more slender bones. Each foot had three toes, the median larger than the others; all premolars except the first were molariform. In many respects *Hyracodon* resembled the contemporary horse, *Mesohippus*. Both had three toes, light bones; both were fleet of foot. But *Hyracodon* died out in the Oligocene, whereas *Mesohippus* left numerous descendants. Why? Apparently because one was a little more successful than the other in eating the most abundant food then available, the grasses of the plains.

According to the latest views, a Lower Eocene mammal similar to *Homogalax* or *Eohippus* was probably the ancestor of the rhinoceroses, the tapirs, and the horses. Some of the descendants of *Eohippus* deserted the forests. In these the teeth became modified by the grinding of dried grasses, the limbs changed as a result of the mechanical reaction to running on hard ground, and the foot was lengthened. Others remained in the forests, and in them the teeth were modified by use in the constant chewing of leaves and twigs, the legs changed as the result of impact on an irregular, relatively soft soil, and the lower arm and leg bones, rather than the feet, were elongated. During the Oligocene the drier climate caused a great reduction of the area of forest and a corresponding increase of grassy plain. Running rhinoceros was forced to compete with running horse. Teeth, not legs, decided the issue.

The oldest true rhinoceros now known is *Trigonias*, a Lower Oligocene animal with four toes on the front feet and three on the hind, the little finger small but functional. Except for the loss of the lower canines, it has the dental formula of the Eocene *Hyrachyus*. The limbs were more massive, but, as in the earlier form, there were no horns. These animals, known as the aceratheres (Fig. 135), persisted in North America till Mid-Pliocene times.

Somewhat nearly allied to them were the first to bear horns, the diceratheres (Fig. 136), which seem to have had small ones on the anterior tips of the nasals. Since the horns are represented merely by rugosities on the bones, it is probable that, as in the living forms, they were composed of densely matted hair rather than of bone or horn. They were placed side by side, not in tandem as in the modern double-horned animal. This group had a brief existence, from Mid-Oligocene to Lower Miocene. Probably most readers have seen the remarkable slabs in various museums showing the bones of these creatures. All of them, containing thousands of limb bones, ribs, vertebrae, and skulls, have been excavated from a single layer in a hill on the ranch of Mr. Harold Cook at Agate Springs, Nebraska.

These were the only double-horned rhinoceroses. The type with the horns on the median line appears to have been chiefly Eurasian. *Teleoceras* (Fig. 135), with a single horn on the nasals, was in America during the late Miocene and Pliocene, but was a

FIG. 135. Above, the short-legged *Teleoceras*, an Old World type of rhinoceros which reached America in Miocene times. Below, a primitive Oligocene acerathere. Redrawn after H. F. Osborn.

short-legged, paludine type which could not have given rise to modern forms. It was also in Europe in Miocene times and may have been of Eurasian ancestry.

Although primarily North American, some rhinocerine group reached the Old World in the Oligocene or earlier. Evolution appears to have proceeded there along several lines, although the records are as yet too scanty to allow connections to be made with either the modern descendants or the American ancestors. Asiatic localities have furnished remains of the largest and most extraordinary of all, the gigantic *Baluchitherium* (Fig. 137) of Baluchistan, Turkestan, and Mongolia. Seventeen feet high at the shoulders, with huge hornless skulls more than five feet long, the baluchitheres must have equaled the giraffes in ability to browse high in the trees. They had but a brief history, their remains being known only from late Oligocene or early Miocene deposits.

When camels are mentioned, one's thoughts turn naturally to distant places; visions arise of the shifting sands and elusive oases of the Sahara or the romantic, almost legendary, caravans which conveyed the luxuries of the East into Europe. If one remembers something of the geography of his school days, the llamas of Andean South America may come to mind; but the name has no association with North America. Nevertheless, this continent was the home of the camels throughout their history. It was the nursery of their infancy, the romping ground of their adolescence, and the Eden whence in the fullness of their strength they were at last expelled, "to labor and sorrow all the days of their lives." Eocene times saw their inception in North America; only recently, probably since the end of the glacial period, did they disappear. Not until the Pliocene, so far as present records tell, did any of them migrate to Asia or South America.

Although they retain some primitive characteristics, the camels are among the most specialized of the artiodactyls. Their position in this respect is comparable to that of the horses among the perissodactyls. The larger camels, it is true, are rather awkward, lumbering animals, not to be compared in fleetness with the antelopes,

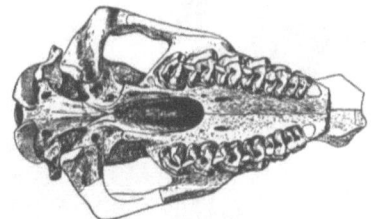

Fig. 136. Palatal view of the skull of a *Diceratherium*, showing the typical lophodont teeth. One-ninth natural size. From O. A. Peterson.

but some of the Mid-Tertiary representatives of the group must have rivaled the gazelles in cursorial powers. The two most obvious specializations of the modern camels are the elongation of the vertebrae of the neck and the growth of one or two "humps." The latter are reservoirs of fat, without any unusual skeletal support. Hence the presence or absence of a hump on any extinct camel cannot be determined. The general impression, however, is that this method of storage of potential energy is a modern development; it is doubtful if any ancient member of the group had a hump. The dentition is reduced to thirty-four by the loss of the first and second upper incisors, one upper and two lower premolars. The outer upper incisors, the canine, and the first premolar are all short, sharp teeth, widely spaced in the elongated muzzle, and also separated from the selenodont cheek teeth by a wide diastema. The molars are high-crowned but not hypsodont. Ulna and fibula are vestigial, the former fused to the radius, the latter represented only by its lower end, which forms the so-called "malleolar" bone. The foot is greatly elongated. There are only two toes, the third and fourth remaining after the loss of the outer ones. They are of equal size, somewhat divergent, supported by pads in an unguligrade position. The corresponding metapodials are long, fused into a cannon bone which has a Λ-shaped notch at the lower end, separating the surfaces with which the proximal segments of the toes articulate.

The oldest "camel" is *Protylopus* (Fig. 138, at left), an animal not larger than a jack rabbit. But how large a jack rabbit may be depends upon the veracity of one's informant; some which have jumped from their hiding places at the approach of the writer have seemed as large as deer. At any rate, the jacks are larger than the cottontail, perhaps as large as a "small dog." The comparisons made by vertebrate paleontologists in their efforts to enable the student to visualize the extinct animals

Fig. 137. Restoration of *Baluchitherium*; height, 17 feet 9 inches at the shoulder. Redrawn after Charles R. Knight, American Museum of Natural History, New York City.

are amusing in their indefiniteness. The largest titanotheres were as large as "small elephants"; the earliest horses were as tall as "large cats"; the oreodonts were as big as "tall pigs." Perhaps, after all, these statements are the best that are possible, for different species of extinct animals varied in size just as individuals of modern species do. The Upper Eocene *Protylopus* may or may not be the direct ancestor of modern camels, but it shows a condition which is primitive in the group. All forty-four teeth are present, in a continuous series. The molars are short-crowned (brachydont) and square, not elongate. The pattern, however, is selenodont. Ulna and fibula are complete, not coössified with their companion bones. The hand has four fingers, the outer ones smaller than the inner, and on the feet are dewclaws, small, vestigial remains of the second and fifth digits.

Even more primitive than *Protylopus* is *Diacodexis* of the Lower Eocene, considered by Matthew to be the most primitive artiodactyl known. It is not, unfortunately, of such structure as to indicate that it was ancestral to all even-toed mammals, but it may represent the stem toward which all selenodonts, namely, oreodonts, camels, deer, and ruminants, converge as their histories are traced further and further backward. *Diacodexis* was not so large as a jack rabbit, but more nearly comparable in stature to "bunny," the semidomesticated rabbit. Not only had it the primitive forty-four teeth, but the cheek series were of the bunodont rather than the selenodont pattern, the molars actually trituberculate. Both front and hind feet retained four toes, although the outer ones of the latter were slender, the only reason for excluding this animal from the proud position of ancestor of all artiodactyls. Its immediate fore-

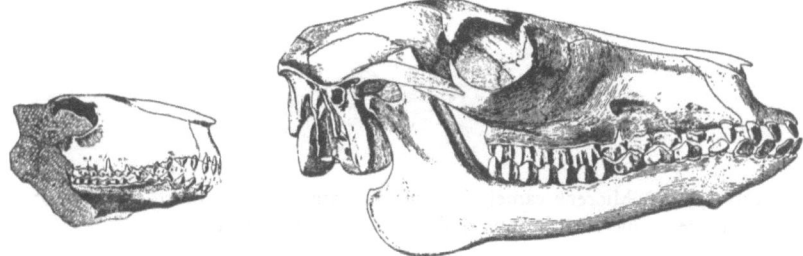

FIG. 138. At left, *Protylopus*, the possible Eocene ancestor of the camels. One-third natural size. At right, the Oligocene *Poebrotherium*. Two-fifths natural size. Both from J. L. Wortman.

bear may have had fully developed second and fifth digits on the hind foot, in which case it would be the connecting link between artiodactyls and creodonts, for *Diacodexis* retains many of the characteristics of the latter group.

The Oligocene *Poebrotherium* (Fig. 138, at right) is truly ancestral to existing camels, whatever may be the ultimate interpretation of the Eocene animals just mentioned. About as large as a sheep, it had a complete dentition, although not in a continuous series, for there are spaces between the anterior teeth. Incisors and canines are all of approximately equal size, the lower incisors erect and sharp, not broad and procumbent as in the modern animals. The molars are selenodont but short-crowned. Skeletons have been recovered from several different levels in Oligocene strata, each higher bed furnishing somewhat taller and larger individuals, but all belong to the same genus. According to W. B. Scott, all were slender, "with small pointed head, long neck and body, and long, very slender limbs and feet . . . the forearm bones were fully coössified, and in the lower leg only the two ends of the fibula remained." Although the lateral toes were reduced to vestiges, only two remaining functional, each with a deerlike hoof, the metapodials were not fused into a cannon bone.

The Oligocene was the period of childhood for this group, but the Miocene was that of their lusty youth, the time during which they reached man's estate. According to Matthew, the Lower Miocene *Oxydactylus*, the Mid-Miocene *Protolabis*, and the Upper Miocene *Procamelus* are the genera which carry the blood in direct line toward the modern *Camelus*. *Oxydactylus* did not differ greatly from *Poebrotherium* except that it was taller, with longer legs, neck, and skull. The molars had high crowns but were not hypsodont, and the feet had deerlike hoofs, without a pad. Its successor, *Protolabis*, still retained the whole set of teeth, and a cannon bone had not yet been formed. *Procamelus* (Fig. 139), however, was as large as a modern llama, had cannon bones, and had lost the first two pairs of upper incisors, although it retained all the premolars. It may have been ancestral to both the true camel and the llama. Not till Upper Miocene times did the molars become hypsodont and the lower incisors procumbent.

FIG. 139. A Miocene camel, *Procamelus*, with dentition much like that of Recent ones. One-twelfth natural size. From Earl Douglass.

The Pliocene camels were racially fully adult, most of them as large as any modern camel, and at least one was larger. The end of the Miocene apparently marked the parting of the ways of two lines, so far as size is concerned. The llamas, migrating to South America, where they now occupy the uplands, failed, perhaps because of the rigorous environment, to reach greater size than *Procamelus*. The toes remained separate, each with its own pad. In contrast to this conservatism, loss of teeth proceeded further than in *Camelus*, two upper incisors, two upper, and two lower premolars being absent from each jaw of the modern llama. The large Pleistocene *Camelops* has the same formula, being in this respect somewhat more specialized than the camel which has survived. *Pliauchenia* is a genus which includes many Pliocene forms. Some of the species are relatively small, and may connect *Procamelus* and the llamas; others are large, and consequently in the line leading to modern Asiatic types. Among these large creatures are some which, apparently for the first time in the history of the group, became adapted to life in deserts. The broad, poorly developed ungual phalanges indicate that, like the modern camels, the large *Pliaucheniae* had large single pads on each foot.

The history of the camel is in the main parallel to that of the horse. The Eocene ancestors of both were small, and there has been a constant increase in size, culminating in the Pliocene. Concomitantly with increase in size there has been an increase in length of limb, accomplished principally by the elongation of the foot. Both have suffered loss of toes but have gained thereby in fleetness. In both the skull has become elongate, with loss of teeth between the incisors and molars. The latter are in successive stages higher and higher crowned till at last they become truly hypsodont, almost rootless. In fact, convergence in the evolution of these two unrelated groups extends further than mere similarity of general trend. A horse and a camel from a formation of any particular stage of the Cenozoic are of about the same size, about the same length of leg, and, in all respects, of about the same degree of specialization. This suggests that evolution was controlled by environment, for in both lines the ancestors were browsers, living on leaves and shoots found in the forests, and the descendants grazers, more and more adapted to exist under the increasingly semiarid climates of the plains on which they lived.

Collateral lines have purposely been avoided in the phylogeny of the camels, but this should not lead to the impression that camel evolution was along one line only or that camels as a whole reached the "top of their form" only when they achieved their greatest size in the Pliocene. As a matter of fact the survivors of the group appear to belong to a strain which was rather backward, conservative in its changes. True grazers did not appear in this line till the Upper Miocene, whereas the Lower Miocene *Protomeryx* seems to have been fully adapted for subsistence on grasses. The contemporary *Stenomylus* (Fig. 134) the "gazelle camel," was better adapted for rapid locomotion than any other member of the group, living or extinct. But these are representatives of phyla which for one reason or another were unable to survive.

FIG. 140. Restoration of *Eohippus*, redrawn after Charles R. Knight's painting in the American Museum of Natural History, New York City.

XXVI

MAMMALS OF YESTERDAY

O! call back yesterday, bid time return.

King Richard the Second, Act III, scene 2

Mammalian history reached its climax just as man was emerging from obscurity. He was confronted in both Eurasia and North America by an extraordinary abundance of beasts, many of them superior to him in strength, speed, and cunning. Some were active, others potential, enemies. But, by and large, abundance of mammals meant abundance of food, provided only that it could be captured. And so man became a hunter.

The North American Pleistocene fauna seems at first view to have been a strange one, for it contained a mixture of creatures now belonging to this and other continents and provinces. Wild camels and horses are now Asiatic in their distribution, elephants are Asiatic and African, lions are African, and tapirs and sloths are Central and South American. All these animals were common in North America during the Pleistocene. Some of them had evolved here and have disappeared recently; others were immigrants during the Pliocene or Pleistocene.

The most conspicuous members of the fauna were the elephants and the mastodon. The latter was the more primitive, the last survivor of a stock which entered North America from Asia as early as the Miocene. One of the chief differences between the two is in the teeth. The cheek tooth of the mastodon is much simpler than that of an elephant, for, although large, it has definite roots and a crown, on which are a few bunodont tubercles. Wear removes the superficial layer of enamel and produces a pattern showing four or five broad areas of dentine alternating with narrow bands of the harder substance. The elephant tooth, on the other hand, is almost rootless (hypsodont) and contains numerous cross-crests, the upper edges of a series of plates in which enamel, dentine, and cement (bone) are arranged in alternating sheets.

The oldest representatives of the elephant tribe (Fig. 141) are *Moeritherium* of the Upper Eocene and Lower Oligocene of the Fayum in Egypt, and *Paleomastodon* and *Phiomia* of the Lower Oligocene of the same region. These animals were less than half the height of modern elephants and had many more teeth. *Moeritherium* had thirty-six, and the others twenty-six. The second upper incisors, which were to become the gigantic tusks of later elephants, were only slightly enlarged in the first, although conspicuous in *Paleomastodon* and *Phiomia*. The molars of all three were of normal bunodont type, foreshadowing those of the mastodons.

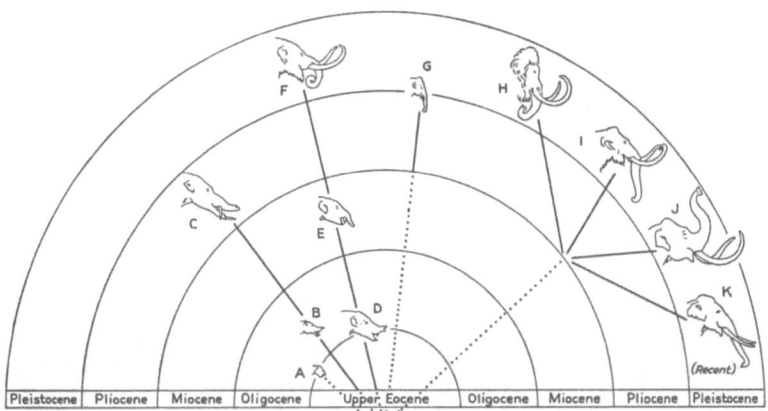

Fig. 141. Diagram to indicate the radial evolution of mastodonts and elephants. Mastodonts: A, *Moeritherium*; B, *Phiomia*; C, *Amebelodon*; D, *Palaeomastodon*; E, *Miomastodon*; F, *Mastodon americanus*; G, *Stegodon*. Elephants: H, *Mammonteus* (= *Mammuthus* = woolly mammoth); I, *Parelephas*; J, *Archidiskodon*; K, *Elephas*. Simplified after H. F. Osborn.

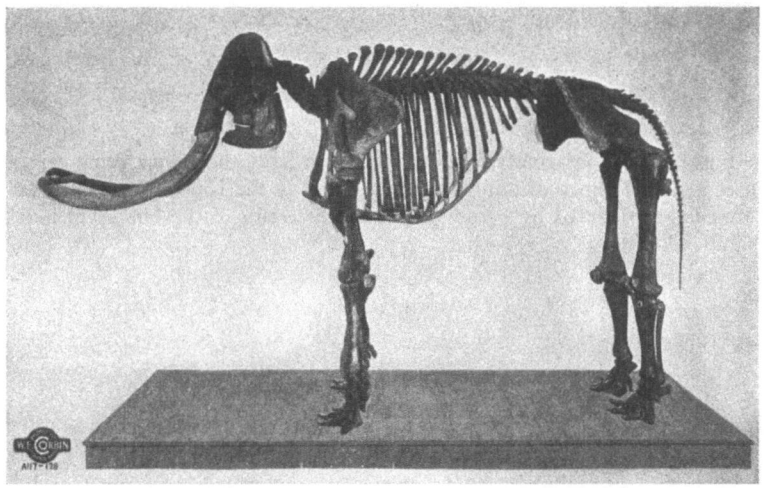

Fig. 142. The skeleton of a Pleistocene elephant (*Parelephas columbi*) in the Amherst College Museum. Photograph by courtesy of W. E. Corbin.

Fig. 142A. An imaginary Siberian Mammoth hunt, the animal being driven over a cliff. From a painting by Ernest Griset in the United States National Museum, reproduced by permission of the Secretary of the Smithsonian Institution.

The Miocene seems to have been the time of greatest mastodont prosperity. Then the group suddenly spread into Europe, Asia, and North America, occupying vast areas never before trodden by elephantine feet. They flourished exceedingly, increased greatly in stature, and lost the useless teeth between tusks and molars. Most remained in the forests and adhered to the ancestral browsing method of feeding. Not till the Pliocene do true elephants, with grinders of the modern type, appear. Just when or where or how they became adapted to life on the plains and to a diet of grasses is still a mystery. The transition probably took place in Asia, but some of the largest and most specialized roamed North America in Pleistocene times.

Least conspicuous of this group was the Siberian or woolly mammoth (*Mammonteus primigenius*), well known in central and western Europe, where it was trapped and pictured by primitive man. It spread across northern North America, occasionally venturing as far south as North Carolina. That its normal habitat was in the cold regions of the north is shown by the dense wool beneath the coarse hair of the frozen carcasses found in Siberia, its native region. It was only nine feet high at the shoulder, but the related Columbian elephant reached eleven feet, about the size of the modern African *Loxodonta*. The Columbian elephant had a more southern range than the true mammoth. The best specimens have been found in the southern and western states, but it seems to have been at home all over what is now the United States and in parts of Mexico. An impressive skeleton is on exhibition in the museum at Amherst College (Fig. 142). Both this and the woolly mammoth have numerous (27–30) thin lamellae in their molar teeth. Largest of all the North American forms was the imperial mammoth of California and other southern states. It was thirteen and a half feet high at the shoulders, rivaling the contemporary straight-tusked elephant (*Palaeoloxodon antiquus*) of southern Europe and northern Africa.

The Pleistocene American mastodon was about as large as the woolly mammoth. It is, perhaps, the most widely distributed of all American vertebrate fossils, its remains having been found at numerous localities from Alaska to Mexico. The legs are comparatively short, and the head lower and more flattened than that of the true elephants. A coat of long coarse hair fitted it, like the Siberian mammoth, for life in cold regions, but it chose wooded areas rather than open plains. It survived the last of the glacial stages and seems to have lived until recent times.

The mastodonts had been established in North America since the Miocene, but the elephants were new arrivals. Other immigrants from Asia during the Pleistocene were the mountain sheep, the Rocky Mountain goat, the musk ox, and the bison. The last were the only ones really to flourish, although musk oxen were once much more numerous and diversified than at present and at times came as far south as Kentucky. Great herds of bison grazed on the western plains, just as did their one survivor, the "buffalo," within the memory of men still living. In earlier days there were numerous species, some including individuals large enough to support horns with a spread of from seven to ten feet. As was mentioned in a previous chapter, deer had reached North America as early as the Miocene, but they were not abundant till the Pleistocene. There is some evidence that there were new immigrations from Asia at that time.

Horses and camels, typical American animals, throve during the glacial age, but after having survived the worst nature could furnish in the way of cold and climatic changes they were blotted out from this continent before the advent of white men. Horses ranged in great herds all the way from Alaska to Mexico. All belonged to the modern genus *Equus*, but neither *Equus caballus*, the domesticated horse, nor its immediate ancestor was among them. There were at least ten species, varying in size from the pygmy horse of Mexico to *Equus giganteus* of Texas, larger than any modern draft horse. There was a forest-dweller at the time, an animal of moderate size, but the favorite habitat was the grassy plains.

Peccaries, the American representatives of the swine, had probably been in this country since the Miocene, but, like some other groups, they became most abundant in the Pleistocene. They still remain in the region from Texas to Brazil, though they have been driven from the more northern part of their range.

Tapirs passed through most of their evolution on the northern continent but are now found only in Central and South America and southern Asia. Some of them came as far north as Pennsylvania during warm interglacial periods. Two species then roamed the United States, a large one now extinct and another smaller animal which seems to be the same as that still existing further south.

Among the most striking of the immigrants are those from South America. Strangely enough, one is the Canadian porcupine, a rodent which has become so firmly associated with the north that knowledge of its origin comes to one as a surprise. Another rodent, the South American water hog, the largest existing member of its order, came north but failed to get a foothold. Even more typically South American were the great glyptodonts and ground sloths, now extinct but a conspicuous part of the fauna of both continents during the Pleistocene. They mark the culmination of the peculiar order Edentata, represented nowadays by the hairy anteaters, tree sloths, and armadillos of subtropical and tropical regions of South and Central America. The armadillos, like the peccaries, are numerous as far north as Texas.

The term "edentate" really applies only to the anteaters, for they alone are toothless. But tooth trouble is characteristic of all members of the order, even to the most distant ancestors. In general, incisors and canines are absent, and the cheek teeth are rootless pegs with no enamel on their surfaces. The number of molars is reduced in some, but, curiously enough, a few have more teeth than the normal placental.

The case seems to be parallel to that of the whales, in which supernumerary teeth were added when the molars became simple and rootless.

The edentates may have originated in North America, for the remains of primitive members of the group have been found in Paleocene and Eocene formations there, but their real evolution and differentiation took place in the southern continent, where they were early arrivals. An origin in North America, however, is not supported by the present opinion of most paleontologists. Following Matthew's idea that the center of placental evolution was somewhere in the Arctic region and that the paths of migration radiate in all directions, it has been suggested that the early Tertiary specimens just mentioned were stragglers left behind as the group pushed

FIG. 143. South American types of edentates which were on the northern continent in Pleistocene times. A *Glyptodon* in foreground, *Nothrotherium* at right, and *Mylodon* in middle. The ground sloths are redrawn after restorations by Charles R. Knight under the direction of Chester Stock; the glyptodont after R. Bruce Horsfall.

southward toward its ultimate home. In this particular case, however, Matthew's idea of radiation does not apply, for the edentates never reached Eurasia or Africa. The Old World pangolin and the aardvark are toothless but are not members of the Edentata.

Whatever may be the ultimate decision on this matter, all will admit that the oldest edentates now known had departed considerably from the primitive placental condition. Only two incisors were present, and the cheek teeth were peglike, with mere vestiges of enamel. The best-preserved specimens are Mid-Eocene in age. Much may yet be learned as Paleocene deposits are explored. Although the North American Eocene specimens retain no traces of bony scutes, their osteology suggests a relationship to the armadillos, and parts of the armor of these creatures have been found in the Eocene of Patagonia.

Modern armadillos do their best to prove that there are notable exceptions to the statement that sluggishness begets armor. They have broad shields of bony plates over shoulder and rump, and the area between is protected by narrow transverse bands, the whole covered with horny scales. Although the animals are almost completely covered by a flexible coat of mail, their short legs are capable of carrying them across the open plains on which they live with great rapidity and of scratching out burrows with equal facility. They are not "choosy" about their food so long as it is tender. Insects, worms, small animals, and "root crops" appeal to them; in short, they are omnivorous. They have been abundant in South America since Eocene times. They reached their maximum size and differentiation in the Pleistocene, culminating in a form as large as a rhinoceros. This swashbuckler invaded North America but, after a short and glorious life, disappeared. Among his numerous relatives still surviving in the old home there is, however, one a yard in length and with fifty pairs of teeth. Hardly an edentulous edentate.

The glyptodonts were the more phlegmatic and more fully armored cousins of the armadillos. Their shells, a mosaic of small polygonal bones fitted edge to edge like the plates of an echinoderm, were solid, inflexible, turtlelike. Some had a bony casque protecting the head, and rings of plates on the tail. A few had spikes on the caudal appendage, reminiscent of those of the stegosaurs. The legs were short, the claws broad, the jaws deep, with numerous ineffective teeth. These creatures could hardly have been burrowers, but on the other hand they could not have been really active. Yet it seems that they found food in abundance, for after their first appearance on the southern continent in the Oligocene they became common in the Miocene and abundant in the Pliocene and Pleistocene. In the latter epoch they reached their maximum size, with a length of nine or ten feet, and their greatest geographic distribution. Their remains are not common in the United States, although some reached the southern states and one adventuresome individual visited Atlantic City. All that has been found of him is a heel bone.

Less is known of the history of the hairy edentates than of their armored relatives. Neither the tree sloths nor the hairy anteaters can trace their ancestry. They are unfortunate in that they have lived in regions where records were not kept. The ground sloths, although now extinct, seem to have belonged to the nobility, for their pedigree extends back to the Oligocene. As with modern noble families, a bit of tradition seems to be mixed with facts, for it is not till the Miocene that the group emerges from obscurity. As in the armored edentates, maximum size, differentiation, and distribution were reached in the Pleistocene. The large creatures are the ones in which we are chiefly interested, for their remains are common in southern United States, and not rare as far north as the states of Washington, Idaho, Kentucky, and Pennsylvania. Thomas Jefferson fathered a fabulous story of a gigantic bear which had once roamed the mountains of the Allegheny plateau; by avocation an amateur paleontologist, he had acquired huge bones from West Virginia which formed the basis of his tale. They were scientifically described as *Megalonyx* by Richard Harlan, who showed that they really belonged to an un-bearlike animal similar to a hairy ground sloth which Cuvier had described from South America in one of his first paleontologic papers.

Most of the specializations of the ground sloths are associated with their ability to assume a semierect posture. The hind legs are short and massive; the feet extremely awkward, large, and flat, with only one clawed toe; the pelvis a huge basin capable of supporting a great visceral mass. The long arms show that the animal was really quadrupedal, but the wrist was so twisted that the weight must have been borne on the knuckles of the outer fingers. All the digits are clawed, useful in drawing branches and leaves to the mouth. The ground sloths were vegetarians and browsers though their teeth were soft and incompetent.

Despite the various absurdities in their construction, nature dealt kindly with these creatures until comparatively recent times. Their giant, *Megatherium*, eighteen feet long, lived in both South and North America during the Pleistocene. Somewhat smaller relatives survived even later, for specimens of *Mylodon* have recently been found in New Mexico with the tendons still undecomposed. Animals of its sort made the famous "human" footprints on the rocks in the prison yard at Carson City, Nevada. Another form (*Glossotherium*) seems to have been confined by man in a cave in Patagonia. Some naturalists have inferred that this animal was domesticated and that its cows were kept for their milk; the numerous droppings indicate that the animals were inhabitants of the cave for a considerable period of time.

All of the Pleistocene creatures so far noted are of inoffensive sorts, herbivorous or omnivorous. Preying upon them in North America were numerous carnivores, some now extinct. They included minks, weasels, martens, raccoons, badgers, bears, foxes, wolves, coyotes, and pumas, which are still with us. Bears and badgers were new arrivals during the late Pliocene or the Pleistocene, having been strictly Eurasian previously. The most prominent of the extinct types were the saber-toothed tigers, repatriated descendants of Oligocene ancestors; they reached their culmination in size just before their extinction. Remains of these great catlike creatures have been found at many localities in the United States, and they were in the Old World as well. True cats, closely allied to the African lion, roamed the southern and western states. They are now gone, but some of the existing pumas are of a size not to be despised. Among the dogs, the "dire wolf" receives the premier place. A little larger than a modern wolf, he seems to have won through his name a reputation somewhat out of proportion to his real importance.

It has not been possible in these few pages to express the full diversity of the Pleistocene mammalian fauna. O. P. Hay states that there were over six hundred species of vertebrates on the continent at the time, and that 60 per cent of them are extinct. I trust the reader has received the impression that North America then had numerous creatures which were newly immigrated, and that during the glacial period or shortly thereafter it lost a vast population. Some of the animals that disappeared were new arrivals, others descendants of ancient stocks. These changes are not easily explained.

Till the Pleistocene most mammalian races increased slowly but continuously in cranial capacity, in size, and in adaptation to life in particular environments. Then many groups disappeared or were relegated to a minor position. It would be interesting to know definitely what were the controlling factors in the blocking of these evolutionary trends. Is there a limit beyond which size cannot increase? Undoubtedly there is, for an animal of great bulk requires more food for its maintenance than a small one, and, although there may be no limit to the amount of food available, there are definite limitations upon the amount which can be ingested and digested during a day. May not the evolution of some mammals and reptiles have stopped because each, according to its own method of feeding, had reached the greatest size possible? This in itself is not a cause for extermination, but it brings animals to a condition in which they are vulnerable. It seems likely, too, that the extinction of some of the great races must be ascribed to geographical changes. In the case of the reptiles the worldwide mountain building which followed the Cretaceous appears to have been responsible for the downfall of the ruling houses of terrestrial animals. In the same way the glacial periods of the Pleistocene may have been, directly or indirectly, responsible for the fall of the mammals.

The influence of the Pleistocene glaciation appears to have been chiefly indirect, for the various advances of the ice could hardly have been so rapid as to cause the extermination of warm-blooded, hairy mammals, capable of withstanding considerable cold and certainly well adapted for migration. Since the glaciation was bipolar, its effect was to drive animals both southward and northward. The effects of the refrigeration were felt far beyond the limits of the areas actually invaded by the ice, the result being a southward shifting of the climatic zones in the northern hemisphere and a northward movement in the southern. Throughout the Pleistocene, an epoch in which there were three or four advances and retreats of the ice that produced alternately cold glacial and warm interglacial stages, animals were forced to follow the shift of the climatic zone to which they were adapted or, if not that, at least one in which it was possible for them to survive. Changing climates altered the vegetation tremendously and consequently affected the food of the herbivorous animals to a marked degree. Since plants, although more sensitive to changes in climate, do not migrate as rapidly as animals do, each stage of advancing ice put a double stress upon the mammals. Their crowding toward the equator greatly increased the population of the tropical and subtropical belts, and forced a spirited competition for food. Regions already fully occupied were called upon to support an enormous number of immigrants. Not only did the incomers have to adapt themselves to a diminishing food supply if they were to survive, but, since the plants themselves were changing, they had also to accept alterations in their diet.

It was unfortunate for North American mammals that dry land becomes narrower and narrower through Mexico and Central America to the bridge at the Isthmus of Panama. Although that bridge probably was higher and wider in Pleistocene times than it is now, it was the bottle neck of migration southward. Still less favorable was it as a route for animals coming northward. Yet it was used to a considerable extent by creatures moving in each direction. During the height of the glacial stages the climate of Pennsylvania and Iowa must have been much like that of Greenland today — habitable in spots, but barely so. Only the most hardy could live there. Of all North America, only Florida, the Gulf States, the Southwest, and Mexico were really hospitable regions. Animals accustomed to life in these areas had to move still further south. Hence it was that pressure from the north forced many sorts of animals across the bridge of Panama.

Throughout the Tertiary, South America was, in effect, an island. Although it was connected at various times with the northern continent, there was comparatively little migration across the isthmus except during the Eocene and Pleistocene. The tropical jungle of Brazil seems to have been almost as effective a barrier as though its area had been submerged. As a result of this isolation, the evolution of mammals in South America was practically independent of that at the north. A few stocks furnished the nucleus for a large and diversified population. We have already seen that some members of these groups traveled northward during the Pleistocene. Let us now list the principal ones which invaded the southern continent.

Immigrant hoofed animals included horses, which apparently throve, for remains are found not only of *Equus* but of genera unknown in the north. Mastodons of various species crossed the isthmus, but, curiously, no elephants. Cloven-hoofed creatures were represented by deer, antelopes, peccaries, and many species of the camel tribe. Among the invading rodents were rats, mice, and rabbits. The descendants of some of them are still in South America; others failed to meet the competition of the native fauna. True carnivores, until then unknown in South America, entered the field and quickly ousted the native marsupial "wolves." In the list are a few bears, which is somewhat strange, for they had only just reached North America from Asia. With them came the saber-toothed tigers, jaguars, and pumas. Among the dogs were various small, foxlike wolves and, curiously, a dog of the same genus (*Cyon*) as the modern dhole of India — a freak of geographic distribution. Smaller carnivores were the weasels, raccoons, skunks, otters, and a few others, but not all the North American fauna. The effect of such a sudden influx of flesh-eaters upon the great native population of herbivorous mammals must have been similar to that following the introduction of a family of weasels into a hen yard.

South America is, from its shape, much better fitted to withstand the rigors of polar glaciations than its northern sister. Its southern end is not only considerably removed from the south pole but it has no great area, as compared with the vast surface of tropical Brazil. Ice covered Patagonia more than once, perhaps thrice, during the Pleistocene, urging mammals northward. They need not have crossed the equatorial belt, so far as we can see; yet some of them did. It may be that a period of glaciation in South America occurred at the same time as a warm interglacial stage on the northern continent, thus inducing a northern migration. Or possibly the influx of North American carnivores into the southern continent may have led to a northward drift in population. At first sight it may seem that the carnivores had much the best of this situation. It mattered little to them if plants disappeared from particular areas; the concentration of herbivores in restricted areas

FIG. 144. Hypothetical restoration of the shovel-tusked mastodont, *Amebelodon*, from the Pliocene of Nebraska. The lower jaws were nearly six and a half feet long. After a sketch by Erwin Hinckley Barbour.

made for good hunting. In the earlier centuries of each ice advance beasts of prey doubtless flourished and acted their part, an important one, in the destruction of various groups of vegetarians. As the cold increased, however, the forests and thickets decreased. Lacking coverts, the pursuers became the pursued, for despite the absence of piercing claws and trenchant teeth, most large herbivores are far from being defenseless, or even of a pacific temperament. Lacking shelter or place of concealment, few of the solitary sorts of carnivores can compete successfully with herds of hoofed creatures. Hence many of the carnivores, particularly those of the cat tribe, became extinct during the Pleistocene. Pack against herd is, however, another matter. Wolves were more successful until they met that defenseless two-legged creature who used weapons to subdue brute force.

In spite of widespread sentiment to the contrary, it cannot truthfully be said that man has "interfered with the processes of evolution." Many writers, still unconsciously under the domination of the theory of the special creation of man, lament the fact that he has interfered with nature. Such and such would have happened if it had not been for the intervention of man. It should be realized that man is just as much a product of nature, and of evolutionary processes, as any other animal. Because through the use of his large brain he has been able to subjugate all other creatures, it does not follow that he is set apart from other animals or that they were created for his use; he is merely one of the factors in the evolutionary forces which control the history of life. Man is unique among animals in that he is able to exercise considerable influence over his environment. Nevertheless, he is not, and never will be, all-powerful. His skill has made many previously hostile regions habitable; on the other hand, some of his practices have brought desolation, through flood and erosion, to formerly fertile regions. Man is still being molded by his environment mentally, just as in Cenozoic times he was molded by it physically.

XXVII

THE ANCESTRY OF MAN

He that overcometh shall inherit all things.
Revelation, xxi, 7

A survey of the chemical and physical properties of inanimate things in their relations to living matter led Professor Lawrence Henderson to the conclusion that all matter is biocentric. Certain it is that, in his egoism, man considers himself the creature to whom all else is subservient, and, with or without reason, has brought himself to the position of regarding the earth as Homocentric. Man's chief curiosity is, therefore, in regard to his own origin.

To write an account of the origin of man at the present time is to write a detective story. Nature, in one of her playful moods, has scattered clues, but, pixie-like, she seems to have delighted in giving just enough information to keep up the suspense. The romance of the search has attracted the keenest of morphologists, anthropologists, and paleontologists. Amateurs and professionals have given freely their time, resources, and mental abilities. Many solutions have been reached: some of them have been proved incorrect; others are strongly supported; but there are probably few students who think that the case is closed. The many discoveries which have been made in recent years encourage one to believe that the story may finally be completed. Interesting as they are, however, the new clues have not been of much help in tracing the ancestry of man. Remains of Tertiary anthropoids, particularly of Miocene and Pliocene age, are badly needed. The world has been ransacked for such material, with meager results, but since there are still large areas in Asia and Africa which have not been explored half so intensively as those of North America there is still abundant opportunity for new discoveries.

The conception that man is supernaturally set apart from all the other animals finds no support in the study of comparative anatomy. In fact, Huxley long ago showed that every bone and every part of man, save three little muscles of hand and foot, are present also in the higher apes, so that on anatomical grounds it is obligatory to class man, apes, and their allies together in a single order, the Primates. This does not mean that man is descended from the apes, as we know them, but that he must have had the same ancestry.

As with horses and camels, it is necessary to understand the skeletal characteristics of man if one is to realize his place in the animal kingdom and understand how it has been reached. In some respects the skeleton is primitive. There are five digits, and the number of phalanges is the same as those of the Eocene mammals. The teeth are short-crowned, with few and low cusps, comparatively little reduction in numbers, and little specialization; they remind one of late Eocene mammals in which quadrangular molars had superseded the primitive triangular ones.

The specializations have to do chiefly with adaptations to an erect posture and with the large size of the brain. The bones of the hind limbs are straight, the pelvis basin-shaped, and the spinal column doubly sinuate, all characteristics which help to support the body in a vertical position. Few appear to realize that in the whole vertebrate group there is but a single species, *Homo sapiens*, with its weight so distributed that the axis is truly vertical. What plants and sessile animals do naturally in response to the stimulus of light has been accomplished but once by vagrant animals. Locomotion by the aid of two or more pairs of appendages has held the body in a more or less horizontal position. Dinosaurs achieved bipedality with the aid of a long tail which balanced the anterior part of the body; short-tailed birds have reached somewhat the same pose, partly because the greater portion of their bulk is concentrated in a relatively short body, but chiefly because the femora extend forward in an approximately horizontal position, bringing the lower legs beneath the center of gravity. The foot in man is specialized, although plantigrade. (In fact, it is a question whether the plantigrade mode of locomotion is primitive. Mammals with feet of this type are rare, and on the whole rather awkward; women realize the latter fact and employ high heels to make themselves pseudodigitigrade, a reversion to the semidigitigrade condition of the early Tertiary mammals.) The heel bone is elongated; the tarsals and metatarsals are so modified as to produce a strong arch; and the axis of the foot is not on the median line but on the inner side, where it passes through the big toe, which is more fully developed than in any other mammal. The outer toes are in process of reduction. The arms are relatively short, but man shares with the higher apes a striking feature, the power of rotating the thumb so that it is opposable to the remaining digits. This, however, is a characteristic of primates in general, probably inherited from the arboreal Paleocene ancestors. In contrast with the other primates man shows a decided loss of hair.

Man has the same number of teeth, thirty-two, as the Old World apes and monkeys ($i\frac{2}{2}$, $c\frac{1}{1}$, $p\frac{2}{2}$, $m\frac{3}{3}$), but all are relatively smaller than those of apes. The canines do not project beyond the others, there is no diastema, and the series forms a semicircle rather than the U-shaped curve of most mammals. That reduction in the number is still in progress is shown by the fact that the first premolars and second incisors are small or may be absent, as are the third molars (wisdom teeth) of some individuals. This reduction in size of the teeth is correlated with the shortening of the muzzle and the retreat of the chin.

These are some of the more obvious specializations of man. The characteristics of the skull deserve special mention because they are among the more striking, are most intimately connected with man's greatest asset, his brain, and because more information is available about their evolution than about that of the other portions of the skeleton. Most conspicuous are the shortness and height, the straight face, high forehead, and large, forward-directed orbits. The brain case is large, bulging upward and backward, and the *foramen magnum* is beneath rather than at the back of the skull. The nasal bones are small and short, the malar arches small, and the eye sockets partitioned from the temporal openings. In apes a very striking feature is the projection of the tooth-bearing portion of the jaws to form a muzzle; in man, particularly in the more civilized and higher races, there is no such prognathism. The shortness of the jaws appears to be correlated with the large size of the brain. As the anterior part of the cerebrum, the seat of the intellect, increased in size, the necessary enlargement of the skull in itself tended to straighten the face (or the facial angle); and, possibly to maintain a balance, or as a result of the changed direction in the pull of the muscles, the jaws shifted backward. Not all these characteristics are exclusively human, however, for some of the monkeys and apes have large brains, and hence large skull caps (calvaria), short faces, and occipital condyles as far forward as those of man.

The order Primates is subdivided into three groups; the Lemuroidea, the Tarsioidea, and the Anthropoidea, the last including the apes and monkeys. In the last suborder there are several families. The Old World anthropoids (catarrhines) are sharply differentiated from those of the New in two ways; they have narrow instead of broad nasal septa, which results in the nostrils' being close together and pointing downward, and they have either nonprehensile tails or none at all. The Simiidae, the highest family of Old World apes, share with the Hominidae the tailless condition and the dental formula, but have longer fore limbs than hind and differ in many other respects. These are, however, matters of proportion, not of the presence or absence of fundamental structures. The New World monkeys (platyrrhines) have flat nostrils, long tails (in some cases prehensile), and three instead of two premolars on each side. They are less closely related to man than the Old World series and probably had an independent evolution after Eocene times.

The lemurs (Fig. 145, at right), which today are confined to the tropical forests of Africa, Madagascar, and southern Asia, are the most primitive primates. Their ancestors are found in Paleocene and Eocene rocks of North America and Europe. These early Tertiary primates have many characteristics in common with the insectivores of their day, as shown by their small size, elongate skulls, numerous teeth — most of them have forty — and by other structures. Some of the Eocene lemuroids are represented by fairly complete material. Perhaps the best-known American form is *Notharctus* (Fig. 145, at left), from the Mid-Eocene. It was a small animal with a skull that was about three inches long, a narrow brain case, and large orbits. There was no partition between the eye sockets and the temporal openings, but a continuous bar of bone behind the eye indicates the initial stage in the production of such a structure. There are forty teeth, the lack of one incisor on either side being the only reduction. The canines are larger and longer than the adjacent teeth, the premolars simple, and the molars of the crushing type, with low tubercles. The molars of the older species of the genus are somewhat triangular, those of the later ones quadrangular. Such teeth suggest omnivorous rather than insectivorous feeding. The feet show that the great toe diverged from the others, an indication of arboreal habits. *Adapis* is a similar but slightly larger and more powerful-jawed animal from the French Eocene. It is probable that the modern lemurs are descendants of animals not unlike these. They have been variously modified in the course of time but retain

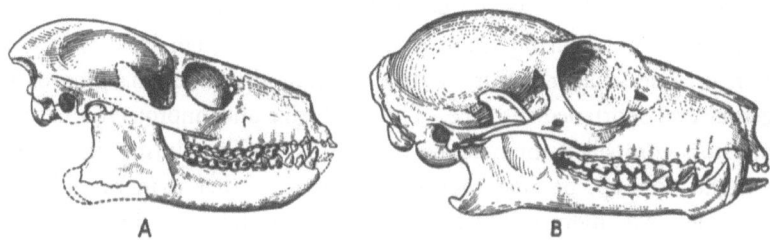

FIG. 145. A, the Eocene lemur, *Notharctus*. One-half natural size. B, a modern lemur, for comparison. One-half natural size. Both redrawn after W. K. Gregory.

the primitive form, the grasping type of foot, and the long muzzle, though they have lost a few teeth, reaching the same dental formula as the platyrrhines.

As early as the Paleocene, however, short-skulled primates made their appearance. They seem to have sprung from the primitive lemuroids and are in many respects so like them that only recently have paleontologists and osteologists recognized them as a special group, the tarsioids. This suborder gets its name from extremely odd animals found on various peninsulas and islands from the Malay peninsula to the Philippines. There are four species of the genus *Tarsius*, all queer little creatures with short, wide heads, big eyes which face almost directly forward, and highly specialized hind legs.

Eocene tarsioids have been found in regions as widely separated as Wyoming, France, and Switzerland. They are somewhat longer-jawed than the modern animal but exhibit the same important characteristics, that is, a V-shaped arrangement of the dental rami and the rodentlike nature of the anterior teeth. Although in general form intermediate between lemurs and monkeys, all known tarsioids are too highly specialized to form connecting links. *Tetonius* (Fig. 146), the best-known American member of the group, is alleged to have been the most "brainy" creature of the

Eocene. The skull is short and wide, rather catlike, but with enormous orbital openings. Its dentition, and that of its relatives, is reduced, in some species at least, to the same formula as in the South American monkeys. Unfortunately *Tetonius* appears to have lost all its lower incisors, the canines taking their place to form rodentlike anterior teeth. The European Eocene tarsioids are somewhat more monkeylike than the American ones, but none has yet been found which could have been ancestral to any of the later anthropoids. Nevertheless, there can be little question but that shortening of the skull and increase in size of the orbits, with concomitant shifting toward the front, took place during Eocene times. Some as yet undiscovered tarsioid must have given rise to the European catarrhine monkeys and their more primitive South American platyrrhine cousins. As the brain increased in size and the sense of hearing became more acute, the breadth of the skull increased, tending to push the orbits into a more frontal position. At the same time the olfactory lobes of the brain, less important as sight and hearing improved, atrophied through lack of use, and the nasal region became smaller. This in turn brought the inner margins of the eyes closer together and shortened the face. The small size and unprotected nature of our distant ancestors compelled them to be timid folk, hiding during the day, abroad chiefly at night. Perhaps because they were arboreal rather than humble terrestrial quadrupeds, sight was more important to them than ability to distinguish scents.

The belief that the tarsioids were ancestors of the anthropoids is strengthened by the study of two little lower jaws, the most important of the few specimens of primates which the reluctant Oligocene strata have as yet yielded. They were found in the Fayum desert southwest of Cairo and have been hailed as the jaws of the ancestors of monkeys, apes, and man. They show the typical tarsioid pattern in the V-shaped joining of the rami of the jaws, but the incisors are erect, not procumbent, and the formula is that of the catarrhines and man, i $\frac{}{2}$, c $\frac{}{1}$, p $\frac{}{2}$, m $\frac{}{3}$. One of these, *Propliopithecus* (Fig. 148 A), is so like the Upper Miocene and Pliocene *Pliopithecus*, and that in turn so like the modern gibbons, that there can be little doubt of its being ancestral to one or more lines of monkeys. The jaw is short, the canines are somewhat larger than the adjacent teeth, the first premolar is bicuspid, and the last molar has five tubercles. The fact that it is a shorter jaw than that of the gibbons, with less prominent canines and less sectorial premolars, shows that the face has become more elongate, more prognathous, during the time since the Oligocene. On another line *Propliopithecus* may have been ancestral to the Miocene *Dryopithecus* and its allies, possibly to the African chimpanzee and gorilla, and possibly also to man. The single little imperfect jaw carries a heavy responsibility. The other Oligocene jaw is better preserved than that of *Propliopithecus*, for it retains both rami and is almost complete. It would be an excellent connecting link between the Tarsioidea and the Anthropoidea if there were only something with which it could be connected. One eminent student of the primates speaks of it indifferently as a tarsioid and a catar-

THE ANCESTRY OF MAN 279

rhine. The teeth are much like those of monkeys and apes, except for the fact that the canine is not enlarged, and there is no suggestion of a diastema. Yet *Parapithecus* (Fig. 147 A) has been, and still is, considered by some as a connecting link between the Eocene lemuroids and tarsioids with large canines, and later monkeys and apes with even larger ones.

During Miocene times platyrrhine monkeys were present in South America, and catarrhine monkeys and tailless apes in Europe and southern Asia. Early in the Pleistocene, or before it, the higher anthropoids disappeared from Europe, although the monkeys remained in restricted areas, as they have to the present day at Gibraltar.

It is not till the Pleistocene that man definitely comes on the stage. The first discovery to attract attention was that of *Pithecanthropus erectus*, found by Dr.

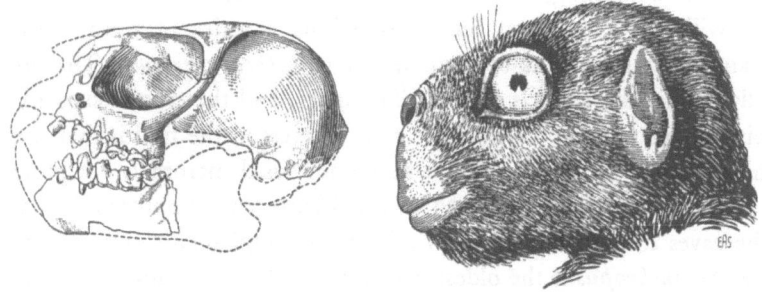

FIG. 146. At left, skull of *Tetonius*, the most advanced Eocene tarsioid, after W. D. Matthew. At right, a restoration of the head of the same, by Edward R. Schmitz.

Eugene Dubois near Trinil in Java in 1891–92. The material consists of the top of the skull, a femur, and three teeth. These bones are probably not those of a single individual, or even of one sort of primate. The femur belonged to a person having an erect posture and of average present-day size, but the skull has been shown by various tests to indicate a creature intermediate in cranial form and capacity between the higher apes and man. If the bones belonged together, this would indicate that in the evolution of man the erect posture and full height were attained before any great development of the skull, which is contrary to the teachings of embryology. The specimens were found associated with species of *Hippopotamus*, *Elephas*, and *Stegodon*, and are now believed to be of Mid-Pleistocene age.

The cranial capacity of any skull in cubic centimeters is readily obtained by the simple process of filling the brain cavity with shot and then measuring the amount required. The average for modern Europeans is about 1450 cubic centimeters, though in exceptional cases it is as high as 2000 and as low as 1200, seldom less than the latter figure in normal healthy individuals. The gorilla has the largest brain among the Simiidae, and although there is a good deal of variation the average is about 500 cc., or one-third that of the normal European. The maximum is 650 cc. The cranial capacity of the skull cap from Java is estimated at 940 cc., intermediate between that of the gorilla and the modern European. Some among the primitive bushmen of Australia, however, have been shown to have as low a rating as 900 cc.; *Pithecanthropus* is definitely on the human side of the line. Lately it has been suggested that "he" was a lady, and that a male of this species may have reached 1100 cc. The calvarium is the most important of the various fragments of *Pithecanthropus*, for it proves that this hominid had a small brain, a markedly retreating forehead, and prominent ridges over the eyes.

Sinanthropus pekingensis (Fig. 148 E) is the latest great addition to the human family. Filled fissures in Ordovician limestones at Chou Kou Tien, thirty-seven miles southwest of Peking, have yielded to the zealous labors of several geologists connected with the Geological Survey of China a series of teeth, well-preserved calvaria, and other fragments of men similar to *Pithecanthropus*. The brow ridges are equally strong, the vault is only a trifle higher, and the cranial capacity but little more, being about 1000 cc. Perhaps the chief interest in the find is that it establishes the Trinil race on a firm basis. Since numerous fossil mammals are present in the fissures, the date can be definitely fixed as early Pleistocene. Another important feature is that the caves contain numerous primitive flint implements and evidences of the use of fire. *Sinanthropus* is the oldest man yet found whose culture is known to have reached this relatively high stage.

Another discovery was made in 1907 in a gravel pit at Mauer, near Heidelberg, Germany, under a thickness of about seventy feet of sand and loess. The associated animals include the straight-tusked elephant, a rhinoceros, and a lion, all African animals, but with them were remains of bear, bison, horse, boar, and ox. This mixture of African and northern species indicates a warm climate and, according to the best evidence, suggests the age to be the second or third interglacial period. The fossil is a lower jaw, with all the teeth preserved (Fig. 147 C). The most striking features are the massive ascending ramus, with a shallow notch at the top, and a retreating chin, both simian characteristics. The teeth, however, tell a different story. They are too small for those of an ape with a jaw of such size, there is no diastema, and the canines do not project beyond the other teeth. The molars likewise show patterns characteristic of primitive human races. The fossil was first named *Homo heidelbergensis*, but it has since been made the type of another genus, *Paleanthropus*. There are no processes for the attachment of the lingual muscles, an indication of poor development of speech.

Next in order of age comes the Sussex man, but it is more convenient to pass on to the Neanderthal type, which is closely allied to the one just mentioned. Neanderthal man (Fig. 148 F) has been known for a long time, the first skull having been found in 1856 in a cavern on the side of the Neander valley near Düsseldorf, Germany. More recently, numerous other remains of this type have been discovered, some of them associated with other fossils. Two skeletons found at Spy, near Namur, Belgium, were in strata containing remains of the hairy mammoth, and such evidence, together with the fact that most of the human relics are found in caves and rock shelters, suggests that the race flourished during the cold period of the last glacial advance.

The abundant material permits an accurate reconstruction of *Homo neanderthalensis*, as he is technically called. Averaging only five feet (range 4'8"–5'3")

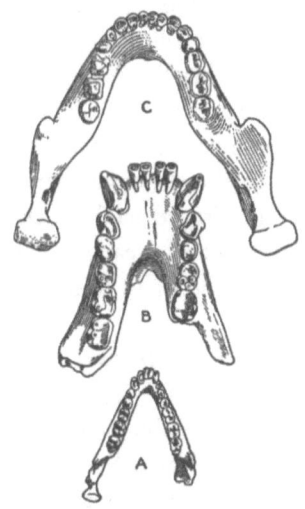

Fig. 147. Jaws of (A) *Parapithecus*, Lower Oligocene; (B) *Dryopithecus*, Upper Miocene; and (C) *Homo heidelbergensis*, Pleistocene; to illustrate the change from the Λ-shaped to the parallel-sided and thence to the Ω-shaped arrangement of the teeth. A, after W. K. Gregory, modified from M. von Schlosser; B, after W. Branco; C, after O. Schoetensack.

in height, he was not so tall as modern man. His posture was not absolutely erect, for the knee joint did not entirely straighten, and his enormous head was thrust forward. The trunk was short and thick; the arms were short, the hands large; the femur was curved, the lower leg short, and the foot clumsy. The skull is that of a brutish man, with heavy ridges over the eyes, a retreating forehead and chin, and large ocular and nasal cavities. The shape of the brain case indicates a high development of the posterior part of the cerebrum but less of that anterior portion which is supposed to be the region connected with the power of thought. In actual capacity the cranial cavity was larger than that of the average European, some skulls measuring 1600 cc.

This type of man appears to have existed for some thousands of years in Europe and other lands bordering the Mediterranean. Associated with his remains are found flint implements of the type known as the Mousterian, which indicate a stage of development that could have been reached only after long practice in the art of working flint. The study of implements has been actively pursued by archaeologists, particularly in France, Belgium, and England. As a result there has been built up a history of man based upon successive "cultures," as indicated by the flints found in those countries. Three broad classes are recognized: eoliths, either unchipped or with only casual working; paleoliths, flints chipped into shape but not polished; and neoliths, tools not only carefully chipped but polished as well. Those of Neanderthal man are of the second class. They are finished on one side only, the other showing the surface resulting from spalling off the flakes. Nevertheless, they are adapted for various uses and prove that the Neanderthals were accomplished in the chase and that they could create the necessary weapons: they are credited with making the first hafted implements. There are also other evidences that they had achieved considerable civilization. Many of the skeletons were given burial by their fellows, and one found at Le Moustier in France had the skull resting on a flint plate of careful workmanship. The interment of weapons with the bodies shows a certain respect for the dead and indicates a belief in some sort of after life, if not (as Lull has inferred) actual immortality.

Unlike all other fossil men is *Eoanthropus*, known from a fragmentary skull and the right half of a lower jaw (Fig. 148 C), with two teeth, the first and second molars, in place. The specimens were obtained by Mr. William Dawson from a small opening by the roadside at Piltdown, Sussex, England, and described by Sir Arthur Smith Woodward. It is difficult to determine their age, for fragments of mammals characteristic of the Pliocene and Pleistocene are mingled in the river-borne gravel. If contemporaneous with the most modern of them, Piltdown man was probably not more recent than the third interglacial stage, since *Hippopotamus* and other subtropical animals occur with it.

The skull is so fragmentary that those who have studied it have been unable to agree as to the proper reconstruction: estimates of its cranial capacity have varied from 1079 cc. to 1500 cc., and an intermediate figure of about 1300 cc. has finally been reached. It is not at all of the Neanderthal type, but has a high forehead like that of modern man. Aside from the fact that the bones are exceedingly thick, it is not peculiar. The jaw, however, is admitted by all to be more like that of a chimpanzee than like that of any man, living or extinct. This was recognized in the original description. The two teeth are like human molars, but the remainder of the jaw affords too much space to be filled by ordinary teeth. Hence, in his restoration of the anterior part, Smith Woodward made the canines large, like those of a chimpanzee, and allowed for a small diastema. The correctness of his view was demonstrated in a striking way the year after publication, when Dawson and Father Teilhard de Chardin, who were resifting the gravel at the spot where the jaw was found, found a large canine. It is twice as large as that of a man and almost exactly like that of a modern chimpanzee. This association seemed to many to be an unnatural one, so the jaw was attributed by some to a species of chimpanzee. The later finding of a few more fragments at a near-by site seems, however, to have convinced most of those

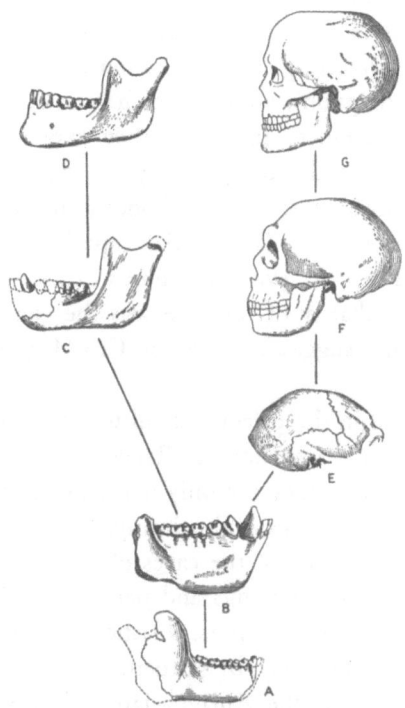

FIG. 148. Sketches illustrating two of the possible lines of descent from *Propliopithecus* to *Homo sapiens*. A, *Propliopithecus*, after W. K. Gregory. B, *Dryopithecus*, after W. Branco. C, *Eoanthropus*, after Sir Arthur Smith Woodward. D, *Homo sapiens*, after W. K. Gregory. E, skull of *Sinanthropus*, lacking face, after A. S. Romer. F, Neanderthal man, after M. Boule. G, *Homo sapiens*, Cro-Magnon race, after R. Verneau.

interested that skull and jaw belong together. *Eoanthropus dawsoni*, then, is to some people the missing link between man and the apes. The forehead is high, the brow ridge insignificant, and the brain large, all features of man, but the chinless jaw has the big canines of an ape.

Toward the end of the last epoch of glaciation, modern man, *Homo sapiens* (Fig. 148 G), came into Europe. Whence he came is not known. The general opinion is that his was one of the waves of immigrants which have followed one another from undiscovered centers in western Asia; a few students look upon Africa as the possible homeland. The species had already differentiated into more than one race; just how many is still a subject of discussion. It is not likely that all came together, for they occupied separated regions. Whether they came while the Neanderthals were still in possession and drove them out or whether they followed a line of least resistance and took possession of an area previously depopulated by "war, pestilence, and famine" is unknown.

All early representatives of our own species differ from the Neanderthals in having higher foreheads, projecting chins, and a more erect posture. The best-known are the tall, or true Cro-Magnons, whose chief habitations were in northern Spain and southern and central France. In stature they may have exceeded present-day Europeans, the average for males being nearly six feet. The forehead was high; brow ridges were practically absent; the thigh bones were straight and the lower limbs long, indicating swift-footedness. The cranial capacity was above the average of today, perhaps as high as 1800 cc., even the women exceeding the average of present-day males. The chief peculiarity of the race was the combination of broad cheek bones with a narrow skull, suggesting that the Cro-Magnons somewhat resembled the modern Eskimo.

These people were still in the paleolithic culture stage, for they used beautifully chipped but unpolished stone implements. They show great progress beyond the Neanderthals, however, in that they not only made use of bone, as well as flint, but carved it, and the carvings, drawings, and paintings ascribed to them are the wonders of archaeological discovery. Many of the caves of central France and the northern Pyrenees bear upon their walls drawings and paintings of hairy mammoths, steppe horses, bisons, reindeer, and other subjects. Plastic art is represented by carvings in bone and soft stone and, more rarely, by statuettes in clay. This art continued over three culture periods known as the Aurignacian, Solutrean, and Magdalenian. It had its beginnings in Aurignacian times near the close of the glacial period with figures incised on bone or on the walls of caves. These pictures are of contemporary animals, crudely drawn, the proportions poor, quadrupeds usually shown with two legs only. Considerable progress was made during Aurignacian times, but during the Solutrean art was more or less in abeyance. The Magdalenian period saw the culmination of prehistoric art, winning for the Cro-Magnons then living the distinction of being called the "Greeks of the Stone Age." Not only were animals depicted in their proper proportions, but lifelike poses were attained, and there was some success in composition. Colors were employed, and a few of the better-preserved paintings show four shades, obtained by the employment of red and yellow ochres and manganese with grease as a medium.

East of the domains of the tall Cro-Magnons, on the plains of Moravia, lived the Brünn or Predmost race. What they knew of art or culture we cannot tell, for they decorated no caves. Many of the remains so far found are burials, deep in the regional loess. One great pit has produced fourteen skeletons, with fragments of six others. Above and below them were fragments of almost a thousand mammoths, a circumstance which has led to the designation of this race as the "mammoth hunters." With the elephant bones were found implements of the "laurel leaf" type which show contemporaneity with the mid-age of the Cro-Magnons. Other stations have produced implements of a simpler type, Aurignacian or older. The Předmost people differed from the Cro-Magnons in being somewhat less tall, about 5 feet 6 inches to 5 feet 7 inches in mean height, in being heavier-boned, and in having a slightly smaller cranial capacity, only about 100 cc. more than that of modern Europeans. Their jaws were strong and somewhat prognathous, but the chin was prominent, not retreating. A neanderthaloid feature is seen in the brow ridges of the males. Some anthropologists believe that this indicates that they may have been connecting links between Neanderthals and sapient men, but similar ridges are shown by some modern Australian bushmen.

Another race, the Grimaldi, may have lived along the northern shores of the Mediterranean in early Aurignacian times. Much has been written about two unusually well-preserved skeletons, mother and son, found in the Grotte des Enfants on the Riviera in 1906. The specimens have certain negroid characteristics which have aroused great interest. Were they connecting links between the Negroes and Caucasians? Or were they wanderers who somehow crossed the Mare Interius and through some service became entitled to honorable burial? Sir Arthur Keith points out in a recent book that not only the Grimaldi but some of the Předmost people showed negroid features. Such seem, in fact, to be characteristic of the whole Cro-Magnon stock, using that term in a broad sense for all the peoples who employed the Aurignacian, Solutrean, and Magdalenian types of implements. Keith is inclined to think that this indicates that Cro-Magnons and negroids had the same ancestry.

With the first appearance of *Homo sapiens*, a nonspecialist must stop and let the anthropologist take over, even though the field of human paleontology is not necessarily so limited.

It is not possible in a chapter of this sort to describe all known human fossils, but a few others will be mentioned in the discussion which follows. We are interested chiefly in seeing what light the present meager evidence throws upon the ancestry of man. Although there are several possible interpretations, the current theories fall under two headings. One view is that man is closely allied to the Simiidae and became separated from that group relatively recently, perhaps during the late Miocene or Pliocene. This is the opinion championed particularly by Professor William K. Gregory, who reached it after an intensive study of the teeth of all modern and extinct primates. Another is that man came from the same ancestors as the anthropoids, the group including monkeys as well as the Simiidae. In concrete terms, this would mean a parting of the ways as early as the Eocene, the lemuroids of that age being the common ancestors. This is the view most acceptable to conservative people, especially those who do not want a monkey for an ancestor. Few scientists, except, perhaps, those in their later years, accept this view in its extreme form, but there are many who believe that the Hominidae and the Simiidae separated from their common ancestor in late Oligocene or early Miocene times.

That man early left the simian line is suggested by the large brain and, particularly, by the great toe, which has lost its primitive divergence from the others, become enlarged, elongated, and the principal bearer of the weight. Sir Ray Lankester showed that this feature indicated that man's ancestors must have left their arboreal homes and become fully terrestrial many eons ago.

Another fact which suggests an early divergence of the human and simian lines is that the young of modern apes are more manlike in appearance than the adults, whereas human babies are more apelike than fully grown individuals. In other words, human and simian infants are much alike but during the youthful stages continually diverge. Interpreted on the basis of the theory of recapitulation, this would mean descent from a common ancestry, not one from the other. Young apes have a big brain, a highly vaulted skull, a relatively high forehead, and no brow ridges. H. Klaatsch, and many others since, have inferred from this that the ancestral simian, although prognathous, was considerably more human-looking than the modern gorilla or chimpanzee. It is possible that from a common ancestor there were two lines of evolution: one in which the brain case became elongate, with heavy superciliary and in some cases sagittal crests, and in which prognathism increased rather than diminished; another in which the brain case increased in size laterally and upward, and the jaw retreated, producing a straight face. Since this idea involves the "biogenic" law, it has, naturally, been severely criticized. One critic has said that to suppose that the remote common ancestor of man and anthropoids had a vertical forehead without brow ridges is to invent an entirely hypothetical group; but in another place the same author has written of *Parapithecus*: "Although only the lower jaw is known, this highly important form must have had the shortened face and swollen brain-case, and probably the large eyes of the small insectivorous tarsioids." The deduction that the ancestor was a short-headed, large-brained, smooth-browed, weak-jawed creature seems to be sustained.

The chief reason for believing that the split did not come till after Lower Oligocene is that this was the time of the first appearance of primates with short heads, the dental formula of apes and men, and lower molars with five cusps. The indications are that *Parapithecus* and *Propliopithecus* sprang from some as yet unknown tarsioid stem, animals with large brains, large and forward-directed eyes, and small canines. There is no evidence of the intervention of a "monkey" stage, in the technical sense. *Parapithecus* or *Propliopithecus*, or both, are generally accepted as ancestors

THE ANCESTRY OF MAN

of both men and apes. It is unfortunate that we know so little about them. Several lines of descent from one or both of them are theoretically possible. One through the Pliocene *Pliopithecus* to the modern gibbons has not been questioned. A second through the Upper Miocene *Palaeosimia* to the Pliocene and recent orangoutangs is reasonable. A third through the Mid-Miocene *Dryopithecus* and Pliocene *Paleopithecus* to the gorilla and a fourth through Miocene and Pliocene species of *Dryopithecus* to the chimpanzee seem probable. A fifth, through some unknown species of *Dryopithecus* to *Homo sapiens*, has been strongly advocated. One route (5a) is through

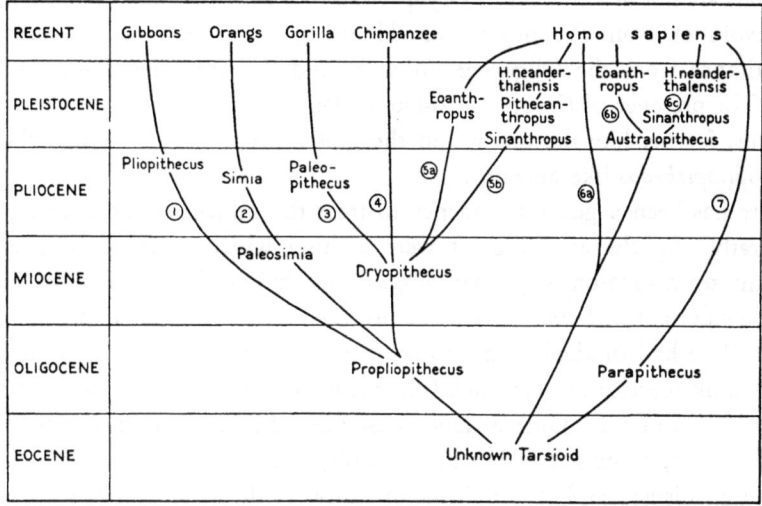

FIG. 149. Diagram showing some of the lines of descent of modern primates which are theoretically possible. It should not be interpreted as suggesting that *Homo sapiens* is polyphyletic.

Eoanthropus with enlarged canines and a high forehead, the other (5b) through his low-browed, small-canined cousins, *Sinanthropus, Pithecanthropus, Paleanthropus,* and *Homo neanderthalensis*. Opinions differ as to whether all these Pleistocene men are in the direct line or whether some or all of them represent lateral branches. A sixth possible line has three variants: (6a) from some tarsioid through unknown descendants with enlarged brains, smooth brows, and decreasing canines to *Homo sapiens*; (6b) the same, with enlargement of canines, to an *Australopithecus*-like ape stage, thence to *Eoanthropus*, and, by decrease in canines, to *Homo sapiens*; (6c) the same, with reduction of canines, through the superciliarily ridged Sinanthropus-Neanderthal line, to *Homo sapiens*; and (7) a line from *Parapithecus*, in which the canines are no larger than the adjacent teeth, to modern man, with no known connecting links. It is obvious that it is impossible to discuss these lines in any detail.

They are listed merely to show how wide a range of speculation the present inadequate evidence permits.

The chief objection to the derivation of man from *Dryopithecus* (line 5) lies in the fact that all known species of that genus — and they range from Mid-Miocene to Pliocene — have large canines, and there is no indication within the genus of any reduction (Fig. 147 B). In fact, the whole trend seems toward increase in size, as is shown by the two known descendants, the chimpanzee and the gorilla. Much has been made of the fact that man and all the post-Miocene simians show the *Dryopithecus* pattern in the molars. Yet it is generally admitted that the orangoutangs and the gibbons are not descendants of *Dryopithecus* but each the result of an independent line of evolution from *Propliopithecus*. Hence the "*Dryopithecus*" pattern is really an inheritance from the Oligocene ancestor, and the argument loses much of its weight. The presence of *Dryopithecus*-like molars in man is no proof of descent from that genus, for he, like the orang and the gibbon, may have inherited them from some *Propliopithecus*-like ancestor.

There has been a general tendency to treat the subject of the reduction of the canines rather lightly, as though it were a thing which might happen almost at will. Some see no reason why *Eoanthropus* (line 5a) of the Mid-Pleistocene should not have evolved into *Homo sapiens* before the end of the glacial period. Perhaps he might, if he had suddenly changed his disposition and gone on a diet of poached eggs and milk toast. But it is doubtful. Remember how long it took the horse to get rid of the extra toes. Once a trend is established and inheritance gets hold of it, a change to an opposite one is difficult. Atrophy can, so far as is known, come about only through disuse; and it is difficult to disuse teeth, as all know who have had occasion to "favor" some sensitive one. Still more difficult to accept is the opinion that canine reduction has passed through two cycles in the history of the primates. Paleocene and Eocene lemuroids with large canines led to Eocene tarsioids, as yet unknown, and they to the Lower Oligocene catarrhines. From its reduced condition in *Propliopithecus* and *Parapithecus* the canine was gradually built up to a fang again, the status it had in Eocene times. Then, after the Miocene, it dwindled away in the line which led to man. Although such a history is not impossible, it is improbable. It would be difficult to find any parallel in the animal kingdom.

The heavy brows of the Pithecanthropus-Neanderthal series (line 5b) have been the chief stumblingblock in the way of their acceptance as ancestors of modern man. It is difficult to imagine so radical a change as that which would be necessary to transform *Homo neanderthalensis* into *Homo sapiens*. Furthermore, there is little doubt that the two races were contemporaneous for a time at least. If *Homo sapiens* is a descendant of a heavy-browed race it must have been a pre-Neanderthal one. Perhaps it was *Paleanthropus*, as Gregory has suggested. It may be that Neanderthal man was derived from an ancestor less brutal-looking than himself, for both the

Galilee and the Ehringsdorf skulls are more highly vaulted than those of the typical members of their race. There is considerable evidence that Ehringsdorf man may have invaded Europe during the warm period which predated the last (Würm) glaciation, and that, although the Galilean neanderthaloids lived during glacial times, their home in Arabia may then have had a moist and comfortable climate. Perhaps the rigors of arctic conditions served to brutalize the later members of the race, though this seems hardly probable, for *Homo rhodesiensis*, who escaped to South Africa, has the strongest brow ridges of all.

On the other hand, some Neanderthal children (Gibraltar and La Quina finds) afford the same sort of evidence as the skulls of young simians do. Brow ridges are absent, and the forehead is moderately high. If this is an inherited and not an adaptive characteristic, it can be interpreted as pointing toward a smooth-browed ancestor for the neanderthaloids.

It is unfortunate that not a single skull or even a calvarium of a Tertiary simian has so far been found. Even *Dryopithecus* is known from jaws only. It is commonly thought that the Miocene and Pliocene members of this group must have had heavy brow ridges, because we naturally infer from the structure of the higher apes that brow ridges and enlarged canines go together, a Cuvierian deduction. But this is not true, for some of the larger-brained South American Cebidae have smooth foreheads and large canines, and the hominids just discussed have brow ridges and small canines.

There can be no question but that everyone in any way connected with the problems of human origin would have been much happier if Piltdown man, *Eoanthropus dawsoni*, had never been discovered. As has already been pointed out, there is small chance that he was ancestral to modern man; yet he has too many human characteristics to be excluded from the family Hominidae. If he is a descendant of *Dryopithecus*, as seems probable, and if *Homo sapiens* and the *Pithecanthropus-Neanderthal* clan are not, as also seems probable, then the family is polyphyletic, and a large-brained man has been evolved independently along two (or more) different lines. This would not disturb us if we were dealing with mere "animals," for no one objects to the derivation of the orangs, the gibbons, and the chimpanzee-gorilla group over three different routes. But the idea that big-brained man, supreme among all earthly beings, should have been evolved twice seems repugnant to scientist and layman alike. No wonder that American paleontologists objected to Smith Woodward's restoration, transferred the canine from the lower to the upper jaw, and made the teeth the type of a new species of chimpanzee!

The description by Professor Raymond Dart of what has come to be known as the Taungs skull is of interest in this connection. It was found in a filled cave near Taungs in Bechuanaland, South Africa, where it was associated with remains of other mammals, including three extinct species of baboons. The age of the deposit is unknown, but it was probably formed during the Pleistocene. When the matrix had been removed from the specimen, there remained a part of the skull retaining the endocranial cast, the face, and the more important parts of the jaws. The teeth showed it to be a young individual, and detailed study of the cast of the cranial cavity proves that it was a simian, allied to the gorilla and chimpanzee. But its brain capacity, 500 cc., is far above that of any of its relatives when at the same age. It also shows such human features as a high forehead, lack of orbital ridges, a parabolic dental arch, small canines, no diastema in the lower jaw, no simian shelf, and an advanced position of the *foramen magnum*, which suggests an upright position (either sitting or standing). As Sir Arthur Keith says, if this skull had been found in a Miocene deposit, it would have been seized upon as an excellent connecting link between simian and human lines. Since it is probably not older than Pleistocene it has to be interpreted as the young of some ape, allied to the gorilla, but with larger brain and other human attributes. Since this paragraph was written, Dr. Broom has found an adult in a near-by region. It is much like a chimpanzee, but the teeth have human characteristics.

To a certain extent, a dual ancestry of Pleistocene man brings us back to Klaatsch's now abandoned theory of two derivations, one from light-boned Asiatic ancestors, the other from heavy-boned Africans. The detailed phylogenies would, however, be entirely different. It is futile, in the present state of knowledge, to continue this discussion. It has been carried thus far only to disturb those who think that the problem of the ancestry of man has been solved.

If we recapitulate briefly what has already been said, it appears that the primates became differentiated from the insectivores in the Paleocene and that some among the late Eocene tarsioids showed a beginning of large cranial capacity. From such stock the Anthropoidea arose in Egypt during Oligocene times. The particular forms which are looked upon as ancestral (*Propliopithecus* and *Parapithecus*) are small, monkeylike creatures. From the first, primates were arboreal in their habits and naturally tended to become fruit and nut eaters, although the early forms were more or less omnivorous and, throughout their history, were ready to add insects, lizards, and other small animals to their diet. The Eocene and Oligocene were in Asia and Africa times of moist and equable climates, and forests were extensive. Under these favorable conditions, the primates increased in size, and, as is known from the considerable number of species found in the Miocene of the Siwalik hills in northern India, Asia became a region in which they were abundant. In that locality lived the various species of *Dryopithecus*, the ancestors of the chimpanzee and gorilla.

During the Miocene, the Himalayas and the mountains of southern Europe began to rise, obliterating the Asiatic part of the ancient Mediterranean; consequently the climate changed, and the present aridity of central Asia was gradually acquired.

Barrell has suggested that this would cause the thinning and final disappearance of the forests from the region north of the Himalayas and that, as a result, any pre-hominids in that region would gradually be forced from the ancestral arboreal life to existence on the ground, with the consequent change to an upright position, which arboreal life had made familiar. Increased aridity meant not only loss of trees but also a decreased supply of food, so that man could no longer depend largely on fruit. He became a hunter, and naturally increased in cunning and swiftness. The use of the teeth changed with the nature of the food. In the Neanderthals and their Heidelbergian ancestors we see the development of the "edge to edge bite," which would be well fitted for gnawing meat, and of enormous jaws with strength enough to crack bones as well. Further, as men on the ground came to use weapons rather than their teeth in fighting, so the canines, if originally large, may have dwindled away. Those anthropoids which were on the southern slopes of the mountains were not deprived of forest conditions. Consequently, they retained their ancient habits of gathering food and fighting with the teeth, and evolved in an entirely different direction from that of their cousins farther north. Fossils to support this theory of Barrell's have yet to be found. Nevertheless it may serve as a working hypothesis in the search for more evidence.

XXVIII

THE IMPORTANCE OF PLANTS

Who can see the green Earth any more,
As she was by the sources of Time?

Matthew Arnold, "The Future"

Was the earth green "by the sources of Time"? Neither geologists nor paleobotanists can answer the question. Probably the continents were dreary, unclothed wastes without visible inhabitants at least till late Cambrian times, possibly even later. All that has been learned from the study of their morphology, and particularly from their methods of fertilization, indicates that terrestrial plants, like terrestrial animals, had aquatic ancestors. This inference is borne out by the record, for few have been found in strata older than the Lower Devonian. It may be that the record is faulty, since no strata which can be proved to be non-marine have yet been found in formations deposited before late Silurian times. Curiously enough, the oldest supposedly estuarine or fresh-water beds contain no terrestrial plants. Woody tissue is characteristic of vegetation on land; until it was formed there was little chance for preservation of such organisms. Since cellulose-bearing fossils occur in no strata, either fresh-water or marine, before the Upper Silurian, it seems probable that there was little "green earth" before that time.

The known pre-Devonian floras are essentially aquatic. Only such simple forms as blue-green, green, red, and brown algae are actually represented by fossils, but the existence of bacteria and diatoms can be inferred from what appears to be compulsive evidence.

Bacteria are animal-like plants, or plantlike animals: their status in one's estimation depends upon one's point of view. They lack the chlorophyll of plants, and most of them feed upon organic matter; hence, by definition, they are animals. But some, despite their lack of green coloring matter, are able to abstract carbon and oxygen from carbon dioxide and its compounds, and so should be classed as plants. Some of them are probably but little changed from the hypothetical plant-animals which are supposed to have been the first living creatures on this planet. Simple as they are in structure, however, bacteria, like all other organisms, have changed (evolved) in the course of time. Many, unfortunately, have become parasitic, their activities resulting in diseases of various plants and animals. Since these affect the human individual directly, the group is more or less in disgrace, and few people seem to realize that the majority of bacteria are beneficent. They are the chief agents in the formation of soils, for without the acids and alkalies they produce the decomposition of rocks would be an exceedingly slow process. They alone among organisms have the power of taking nitrogen from its solution in water and building it into compounds available to plants. They are the agents of putrefaction and fermentation. Without them, agriculture would be impossible; in fact, one might go so far as to say that, were it not for them, terrestrial vegetation would never have become abundant. Sediments of pre-Cambrian age show the same physical characteristics as those of later times; hence it may be inferred that bacteria have existed since the earliest periods of earth history. The oldest actual records of their presence are furnished by certain late Devonian ostracoderm plates in which the loss of the original structure appears to be due to these organisms.

Closely allied to the bacteria are the unicellular blue-green algae. Their principal importance lies in the secretion of calcium carbonate, and their early history has already been discussed in the chapter on the Pre-Cambrian. Of the modern seaweeds the most conspicuous and most highly organized are the kelp, the brown algae. They are restricted almost entirely to marine situations, growing abundantly on rocky coasts. Like other algae, they have no roots or leaves, and take their food from the dissolved gases and chemical compounds in the medium in which they live. Probably they have been in existence since the Pre-Cambrian, but as they form neither woody tissue nor calcareous deposits they are ill adapted for preservation, and their history is obscure. More important from a geological standpoint are the green algae, forms in which the color of the green chlorophyll is not obscured by other pigments. They live in both salt and fresh water, are less specialized than their brown relatives, and have two claims upon our interest. Some secrete calcium carbonate, contributing greatly to the substance of coral reefs; others form the marl of inland lakes and bogs, a substance much used by farmers to "lime" their lands. Still others, more simple, were the probable ancestors of terrestrial plants. As for the red algae, everyone who visits the seashore knows the pink or white encrustations they form on shells and rocks. Their contribution to modern coral reefs is great; in fact, it is doubtful if, without the protecting and binding qualities of *Lithothamnion*, corals would be able to build reefs. Since these algae cause the deposition of both calcium and magnesium carbonates, they have also contributed greatly to the formation of limestone and dolomite.

Whatever their good qualities, algae are not important in supplying food. That rôle is now played by the diatoms, minute unicellular plants which are the fundamental source of food for marine animals. Although they are microscopic in size, even the national debt seems small as compared with the number present in a quart of sea water during the reproductive season. Innumerable billions of them are present at all times in the upper layers of the oceans, particularly in cold water, and during the so-called "flowering" periods, from March to September, they are so abundant that the water has been called a "vegetable soup." Fortunately they are present in fresh water as well as in the sea, for all small aquatic animals are primarily dependent upon them. With them in surface waters are protozoans and the larvae of all sorts of animals, both benthonic and nectonic, and it is the fate of these minute plants to serve as food for their associates. Since many of the tiny animals are in their turn eaten by larger ones, there are regular cycles in the building up of larger and larger creatures. Small planktonic crustaceans, copepods, eat the diatoms; small fishes live on the crustaceans; larger fishes eat the small ones. (Naturally, the larger the individuals, the fewer their numbers.) This cycle is but one of many. Although they are particularly abundant in the plankton, diatoms are also numerous on the floors of seas and lakes, and on the surfaces of aquatic plants, where mollusks, worms, and other benthonic animals find them.

Unfortunately, the geological record of this group is incomplete. There are doubtful records from the Paleozoic; not till the Jurassic are there deposits with identifiable specimens. But there are two good reasons why this should be so. In the first place, the plants are extremely minute, mensurable only in fractions of millimeters; in the second, their skeletons are exceedingly fragile. Their tests, unlike those of other plants, are siliceous and consist of two shallow "pans," one of which fits into the other like a pillbox into its lid. Even the lightest pressure reduces them to powder of unrecognizable components, so fine that some of the most delicate polishing media are made from diatomaceous deposits of Tertiary age.

Just as there are two explanations for the failure of these plants to be preserved as fossils in the older rocks, so there are two reasons for believing that they were in existence even in pre-Cambrian times. The more obvious is that there were pre-Cambrian animals, which must have had vegetable food. Calcareous algae were too well protected to furnish it in any quantity; brown algae were too large and tough, and too completely confined to the strand to be of any great use to the minute animals of the plankton. Hence, diatoms, or their naked ancestors, must have been present. Secondly, it will be remembered that our discussion of the conditions of early pre-Cambrian time led to the suggestion that the original oceans were somewhat acid, compelling the earliest animals to secrete chitinous or siliceous skeletons, if any. Perhaps the diatoms began to form hard parts at the same time as the siliceous sponges and radiolarians.

Mr. William C. Darrah has recently discovered spores of terrestrial plants in Swedish Upper Cambrian shales, but the nature of the vegetative shoots which bore them is as yet unknown. The oldest recognizable stems are found associated with graptolites in the Upper Silurian of Victoria, Australia. They are similar to forms which have long been known from Lower and Mid-Devonian strata in North America and Europe. All are small spore-bearers, and hence allied to the ferns, although they have a simpler foliage, leaves being absent or short and linear. The roots are poorly developed, but the underground stems are extensive (Fig. 150). Such plants as these were probably in existence as early as Ordovician times, for they fit in well as connecting links between aquatic and terrestrial types. The former needed no roots or leaves, for they could take water, oxygen, carbon dioxide, and other food directly from the medium in which they lived. The latter depend on their roots for water and foods dissolved in it, and upon their leaves for "breathing" and for the transformation of raw materials into food.

Professor Douglas Campbell has developed the most satisfactory theory of the evolution of terrestrial from marine plants. He thinks that the ancestors were green algae, for they alone lack other sorts of pigment, and that their first migration was from the sea into fresh water. At times of low water in rivers and pools, the small weeds came into contact with the mud. The first reaction probably was longitudinal growth; the second, the protrusion of delicate, hair-sized rootlets to keep connection with water beneath the surface. Finally, in response to the influence of light, upright shoots were formed. The production of the first land plants was so important an event that the processes involved seem wonderful and mysterious. Yet the results, and probably the processes, were less remarkable than those physiological-chemical reactions which take place in our gardens every season. The really extraordinary thing is that new terrestrial forms are not constantly being evolved from simple green algae.

The situation is parallel to that of the evolution of terrestrial animals. Probably plants, snails, insects, arachnids, and amphibians emerged from the aquatic environment at about the same time. Only in the case of the amphibians is the record sufficiently full to allow reasonable inferences to be drawn. In that case the result seems to have been due to the fact that certain geographical changes happened to coincide with a particular stage of the evolution of a particular group of fishes. Changing physical conditions repeat themselves at intervals of varying lengths, but evolution in organisms is a cumulative process; hence there can be no exact repetition. No matter how many times particular environmental conditions are repeated, there is only one time in the history of any group of organisms at which it can respond to external influences in one particular way.

The state of evolution of the late Silurian and early Devonian flora supports Campbell's theory. The objection has been raised that the nearly rootless condition of these early forms was a secondary xerophytic adaptation, for it may be that they lived in bogs. The criticism is probably just, but the inference therefrom is misleading. Bogs, seasonally dry, are ideal locations for transitions from semiaquatic to semiterrestrial life. Characteristics which were primitive in the early days may appear later as secondary features. Roots may come and roots may go. Moreover, at least one genus prominent at this time, *Psilophyton*, is so widespread geographically that it seems unlikely that it was confined to swamps.

The Mid-Devonian witnessed one of those sudden influxes of diversified plants which have led to the idea of "explosive" evolution. After countless ages terrestrial floras came suddenly upon the scene. There really is no great mystery here, merely lack of information. When more pre-Devonian terrestrial deposits are found, their fossils will reveal more about the evolution of plants.

Ferns have been in existence longer than any other plants familiar to us today. After their modest beginnings in the Mid-Devonian, their progress was rapid, reaching its culmination in the Carboniferous. Throughout the late Paleozoic and Mesozoic they formed the undergrowth in moist, shady places, just as their descendants do today. With favorable conditions tree ferns have from time to time risen above the general level, but they have never been dominant. Their beautiful foliage makes ferns popular favorites, but quite as fascinating is their Paleozoic habit of spreading by means of underground stems. To this, and to the fact that they seem never to have been particularly acceptable as food, they probably owe their survival.

The great trees of the late Paleozoic forests (Fig. 151) were also sporebearers, relatives of the modern ground pines and horsetail rushes, now comparatively insignificant plants. Most conspicuous were the lepidodendrons and sigillarias, with straight stems reaching heights of a hundred feet or more. *Lepidodendron* branched freely and had short, grasslike leaves and a trunk ornamented by diamond-shaped, diagonally-arranged leaf scars. *Sigillaria* had few branches, much longer leaves, and leaf scars in vertical rows. The fructifications of both included large, almost seed-like megaspores, more highly organized than those of their later relatives. Both trees spread laterally by means of bifurcating underground stems, commonly called *Stigmaria*. These served not only for the production of new trunks, as in the ferns, but also for the support of the trees. This fact and the relatively small size of the roots indicate that the soil in which they grew was swampy. *Calamites* was the late Paleozoic representative of the rushes. The vertical fluting of the trunk, its nodes, from which whorls of branches emerge, and its radially arranged leaves are so like those of modern forms that no one would fail to recognize its affinities. The leaves, known as *Annularia*, are longer and more graceful than those of living species, and the height attained by some, sixty or seventy feet, is not reached even by modern tropical representatives of the group.

Although spore-bearers dominated the late Paleozoic vegetation, seed plants appeared as early as the Mid-Devonian. The oldest are those with fernlike leaves, the pteridosperms (Fig. 152). Since they have the foliage of their putative ancestors, the spore-bearing ferns, they are believed to be the primitive members of their group. During the Carboniferous they were abundant, but it was their fate to disappear at the end of the Paleozoic. There is considerable evidence to show that they were ancestors of various other seed-producing groups. Some were vinelike in habit, others similar to tree ferns. It is probable that they produced many of the nutlike

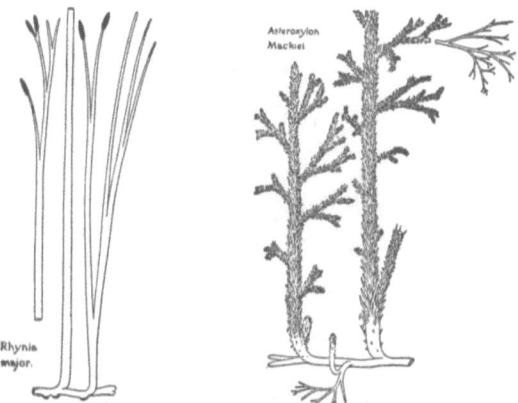

Fig. 150. At left, *Rhynia*, at right, *Asteroxylon*, primitive Mid-Devonian spore-bearers. Note the absence of leaves from the former, their primitive condition on the latter. From Kidston and Lang.

Fig. 151. Restoration of a Carboniferous forest, showing *Lepidodendron*, *Sigillaria*, ferns, and fernlike plants. Photograph by courtesy of the Field Museum of Natural History, Chicago.

FIG. 152. A pteridosperm, *Emplectopteris*, with seeds in place upon the pinnules. An original drawing, contributed by Mr. William C. Darrah.

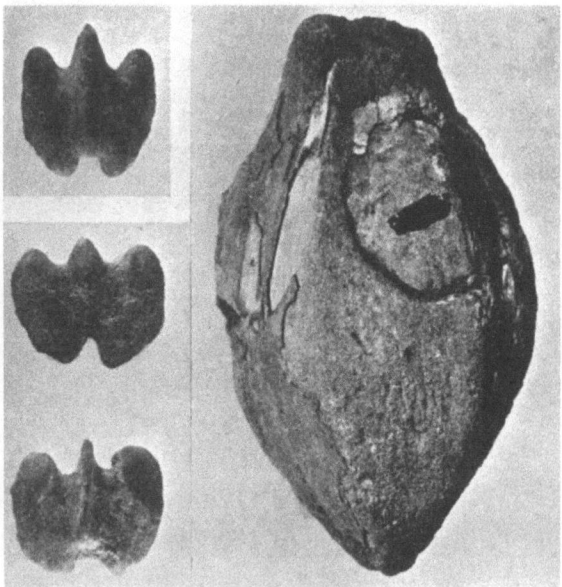

FIG. 153. At left, casts of three hickory-nut meats from the Oligocene of Nebraska. At right, a coconut from the Eocene of Belgium. Photograph of the latter through the courtesy of W. C. Darrah.

seeds so common in Carboniferous rocks, and for long a puzzle to paleobotanists.

True gymnosperms, with naked seeds and simple foliage, also came on the scene during the Devonian. The earliest, and the most abundant during the later part of the Paleozoic, had long, strap-shaped leaves with a parallel venation similar to that of the modern maize. Best known of these is *Cordaites*, commonly called the first "fruit tree," for its yew-like seeds appear to have been enclosed in a fleshy capsule. Short-leaved conifers, much like modern evergreens, did not appear until the Permian.

Paleozoic terrestrial floras were of geological importance in producing ancestors for later ones and in supplying materials for the formation of coal. But as a source of food they were negligible. Terrestrial animals starved in a land of plenty. Amphibians inherited their food habits, along with their teeth, from carnivorous fishes and transmitted both to the early reptiles. In Carboniferous times the food cycle, even of the air-breathers, remained much as it had been in the olden days. Small invertebrates and fish ate the unicellular plants in the rivers, small amphibians and reptiles ate the small invertebrates and fish, and larger fish and tetrapods fed upon all sorts of defenseless creatures. The minute plants in the water were still the primary source of food. A strictly carnivorous population, on land, has its limitations. So long as its primary source was in the fresh waters food was scarce, a fact which probably accounts for the small size of most Paleozoic amphibians and reptiles. A few giants were seven or eight feet long, but the average size was nearer seven or eight inches. All were semiaquatic, and dwelt beside the sluggish rivers on the coastal plains bordering the seas. Probably, in season, they came upon seeds or the tasty fruits of *Cordaites*, and they may have gulped some of them. Possibly they even cast speculative eyes on the source of supply. But they were not built for climbing; it was as much as they could do to drag themselves along on their bellies. They had no real taste for plant food. Millions of years were to pass, and a much more attractive table was to be set, before vertebrates really became vegetarians. Few reptiles and almost no amphibians ever became reconciled to a diet of vegetables. They would rather die than eat the stuff.

Nevertheless, the increasing amount of vegetation began to make its influence felt, although indirectly at first. Decaying plants furnished food for various terrestrial arthropods and their larvae — that is, for diplopods, centipedes, and many insects, all groups which first appeared or which first became abundant during the later part of the Carboniferous. Their presence probably first turned the attention of land animals away from the water, and grubbing for them may have provided the first vegetarian dinner.

The world-wide uplift of the continents at the end of the Paleozoic drained the Carboniferous swamps and extinguished the gigantic spore-bearers and the seed-bearing fernlike plants. Increase in the size of the lands gave new opportunities for the floras to expand, but under new conditions. Plants had to dig for water. It was a time of development of root systems rather than of upright shoots. In a sense, plants had to start over again. As a result the floras of the Triassic and Jurassic were not particularly luxuriant and contained few lofty trees. They have been described as "scrub." Tree ferns, small evergreens, monkey puzzles, ginkgoes, and cycadoids dominated the scene, with ferns and small lycopods in the undergrowth.

This was not only the real age of gymnosperms but the time when reptiles got their feet solidly planted on dry land. Much has been written about the amnion and the allantois, one essentially a water-cushion for the embyro, the other intimately connected with the shell of the egg, providing a means of respiration in the unhatched state. The presence or absence of such structures is said to be a fundamental difference between reptiles and amphibians. But was it in the early days? The geological record suggests that the early reptiles, like Amphibia, laid their eggs in water, and that somehow, as the lands became dryer, there gradually evolved in the reptiles a type of egg which could be hatched on land.

"Propinquity leads to love and marriage." Why not to changes in food habits? Vegetable food was plentiful in early Mesozoic times not only along the streams but on the uplands. Reptiles gradually acquired a taste for it, and throve; the old-fashioned stegocephalians stuck to the rivers, and perished. Unfortunately, vegetarianism did not help vegetarians much in Mesozoic times; they fattened themselves merely to fall prey to the carnivores. Not until the Cretaceous did the herbivorous dinosaurs become abundant. Their conspicuous evolution coincides with the sudden appearance in North America, Europe, and New Zealand of the first plants with conspicuous flowers, the angiosperms.

These plants, dominant today, made a spectacular entry upon the world stage as great trees rather than as modest herbs, and paleobotanists are still ransacking the earth in search of their ancestors. The oldest known are singularly like forms now living. Their wood, leaves, flowers, and fructifications are of modern type. The two great groups, monocotyledons, with parallel venation (palms), and dicotyledons, with reticulate venation (common hardwoods), had already been differentiated. At one time or another during the Cretaceous, forests much like modern ones were established. Many of the trees were sorts familiar today. Willows, birches, magnolias, tulips, sycamores, cottonwoods, sassafrasses, and viburnums were common, but more important sources of food were the figs, persimmons, breadfruits, oaks, beeches, walnuts, and palms. This rapid increase in the supply of seeds, fruits, nuts (Fig. 153), and edible leaves had a profound influence upon the evolution of arboreal reptiles, birds, and mammals, as has been related in other chapters.

The problem of the herbivorous dinosaurs requires a moment's further consideration. Their first abundance coincided with the arrival of the angiosperms. But is there really any connection between the two events? As has been pointed out, there was a noticeable lag in the adaptation of reptiles to vegetable food. It is probable that not till about the end of the Jurassic did they become used to feeding upon the various gymnosperms surrounding them. Once they were so accustomed, their opportunity had arrived, and they increased and multiplied. Their bills and cheek teeth were well fitted for plucking and slicing the coarse trunks and leaves of the cycadoids, plants which reached their maximum in the late Jurassic and early Cretaceous, and then gradually faded from the scene as the angiosperms replaced them in late Cretaceous and early Tertiary times. It may be that the herbivorous dinosaurs were so restricted in their diet that the coming of the flowering plants was no advantage to them. It is even possible that it caused their extinction.

The Paleocene flora much resembled that of the late Cretaceous, though it included such new arrivals as the maples, with winged seeds, and, more important, the grasses. The Oligocene and Miocene, however, witnessed geographic changes in various parts of the world. Gradual uplift brought the Great Plains of western North America to high altitudes, and concomitant erosion of the rejuvenated Rockies flooded them with sediments which raised their surfaces still further. The two processes united to destroy forests and to produce a vast area occupied by grasses and herbs. Still later, orogeny brought up the coast ranges which now catch most of the Pacific moisture on their western slopes. This reduced the high plains to the status of a semiarid region and gave great opportunity for the spread of grasses. The effect of this on the evolution of mammals was all-important.

The general uplift of continents which began in the Miocene culminated in the Pleistocene, a time when continental glaciation destroyed all vegetation over vast areas, later to be repopulated. Trees are not the first plants to occupy a previously devastated area. Ahead of them march a host of lowly angiosperms, grasses, and herbs, which can exist under unfavorable conditions. It is to them and their even more humble associates, bacteria, fungi, and ferns, that trees owe the preparation of the soils which permits them to advance.

It is evident, in short, from all that we have said, that nature, as a food provider, has been somewhat slothful. All animal life is dependent upon plant food, but till Mesozoic times few if any vertebrates ate terrestrial vegetation. Larger animals fed upon smaller ones in a descending series to the smallest, which, perforce, consumed unicellular plants, for there was nothing else they could ingest. Carnivores dominated the world. Direct use of vegetable food, to which must be ascribed the manifold changes that have taken place in birds and mammals, did not begin until the Mesozoic, and did not increase greatly until the late Cretaceous and Tertiary, when at last the angiosperms, the "life-givers," became the dominant plants. Humble servitors of the animal kingdom, they deserve an even deeper gratitude than that expressed in the admiration aroused by their beauty.

XXIX
RETROSPECT AND PROSPECT

The Present is the living sum-total of the whole Past.
Carlyle, "Characteristics"

Carlyle's words are those of a writer who dealt with a limited period, but they are as true for the paleontologist for whom history began two billion years ago as they are for the historian for whom it began seven or eight thousand years ago. The viewpoints of the neo-historian and the paleo-historian differ widely. Carlyle and his predecessors, contemporaries, and successors give the impression that history began when man learned to write and that it ceased when they themselves laid down their pens. They cannot realize that the years which have been the subject of their study are so short a part of past time that the geologist has no unit in his calendar small enough to record them. Three or four thousand years seem a long time to one who has trouble finding out what happened a century ago. The paleontologist, like the astronomer, after years of study eventually acquires a sense of time which cannot be transmitted to others by the mere telling. He is therefore less prone than the ordinary historian to stress the present, and less apt to be disturbed by those contemporaneous happenings which seem unfavorable to the progress of civilization. Millions, not hundreds or even thousands, of years must pass before the fate of a race is determined.

At the present moment man is but a few generations removed from the brute. That does not mean a few generations from the monkey, but a few from ancestors as tall, as big-brained, and as highly developed physically as ourselves. Self-preservation was not only the first law but practically the only law of man a few centuries ago, so few, in fact, that, geologically speaking, it was no time at all. Man is just beginning to learn from experience. Perhaps after a few hundred generations the accumulated knowledge of what has happened in the past may lead to a happier period for those who spend their brief lives on this planet.

Homo sapiens appears to have been one of the latest productions of nature, for there is no record of his presence till near the close of the Pleistocene, seven to twenty thousand years ago — a moment as compared with the supposed billion and a half years since the time when life may have appeared on the earth. Since man's span of existence probably will be comparable to that of other animals, he has a good "expectation of life."

If one looks at the history of other groups one finds that, in general, after an inconspicuous beginning each experienced a period of great differentiation and expansion, followed by a gradual or, more rarely, sudden decline in importance. The cystids furnish an excellent example. A narrow line would suffice to indicate the traces of their existence from the Lower Cambrian to the Lower Ordovician, the time when they suddenly became abundant; after the Mid-Ordovician they gradually declined in numbers till their extinction in the Mid-Devonian. Crinoids have the same history, although they first appeared in the Mid-Ordovician, reached their culmination in the early Mississippian, and have lingered on to the present. The old-fashioned tetracorals have a vague, almost traditional ancestry in the Mid-Cambrian, reached their high point in the Mid-Silurian and Mid-Devonian, and then faded from the scene in the Permian. Among the vertebrates the amphibians, after a feeble beginning in the Devonian, blossomed forth in the Permian, then declined to their present lowly status. Reduced to a diagram, the history of the reptiles shows the same figure, starting in the Pennsylvanian and culminating in the late Mesozoic, since when their variety has remained relatively small. The story of the mammals is similar, a poor start in the late Triassic leading to the extraordinary differentiation in the Tertiary, to be followed by the beginnings of restriction in late Pleistocene and Recent times.

These examples have been drawn from large groups, but the same "law" holds true in lesser ones, whether they be orders, families, or genera. To cite examples, however, it would be necessary to descend to technicalities that would interest the specialist only. There are, in fact, so many such histories that paleontologists have come to accept them as examples of the usual course of events. Hence one infers that, since man is in the initial stages of his phylogeny, the probabilities are that his line will continue far into the future.

As with all rules, there are exceptions, although not many. Those of one class are on the optimistic, those of the other on the pessimistic side, so far as predictions about the longevity of man are concerned. Examples of the former are the gastropods and pelecypods. Starting from scratch in the Lower Cambrian and Mid-Ordovician respectively, both groups have enjoyed continuing prosperity, reaching their greatest variety at the present day. It is possible that they have not yet attained the height of their evolution, even though man has eliminated many terrestrial forms in recent years. Insects show the same increasingly rapid upward swing; man's campaign against the more noxious of them has been only moderately successful. Such examples as these suggest the possibility of almost eternal life for man, but there are others which present a darker picture. One is that of the blastoids, whose brief story has already been recounted in the chapter dealing with the radiates. Their unimportant beginnings in the Ordovician and Silurian are comparable to those of the groups already discussed, but their time of extraordinary abundance in the late Mississippian was followed by almost complete extinction in the early Pennsylvanian. Several groups of crinoids ran against similar walls early in the Mississippian, and there are other orders or families — for example, the ammonites and the ceratopsian dinosaurs — which were extinguished at the height of their evolution. Man may have a similar fate, but the chances are against it.

One of the principal reasons for so thinking is that, although man likes to believe that he sits on the topmost branch of the family tree, he really does not. As has already been shown, he is rather a primitive mammal, and if he has any regard for the future he should be glad of it. Edward Drinker Cope long ago pointed out in his "doctrine of the unspecialized" that simple organisms appear to attain greater geological longevity than more elaborately constructed ones. Dr. Rudolf Ruedemann showed some years ago that forty-five Paleozoic genera have living species and that ten of them have survived since the Ordovician. All are invertebrates, and all are members of the more lowly classes: foraminifera, worms, brachiopods, ostracods, and the like. All have simple skeletons — so simple, in fact, that it is a question whether the genera can be positively identified. It is probable that if the soft parts were known it would be found that those of modern representatives of these long-lived genera were considerably different from those of their Paleozoic namesakes, and students of the particular groups to which they belong would be loath to unite the fossils with their supposed modern representatives. In fact, only two of the ten Ordovician "immortals" have well-supported claims to their titles. One is the geologically famous inarticulate brachiopod, *Lingula*. Fossils have been found which show the mold of the fleshy pedicle, and others which retain imprints of all the numerous muscles of the modern form. The other is the worm-tube, *Spirorbis*. Although there is less evidence in this case, the size, form, direction, and rate of coiling are so nearly identical in Ordovician and recent specimens that there is no way of distinguishing between them. Both these animals appear to have survived because they could live under a variety of conditions. All that they needed was plenty of water. Both are found in marine, brackish water and in what seem to be fresh-water deposits. Attached to floating objects, they drifted far out to sea, but they were equally successful in withstanding the attacks of the surf upon rocky headlands or the shifting sands of beaches. *Spirorbis* is just as common on the late Paleozoic plants of the coal swamps as it is on the brown seaweeds of the present day. *Lingula* probably got into fresh water less often but still lives in estuaries under almost incredibly adverse conditions.

The chief interest in the survival of such an animal as *Lingula* lies in the fact that it is one of the most primitive of all known brachiopods, and close to the central stock from which all the other members of that highly diversified phylum sprang. It has many descendants, some of them so different from itself that the connecting links have not yet been found. A second example of the survival of a primitive stock may be drawn from another family of brachiopods, an extraordinarily interesting group which has, for lack of space, been neglected in this book. The ones referred to are the spine-bearers, productids, most widely distributed and abundant animals of the second half of the Paleozoic. Their progenitor, *Chonetes*, was derived from an earlier spineless form in the late Ordovician. *Chonetes* has spines along the posterior margin of one of the two shells (valves) only. They appear to have served at first as organs of temporary fixation, but the sessile habit, thus initiated, was

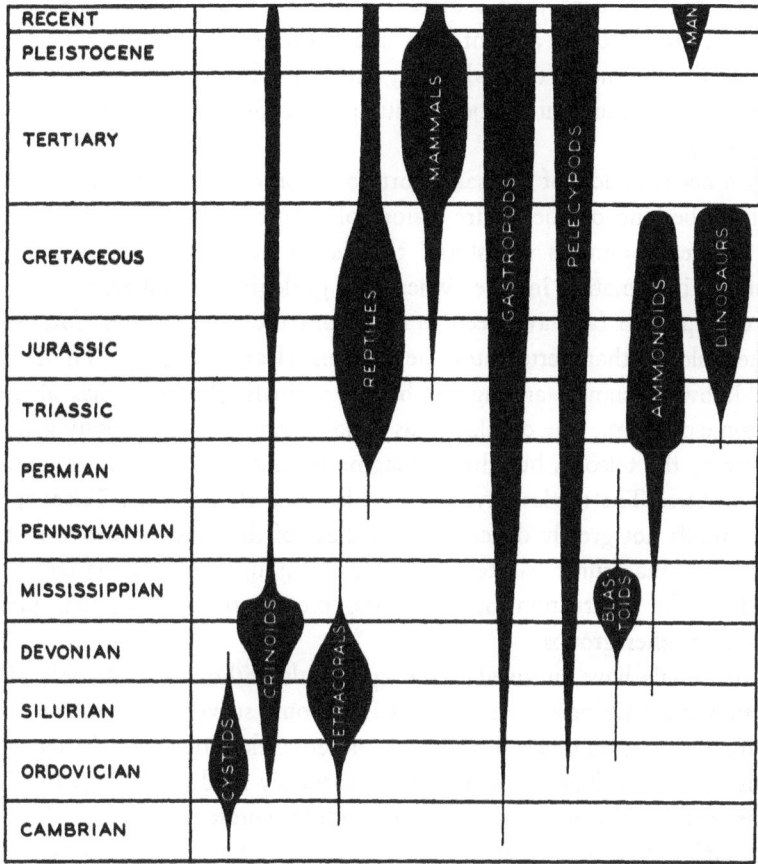

Fig. 154. Chart to show the times of culmination for certain groups, with a suggestion of their previous and subsequent history.

quickly accepted. The Devonian *Strophalosia*, for instance, soon became fully spinose, and lost its freedom and the original symmetry of the shell. A somewhat more conservative race, the productellas of the Devonian, were attached to foreign objects by spines which grew from the larger (convex) valve only. They in turn seem to have been the ancestors of the cosmopolitan *Productus*, with spines on both valves. Although firmly anchored in the adult state, and "degenerate" in that they had lost the pedicle and various other structures of a typical articulate brachiopod, the productids were the most successful members of their phylum. They produced the most numerous, the most varied, the most widespread, and the largest brachiopods of the late Paleozoic. It is probable that they gave rise to the Permian *Richthofenia,* an entirely sessile form attached by the apex of one shell only. Although *Chonetes* seems to have been the ancestor of all these spine-bearers, it survived as long as any of the descendants, longer indeed than some of them. Nor was its tribe one to be pitied as the feeble remnant of an ancient race. New species continued to spring from the central stock till the end of the Paleozoic, and individuals were just as numerous as the extraordinarily abundant Carboniferous productids, although less varied and of much smaller size.

Many other instances of the same sort could be cited. They are important not only for their bearing on the future history of man but also as justification for our belief in "contemporaneous ancestors," that is, in the survival of primitive types, which furnish information in cases where the geological record fails.

The examples so far have been drawn from the invertebrates, for, since they change more slowly than vertebrates, they present clearer evidence. There are, however, well-known instances among the higher animals. Attention has already been called to some of them. The cartilaginous ganoids gave rise to the bony ganoids, and they in turn to the teleosts, but the cartilaginous ones are still with us, although in reduced numbers. The turtles have changed but slightly since the Triassic; some of them have skulls not greatly different from those of the most primitive reptiles, the cotylosaurs. Old man turtle has kept paddling along, ignoring the changes in fashion which led to the destruction of dinosaurs, pterosaurs, ichthyosaurs, plesiosaurs, mosasaurs, and other groups.

Few mammals have survived for long periods without considerable alteration. The opossum is a striking example, but it is the only surviving primitive marsupial. Some of the insectivores retain a primitive placental dentition, which harks back to Cretaceous times, but their limbs have been variously modified. Dogs represent the central line of the carnivores, although their modern representatives exceed the Eocene stock in complexity of brain and in size. Tapirs are often referred to as "living fossils." In some respects they are like the Eocene ancestors of various odd- and even-toed vegetarians. Nevertheless, although they retain some primitive characteristics, they can be called primitive only by comparison with such highly specialized relatives as the modern horses, rhinoceroses, and camels.

Relict animals or plants are those which retain primitive characteristics despite the vicissitudes of the ages. First on the list are *Lingula, Spirorbis,* and the like, animals seemingly able to withstand all sorts of conditions within their own milieu, in this case the aquatic one. Turtles have done a little better, for they produced the terrestrial tortoises. Omnivorous mammals, such as the opossums and insectivores,

RETROSPECT AND PROSPECT

belong in this group. Tapirs typify a second series, composed of organisms which have survived only because of an ignominious retreat to that part of the world where life is easiest, because food is most abundant. The arctic, sub-arctic, and cold-temperate zones of the present day contain no relicts of this type. Only vital groups can live there. It is in the warmer belt alone that such out-of-date animals as apes, monkeys, elephants, rhinoceroses, tapirs, lions, tigers, and the like survive. Man is not a relict, but an active, primitive creature. Like *Lingula*, he is able to live under all sorts of conditions.

In this discussion, emphasis has been placed upon conservative tendencies in evolutionary processes. On the other hand, evolution has been described, as indicated by the real meaning of the word, as a process of unfolding. The fact that primitive creatures do survive shows that, if there be an inherent tendency toward differentiation, it can be inhibited, but the "unfolding" idea is to some extent supported by the facts of geological history. Each group shows a multiplicity of forms. There are, or have been, crawling reptiles, walking reptiles, flying reptiles, gliding reptiles, and swimming reptiles. Mammals show the same differentiation into forms which, externally at least, closely resemble the reptiles occupying corresponding habitats. In fact, animals of all sorts have a tendency to occupy every zone and habit of life. This is what H. F. Osborn called the "Law of Adaptive Radiation." All animals and plants show it, the higher groups more fully than the lower. The sponges, for example, are mostly sessile; a few float, but none crawls or swims, and all have the same method of feeding. The coelenterates have been somewhat more successful, for, although dominantly sessile, some float, others swim, and at least two of them crawl. Fish have done still better, for they swim, float, crawl, jump, and almost fly. Their food habits are more diversified than those of other aquatic animals. Most are carnivorous, but many are herbivorous. The carnivores differ greatly in their diets. Some have piercing, some grasping, some cutting, and others crushing teeth. Terrestrial animals have greater opportunities than aquatic ones, but the earliest of them, the amphibians, show less diversification than the fishes. The first invertebrates to get onto the land, the diplopods, likewise failed to live up to their possibilities, but the success of their cousins, the insects, is well known. No greater radiation than theirs is imaginable; only at sea have they been dilatory.

Radial evolution does seem to be a sort of unfolding, but not in the sense of a direct upward trend. In fact, much of the movement has been lateral or backward, so far as progress toward higher animals is concerned. Reptiles did not evolve to their maximum and then produce mammals (Fig. 155). Mammal-like reptiles were a side line, doomed to expire before the end of the Triassic; even they did not culminate when producing mammals. Their "great men" were big, foolish-looking creatures which managed to thrive in Africa, Europe, Asia, North and South America till the dinosaurs usurped their position at the end of the Triassic. Like the Israelites in their journey, the South American representatives of the group got across the Red

Sea of North America without many casualties, whereas Pharaoh's pursuing army of dinosaurs was properly bogged down. Small, inconspicuous, and lowly were the members of the therapsids which gave rise to the mammals. We are not yet sure that we have identified them.

The culmination of the reptiles was not in the mammals but in the snakes, the most specialized of all. As has been pointed out, the whales are the highest of the mammals, and the birds are at the summit of the whole animal kingdom — not man.

A vase is begun; why, as the wheel goes around, does it turn out a pitcher?

Horace grasped, to a certain extent, the seeming casualness of evolution. If there has been any purposefulness, any inherent direction, it is obscured by radiative adaptation. If nature tried to produce in man the perfect vase, all she achieved was a pitcher.

Most chroniclers of the history of life begin or end with a "family tree," with roots in the protozoans, a trunk of invertebrates, and a series of modern animals ornamenting the branches. I have only an elementary knowledge of botany, but I doubt if there ever was such a plant. A truer idea of the paths of evolutionary change may perhaps be gained from the accompanying diagram (Fig. 156), in which is expressed the idea that although on the whole there has been progress in an upward direction, it has not been straight forward but is the result of the response of animals to all conditions of environment. In other words, progress has been the more or less fortuitous result of radiative adaptation.

Men like to think that the evolution of the earth and its inhabitants was purposeful, which probably accounts for the popularity of the theory of orthogenesis, or "straight-line" evolution. In dealing with the histories of the horse and the camel we followed to the present day lines of descent which showed persistent, although not continuous, change in the direction of lengthening of the legs and feet, reduction in the number of toes, and increasing complexity of teeth. On the other hand, the central, direct line of the rhinoceroses perished in the Pliocene; it did not lead to the modern forms, which must have been derived from some lateral branch whose family records have not yet been discovered. And, as was intimated in the same chapter, there were various other lineages of horses and camels than the ones described. That is, evolution was radial within each of these groups. As we look still further back, we find that the stock from which the horse arose also gave rise, on the one hand, to the tapirs, the rhinoceroses, and the titanotheres, and, on the other, to various lines of even-toed mammals. Radiation was superimposed upon radiation. In other words, an orthogenetic line is one that is man-picked. Every living individual has an ancestry extending straight from himself to the first living protoplasmic particle. Each living creature, coral, butterfly, fern, king, or criminal, is the culmination of an orthogenetic series. Whether a man be a king or a criminal (there have been plenty of instances where the two were combined in one person) depends upon two things, heredity and environment. It is still a question which is the more important. The paleontologist, partly from training, partly from his knowledge of ancient history, is apt to stress the latter.

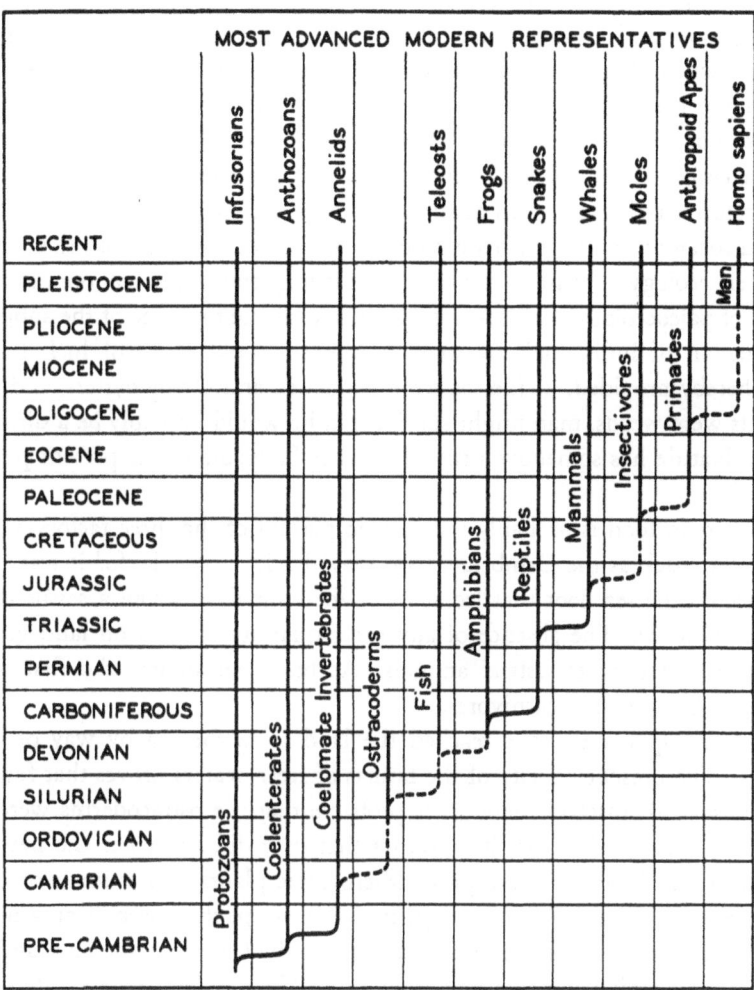

Fig. 155. A diagram showing the geological ranges of certain groups, to emphasize the fact that each line has evolved in its own way, and not in the direction of the next more highly organized group. As indicated, higher groups arise at the beginning, not late in the history of any phylum. The term Coelomate Invertebrates as used here refers to the primitive ones only. The chart was not intended to show all the lines of invertebrates.

The times of greatest change in plants and animals — that is, of most rapid evolution — have coincided with the great physical events in earth history, those which have drastically altered the distribution of land and sea, raised mountains, or partially submerged continents. Quiet intervals have in general been periods of relative stagnation for organisms. Not that change ever absolutely ceases, for, no matter how uniform the conditions under which animals and plants live, the "struggle for existence" is perpetual. The three important factors of environment are geographic conditions, food, and ecological relationships. The most striking changes in organisms occur when there are marked alterations of the first, for the others are dependent upon them. But even though there be no earth movements, physical conditions are not always the same. There are periods of heat and cold, moisture and drought, changes in the salinity of the seas, in position of oceanic currents, et cetera. Even extraterrestrial forces, such as sun spots, have their influence.

The associations of organisms with each other in connection with their habitat — that is, their ecology — must have been of great importance in evolution, but so little is known of paleoecology that it is impossible to evaluate that side of the subject. The overproduction of individuals by some groups is merely a bit of luck for others. Food is thereby provided. If the sea were populated entirely by oysters, and the only land plants were weeds, man might survive, but he would probably be a discontented mammal. Nature has somehow struck a balance favorable to the present proprietors of the earth.

For animals, food is probably the most important of the three factors mentioned above. As far back as the Cambrian, food played a leading rôle, for it was evidently its presence on the sea floor that led some of the animals to adopt a sessile or motile benthonic existence. The absence of any great quantity of food on land greatly delayed the evolution of terrestrial animals. As has been shown, the first tetrapods were not herbivores, but carnivores.

Organisms appear to have reached the land from the sea by way of estuaries and rivers. The present paucity of the faunas of fresh waters proves that few groups have established themselves there. It is probable that the ostracoderms were among the first, and their jawless condition shows that they did not enter the new habitat in pursuit of any large or active prey. Microscopic plants and animals or decaying green algae probably satisfied their appetites. All except the anaspids appear to have been mud-grubbers. The earliest gnathostomes are the Lower Devonian acanthodians, and the Mid-Devonian ganoids and lung fish. Their oldest representatives are so highly organized that it seems probable that these groups originated during Silurian times. Perhaps the first real jaws were those which chewed graptolites and soft-shelled trilobites. The Paleozoic amphibians were carnivores, dependent chiefly upon such food as they could obtain from the waters of streams, lakes, and swamps. Their somewhat more terrestrial offspring, the primitive reptiles, had much the same habits, although there were among them a few which seem to have been satisfied with a diet of insects, insect larvae, worms, and other inhabitants of the land. Apparently the higher animals were reluctant to try a vegetable diet until necessity forced it upon them in Mesozoic and Tertiary times. So long as vertebrates lived in the water, the scheme seems to have been to expect the more lowly invertebrates to feed upon simple aquatic plants, for fish to eat the invertebrates, and for the lordly terrestrial vertebrates to devour the fish, or each other. Until recent times "Devil take the hindermost" was the general custom. No wonder that this rule dominates so large a proportion of human beings.

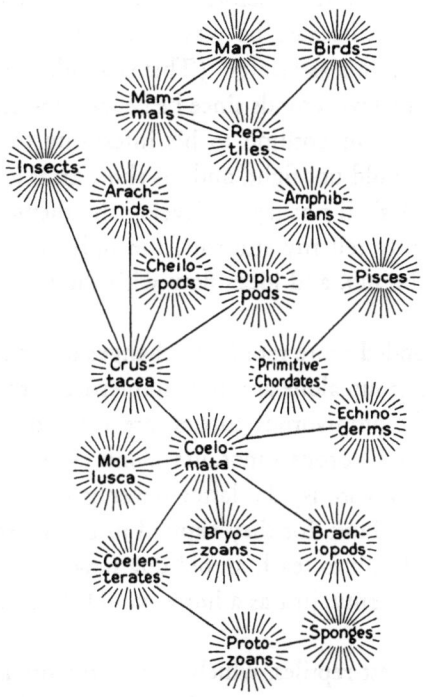

Fig. 156. A diagram showing the probable relationships of the important groups of animals, in the progress upward from the Protozoa. The radial lines about the names indicate that in each group there has been evolution in all possible directions, but only a few lines have led to progress toward a higher group. Coelomata should be read as "Primitive Coelomata." This an attempt to replace the familiar but misleading "family tree."

The beginnings of vegetarianism among the higher animals must have long antedated the time at which true vegetarians appear in the geological record. It seems probable that the sequence was in each case from carnivore through omnivore to herbivore. This was the history of the herbivorous mammals, and it was probably that of the other groups. As has already been noted, certain Pennsylvanian gymnosperms, *Cordaites,* produced tempting fruit, and Triassic and Jurassic members of the same group bore seeds and fruits. These inducements led various animals to essay arboreal life, which culminated in the evolution of the birds, pterosaurs, early arboreal insectivores, and bats. But the climbers carried with them into the trees a taste for flesh, and their new habitat provided abundant food of this kind, larval and adult insects, eggs and youthful progeny of birds, and various other odds and ends. Hence most arboreal creatures are naturally more or less omnivorous. In general, arboreal life and its resultant diet has led to loss of teeth. For some reason this has affected mammals less than birds or reptiles. Fortunately man got down, or was driven down, from the trees before he became a monkey — "driven down," not by climatic changes, but because he was a more primitive, less competent, and less pugnacious animal than his associates. (This is an idea new to the writer, having occurred to him during the few seconds since he began to write the previous sentence. Perhaps man owes his present position to his inferiority to the other anthropoids: without large canines he could not fight, and his short arms and fingers made him an unsuccessful brachiator. I shall not try to develop the theme, but merely list it as a possibility. As I frequently tell my students, the only way to orient oneself is by writing. Merely guide the nib, and ideas flow with the ink — not that the ideas are all good.)

The mammals responded more rapidly to changes in the supply of food than any other vertebrates; in this group alone are there numerous herbivores. Among modern reptiles all, except for the omnivorous turtles, are carnivores. Among extinct ones the sauropods and the herbivorous dinosaurs alone essayed a different food. The former were probably omnivorous; the latter never got beyond the browsing stage. This was not entirely their fault, for they expired just as grasses came into the fields. It is well said that it is difficult to teach an old dog new tricks. The ceratopsians may have died in the midst of plenty, just as a horse would if turned loose in a coniferous forest.

In contrast to the earliest reptiles, which were carnivorous, those early mammals which left descendants were omnivorous. They merely followed the paths of least resistance when they accepted whatever food nature offered them. It can hardly be said that certain mammals were predestined to certain diets, and that they followed routes of evolution already picked out for them. The evidence seems to be clear that the habits of the animals changed with the ever-varying nature of the food. The placental mammals were a new and plastic line in late Cretaceous and early Tertiary times. Their habits had not been fixed, as were those of the reptiles at the dawn of the Mesozoic. The group was still in its infancy, still capable of variation and adaptation. Man has inherited this primitive condition; all is grist which comes to his mill. He thrives, despite fads or fancies.

The geological record seems to show that the progress of evolution has been governed largely by the physical environment. From the beginning, life has depended on external circumstances and more or less accidental coincidences. As Lawrence Henderson has shown in his book, *The Fitness of the Environment*, life as we know it could not exist but for the fact that the most abundant liquid on the earth is water, the dominant gas, oxygen, diluted in proper proportion by inert nitrogen, and that carbon dioxide is present.

Nothing is really known about the origin of life, but it seems probable that its inception was the result of a huge chemical experiment, during which conditions were such as had never previously obtained on this earth and will never be repeated. Whether one believes in a hot nebular or a cold planetesimal origin for the earth, there must have been a period when the outer zone was molten. During the period of cooling, chemical combinations may have been brought about which cannot be duplicated on the small scale of the modern experimental laboratory. Once life came into existence, it might have gone on forever in lowly forms had it not been for the constant driving of external forces.

Evolutionary change seems to have been slow while animals remained in the water, a more or less static environment. As has been pointed out, it appears to have been speeded up when organisms reached the land, and to have proceeded most rapidly during the Tertiary. Nevertheless, there has been no definite rate of acceleration. Loomis tells us that the change from *Eohippus* to *Orohippus*, involving only the addition of a cusp on the third premolar, is spread over a period of two or three million years. The loss of the fourth digit of the front foot of the horses was accomplished in about ten million years. Yet *Merychippus* arose from *Parahippus*, and in its turn gave rise to *Hipparion, Pliohippus,* and *Protohippus* within a brief period in the early Miocene. The rate varies with external circumstances.

One of the most striking features of evolution from the end of the Mesozoic onward has been the general increase in the relative size of the brain of all vertebrates which have survived. Education appears here, either as cause or effect; it is probable that it was the former.

During the Paleozoic, when fertilization was largely external, the mother laid her eggs in the water or on the sand, and, with that, responsibility ended. Whether impregnation occurred before or after the extrusion of the eggs, the father was not interested in the outcome. In the course of time, however, some animals began to protect their eggs, and there was gradually evolved the habit of retaining them within the body of the mother until the time of hatching. Thus arose what is called the mother instinct. Just when this began is not known. Probably not before Mesozoic times. (Only one of the many trillions of trillions of eggs laid during the Paleozoic is now known. It occupies an inconspicuous position in the Museum of Comparative Zoölogy at Harvard. Perhaps it is the only one which failed to hatch. It had lain buried in the muds of the Texan Permian till Llewellyn Price routed it out.) The mothers among the higher reptiles, birds, and mammals feed and train their children during the early days, weeks, or months of their lives, giving them a decided advantage over the hosts of more primitive creatures left by their parents to learn by trial and error, if at all. As Professor Hervey W. Shimer has pointed out, such education led to the keeping together of the family and so to the formation of societies, with social (communal) behavior.

What does the history of life teach? Simply this: that man is a part of nature, physically governed by forces which are as yet only partially within his control. Real progress will be made only when facts are faced. Man is an animal, and a badly assembled one at that. Nevertheless he differs from others in that he is articulate and can be made literate. He alone is conscious of his position in the world, and that largely because of written records. Cats and dogs and birds and innumerable other animals teach their young enough to enable them to get along more or less successfully, but they lack foresight because they have but a limited hindsight. Man will, like the dinosaurs, follow a royal road to destruction unless he heeds the warnings of past history. Just at present he appears to be heading toward another dark period, like that of the Middle Ages. Despotism seems to be coming back into fashion. Let us hope it is nothing more than a threat, for with it would come all the evils of the feudal system. Like any other animal, man prefers to be ruled, bossed, and herded, because it saves him the trouble of thinking. Thousands of years may pass before he learns to use his single unique organ, the specialized brain. But I do not wish to be unduly pessimistic. There is still abundant time to change. At the worst, we know that dark ages have come and gone. Men have built new structures on the ruins of the past. Records of some sort survive, and much is learned from them. There will come a time when there will be only one language, when narrow patriotisms born of isolation will be forgotten, and when increase in population will be adjusted to food supply. Will that be the millennium? Maybe, but only if man lives up to his possibilities. I'll never know, nor will you. Let us hope that man will learn to direct his future by his knowledge of the past.

INDEX

INDEX

INDEX

Abbott, G., 31
Abel, Othenio, 138, 169, 191
Acanthodes, 92
Acanthodians, 89, 91–93, 106, Figs. 49, 52
Aceratheres, 257, Fig. 135
Adapis, 277
Adelospondyli, 112, 115
Aepyornis, 183, Fig. 92
Aglaspis, 55, 56, 58
Aigialosaurs, 159
Air-breathing, initiated, in the arthropods, 60, 103; in the vertebrates, 99, 103
Aistopoda, 114, Fig. 63
Alder flies, 204
Algae, 25; blue-green, 29, 30, 32, 293; brown, 32, 293; green, 293; red, 25, 32, 293
Alligators, 125, 129, 144, 164
Allosaurus, 132, 141
Allotheria, described, 213, 214, 222, 223, Fig. 108
Amblypods, 223, 225, 226, Fig. 116
Amebelodon, Figs. 141, 144
Ammonoids, described, 194–199, Figs. 98–101; range, 302, Fig. 154
Ammonites, *see* Ammonoids
Amoeba, 5
Amphibians, 22; characteristics, 108, 109; crossopterygian characteristics of, 103, 104; described, 108–118; footprints, 109; metamorphosis, 108; oldest, 109; origin, 97–106; range, 301
Amphinome, Fig. 9
Amphioxus, 84, 85, 95; theory of origin of vertebrates, 84
Amphitherium, 216
Anapsids, 125, Fig. 66A
Anaspids, described, 80, 95, Figs. 45, 49; theory of origin of vertebrates, 88–93, 95
Anchisaurus, 131
Ancyloceracones, 198, Fig. 100
Andrews, Roy Chapman, 142, 217
Andrewsarchus, 229
Andrias scheuchzeri, 116
Angiosperms, 298, 299
Animals and plants, differences between, 4

Annelids, 32, Figs. 7, 9; larva, 86, Fig. 48A; theory of origin of vertebrates, 86, 95
Annularia, 296
Annulata, 20, 24, 28, Figs. 7, 9
Anteaters, 266
Antelopes, 251, 252
Anthropoids, 276, 278
Antiarcha, 87, 88, 106; described, 80, 81, Fig. 44
Antilocapridae, 252
Ants, 208
Anura, 108, 116, 118
Apterygota, 202
Apteryx, 183
Arachnids, 22; described, 55–61
Archaeocyathinids, 24, 36, Fig. 8
Archaeopteryx, 173, 176, 183, 184, 188, Figs. 91, 93, 95
Archaeornis, 174, 176, 183, 184, 186–188, Fig. 88
Archelon, 160, 161, Fig. 84
Archidiskodon, Fig. 141
Arctodus, 232
Arctotherium, 232
Armadillos, 226; described, 266–268
Arthrodires, 88; described, 81–83, Figs. 46, 47
Arthropods, 21, 23, 27, 28, 33, 47, 55; theories of origin of vertebrates, 87, 88
Artiodactyls, 226, 239; described, 243–252, Figs. 123A,D; evolutionary trends of, 243, 244
Asteroxylon, Fig. 150
Astrapotheria, 223
Atikokania, 31
Auchenaspis, Fig. 45
Australopithecus, 287
Aves, *see* Birds
Axonolipa, 43, 45, Figs. 15, 17, 19C,D,E
Axonophora, 43, 45, 46, Figs. 18, 19F,G

Bacteria, 25, 30; importance, 292, 293
Bactriticones, 198, Fig. 100
Baculiticones, 198, Fig. 100
Balanoglossus, 84–86, Fig. 48; larva, 85, 86, 94, Fig. 48B
Baluchitherium, 243, 258, Fig. 137

Bark lice, 205
Barnacles, 21, 40
Barrande, Joachim, 22, 53
Barrell, Joseph, 101–103, 291
Barylambda, Fig. 116
Bather, Francis A., 72
Bats, 223, 224
Bears, 229, 231, 232, Fig. 119
Beebe, William, 184
Beecher, Charles E., 123, 127
Bees, 208
Beetles, 205, 207
Belemnites, 190, 191, Figs. 96, 97
Beltina danai, 31, 35
Birds, 22; ancestors, 183–188; characteristics, 174, 175; rare as fossils, 174; reptilian characteristics of, 173, 177; toothed, 14, 173–180; flight, origin of: cursorial theory, 183, 184; gliding theory, 185; tetrapteryx theory, 184, 185
Bison, 265
Blastoids, 20, 40, 62, 70, Fig. 37; range, 301, Fig. 154
Blastomeryx, 250
Blattaria, 202
Bothriolepis, 80, 81, 87, 88, Fig. 44
Bovidae, 252
Brachiopods, 21–23, 27, 28, 33, 36, 42; articulate, 21, Fig. 12A,B; inarticulate, 21, Fig. 11
Brachiosaurs, 136
Brain, regions of, 75
Branchiosaurs, 110, 112, 113, Figs. 57, 59, 61, 62; metamorphosis, 113
Brittle stars, 20, 68, Fig. 36
Brontosaurus, 18, 132, 148; described, 135, 136
Brooks, W. K., 34, 35
Broom, Robert, 98, 102, 103, 145, 187, 211, 212, 290
Brown, Barnum, 138
Brues, C. T., 208
Bryant, W. L., 77, 98, 105
Bryozoans, 20, 22, 23, 27, 40, Figs. 9, 12
Bugs, 207
Bulman, O. M. B., 42
Bunodonts, described, 244–246
Butterflies, 207

Cacops, 112, Fig. 60
Caddis flies, 207
Calamites, 296
Camels, 227; evolution, 258–263; foot, Fig. 123A

Campbell, Douglas, 295
Camptosaurus, 137, Fig. 77
Canis familiaris, 230, 231
Canis lupus, 230, 231
Carnivora, 223; described, 228–237
Carpenter, Frank M., 200, 202, 208
Case, E. C., 127
Casts, 6, Figs. 1, 3
Catarrhines, 276, 278
Cats, 229, 230, 233–236, Figs. 120, 121
Caudata, 108
Cavicornia, 249; described, 251, 252
Cayeux, L., 30, 31
Cephalaspids, 78, 79, 81, 88, 94, 106, Figs. 43, 45
Cephalaspis, 79, Fig. 43
Cephalopods, 21, 37, 40; described, 189–199, Figs. 96–101; phylogeny, 198, Fig. 101; shell forms, Fig. 100
Ceratopsia, 130, 146, 150, 162; described, 141–143
Cervicornia, described, 249, 250, Fig. 130
Chaetopoda, 24, 33, Figs. 7, 9
Chalicotheres, described, 242, 243, Fig. 123C; exceptions to Cuvier's law, 242
Cheirocrinus, Fig. 35
Chilopoda, 22
Chimpanzee, 278, 283, 286–288, 290
Chitons, 40
Choeromorus, Fig. 126
Chonetes, 303, 304
Chordata, 11, 22, 23, 25, 27, 41, 73, 84, 86
Civets, 229
Clark, A. H., 66
Clark, H. L., 66
Clathrotitan, Fig. 105
Cleland, H. F., 45
Climatius, 92
Cobb, Irvin S., 39
Coccosteus, 81, 82, Fig. 46
Cockerell, T. D. A., 208
Cockroaches, 202, 203
Coconuts, Fig. 153
Codfish, Fig. 50
Coelenterates, 20, 23, 24, 26–28, 32, 33, 40, 62
Coelomates, 26, 27, Figs. 13E,F
Coelurosaurs, 134, 145, 146, 150, 186
Coleoptera, 204, 205
Collembola, 202, 204, Fig. 103
Compsognathus, 133, 186
Comstock, J. H., 201
Condylarths, 223, 225, Fig. 114

INDEX

Cook, Harold, 258
Cope, Edward Drinker, 17, 120, 124, 127, 302
Corals, 20, 33, 40, Fig. 12D; described, 62; extinction of, 65; formation of skeleton, 62, 63; history, 65; larva, Fig. 32B; symmetry, 63, 64
Cordaites, 297, 309
Corythosaurus, 139
Cotylosaurs, 121-124, 126, Figs. 64, 65
Credner, H., 113
Creodonts, 225, 229, 230
Crinoids, 20, 40, 62; camerate, 69, Fig. 34; described, 68; free-swimming, Fig. 33; history, 69; range, 301, Fig. 194; stalked, Fig. 33
Crioceracones, 198, Fig. 100
Crocodiles, 125, 129, 144, 145, 149, 162; marine, 160
Cro-Magnons, 284, 285
Crossopterygians, 97, 102, 106, Figs. 53-56; amphibian characteristics of, 103, 104
Crustaceans, 22, 23, 32, 37, 55, Fig. 10
Cryptolithus, 51-53, Fig. 21
Cryptozoön, 30, 32
Cuttlefish, 189
Cuvier, G. L. C. F. D., 128, 168, 242, 269
Cynodictis, 231, Fig. 118
Cynodonts, 212, 214, 215, 219
Cynognathus, 211, 214, 219
Cyon, 271
Cyrtoceracones, 194, Fig. 100
Cystids, 20, 23, 24, 33, 36, 62; as ancestors, 71; described, 70, 71, Fig. 35; range, 301, Fig. 154
Cystoids, *see* Cystids

Daly, Reginald A., 33, 34, 36, 65
Daphaenus, see Daphoenus
Daphoenus, 232-233
Darrah, William C., 294
Dart, Raymond, 289
Darwin, Charles, 253
David, Sir Edgeworth, 32
Dawson, William, 282
Deane, James, 128, 129
De Chardin, Father Teilhard, 282
Deer, 227; history, 249, 250
Deinosauria, *see* Dinosaurs
Dendroidea, 43-45, Figs. 17, 19A,B
Dermaptera, 207
Dermoptera, 223, 225
Diacodexis, 261

Diacodon, Fig. 115
Diapsids, 125, 126, Fig. 67C,D
Diatoms, 25; as food, 293, 294
Diatryma, 180, 181, Fig. 90
Dibranchiata, 189, 190, Fig. 96
Diceratheres, 257, Fig. 136
Diceratops, 142
Dicotyledons, 298
Dictyonema, Fig. 17
Dimetrodon, 127
Dimorphodon, 171
Dinichthys, 81-83, 106, Fig. 47
Dinictis, 234, 236, Fig. 120A
Dinohyus, 245, Fig. 127
Dinornis maximus, 182
Dinosaurs, 125; ancestry, 145; bones in quarry, Fig. 5; brains, 148, 149; causes of extinction, 148-152, 298, 299; classification, 130; described, 128-152; food, 150; origin of bipedality, 146-148; possible ancestors of birds, 185, 186; range, 302, Fig. 154; sacral brain, 149; tracks, 128, 129, Fig. 1; where found, 16
Diplodocus, 18; described, 135, 136, 148, Fig. 72
Diplopoda, 22, 103
Diplovertebron, 123, Figs. 57, 58
Dipnoi, 97, 100
Diprotodonta, 215
Diptera, 201, 207
Dogs, 229-231; genealogy, Fig. 119
Dolichosoma, 115, Fig. 63
Dollo, L., 182
Draco, 164
Dragonflies, gigantic, 203; true, 204, 207, Fig. 106
Drepanaspis, 76-78, Fig. 42
Dromatherium, 214, Fig. 109
Dromomeryx, 250, Fig. 130
Dryopithecus, 278, 287-290, Figs. 147B, 148B
Dubois, Eugene, 279

Earwigs, 207
Eaton, George, 168
Echinoderms, 20, 23, 27, 28, 32, 40, 62; as ancestors of vertebrates, 93, 94; described, 66-72
Edaphosaurus, 127, Fig. 70
Edentates, 223, 226; described, 266-269, Fig. 143
Edrioasteroids, 20, 23, 24, 33, 71, Fig. 36
Edwardsia, 66, Fig. 32A

INDEX

Elasmobranchs, 74, 78, Fig. 39
Elephants, 227; history, 264, 265, Fig. 141
Elephas, 279, Fig. 141
Elginia, 123, Fig. 65A
Embiaria, 204
Embolomeri, 111, 123, Figs. 57, 58, 59A,B
Emperoceras, 198
Emplectopteris, Fig. 152
Entelodonts, described, 245, 246, Fig. 127
Eoanthropus dawsoni, 282, 283, 287-289, Fig. 148C
Eodelphis, 217, Fig. 111
Eogyrinus, 105
Eohippus, 254-257, 311, Figs. 131, 140
Eoliths, 282
Eosauravus, 120, Fig. 69
Eozoön, 32
Epihippus, 254
Eporeodon, 246
Equus, 254, 255, 266, 271, Fig. 131; *E. caballus*, 266; *E. giganteus*, 266
Eryops, 111, Fig. 64A
Eudendrium, Fig. 16
Euparkeria, 145, 187, Figs. 93, 94
Eurypterids, 32, 40, 87, 88; described, 57, 58; habitat, 58, 59
Eurypterus, 31, 57, Fig. 28
Eusarcus, 58, 60, Fig. 30
Eusthenopteron, 98, 102, 104, Figs. 53, 54, 56
Euthacanthus, 92
Evolutionary change, times of, 307, 308; rate of, 25-28, 311

Favosites, 64, 66, Fig. 38
Felidae, see Cats
Felis, 10, 234, Fig. 120C; *F. atrox*, 235; *F. catus*, 10; *F. leo*, 10; *F. tigris*, 10
Ferns, 296, Fig. 4
Fish, 14; bony, 74, Figs. 39-41; fins, 75; scales, 74; skeleton, 73, 74
Fissipedia, 229, 230
Flies, 201, 207
Food, as a factor in evolution, 308-310
Foraminifera, 19, 24, 30, 33, Fig. 6
Fossils, collecting, 12-18; definition, 8; names, 9-11; processes of preservation, 5; states of preservation, 7; study, 8; submarine, 13; where found, 12

Ganoids, 74, 75, Fig. 39
Gaskell, W. H., 87
Gastroliths, 149

Gastropods, 21, 23, 32, 33, 36, 40; range, 301, Fig. 154
Geological timetable, xi
Gibbons, 278
Gilmore, Charles W., 136, 137
Giraffes, 249, 250
Glyptodonts, 266, 268, Fig. 143
Goats, 252
Goniatites, 196, Fig. 98
Gorgosaurus, 134, Fig. 76
Gorillas, 278, 286, 288, 290
Grabau, A. W., 58
Granger, Walter, 142
Graptolites, 20, 33, 40; described, 42, Figs. 15, 17-19; growth of colonies, 43, 44; habitat, 44; history, 45, 46
Graptolithus, 42
Grasshoppers, 203
Gregory, William K., 142, 184, 185, 215, 218
Gymnosperms, 297, 298
Gyroceracones, 194, Fig. 100

Hallopus, 132
Halysites, 64
Handlirsch, Anton, 200
Harrimania, 85
Harmer, S. F., 85
Hay, O. P., 269
Hedgehogs, 221, 223
Heilmann, Gerhard, 177, 185-188
Heintz, Anton, 82
Hemichordata, 85
Hemicyon, 232
Henderson, Lawrence, 310
Heptodon, Fig. 125A
Hesperornis, 177-180, 182, Fig. 89
Heteroptera, 207
Hexacorals, 63-66
Hickory nuts, Fig. 153
Hipparion, 311
Hippopotamus, 244, 279
Hitchcock, Edward, 128, 129
Holm, Gerhard, 42, 59
Holtedahl, Olaf, 30, 31
Hominidae, 11, 276, 286
Homo heidelbergensis, 280, Fig. 147C
Homo neanderthalensis, see Neanderthal man
Homo rhodesiensis, 289
Homo sapiens, 11, 208, 256, 275, 283, 288, 300, Figs. 148D,G
Homogalax, 257

Homoptera, 204, 205
Hoplophoneus, 236, Fig. 121A
Horse, evolution, 253–256
Horseshoe crab, 55, 56
Huntington, Ellsworth, 147
Huxley, Thomas Henry, 274
Hyaenarctos, 232
Hydra, 42
Hydroids, 20, 42, Fig. 16
Hydrozoans, 33, 37
Hyenas, 229
Hymenoptera, 207, 208
Hyolithes, Fig. 10
Hyolithids, 33
Hypsilophodon, 138
Hyrachyus, 256, 257, Fig. 123B
Hyracodon, 257
Hyracodonts, 256
Hyracotherium, 256
Hysterogenicones, 198, Fig. 100

Ichnology, 129
Ichthyornis, 177–179, 182, Fig. 92
Ichthyosaurs, 125, 160; described, 153–155, Figs. 79, 81
Ichthyostega, 109
Ichthyostegopsis, 109, Fig. 55B
Ictidosaurus, 212
Iguanodon, 137, Fig. 73
Implements of primitive man, 282
Insectivores, 215, 216, 218–220, 222, 229, 304, Figs. 112, 115; ancestors probably arboreal, 224
Insects, 41; characteristics, 200, 201; described, 200–208; effects of Permian glaciation on, 204, 207; in amber, 208, Fig. 1; oldest, 202; origin of metamorphosis, 207; origin of wings, 205; venation of wings, 201, 202, Fig. 104
Isograptus, Fig. 15

Jackson, Robert T., 126
Jefferson, Thomas, 268
Jellyfish, 20, 23, 24, 33, 36

Keith, Sir Arthur, 285, 290
Kiaer, Johan, 80, 91
Klaatsch, H., 286, 290
Knowlton, T. H., 147
Koch, Lauge, 109

Labyrinthodonts, 117, 121; described, 110–112

Lacewings, 204
Lakes, Arthur, 16
Lambe, Lawrence M., 142
Lameere, A., 201
Lanarkia, 77, 78
Lane, Alfred C., 33, 35
Lankester, Sir Ray, 60, 286
Lapworth, Charles, 44
Lasanius, 89–93, Fig. 51
Leidy, Joseph, 248
Lemmatophora, 206, Fig. 104
Lemuroidea, see Lemurs
Lemurs, 224, 288; described, 276, 277, Fig. 145
Lepidodendron, 296, Figs. 4, 151
Lepidoptera, 207
Lepidosiren, 99
Lepospondyli, 113
Limulus, 55–57, Fig. 26
Lingula, 21, 302, 304, 305, Fig. 11
Linnaeus, Carolus (Carl von Linnè), 9, 11, 42
Lithothamnion, 293
Litopterna, 223
Lituiticones, 194, Fig. 100
Lizards, 159, 162, 164
Llamas, 262
Lobe fins, 97, 101, 106
Logan, Sir William, 29
Loomis, Frederick B., 231, 244, 249, 250
Lophiodon, Fig. 124
Loxodonta, 265
Lucas, F. A., 183
Lucas, O., 17
Lull, Richard Swann, 147, 149
Lungfish, 97, 102, 106; habits of, 99
Lysorophus, 115

MacBride, E. W., 37
Machairodontidae, see Saber-toothed tigers
Machairodus, 235, Fig. 121B
Mackenzia, 66, Fig. 32C
Mammals, 22; archaic, 209–219; characteristics, 209–211; Cretaceous, 217–219; fossil, where found, 15; Jurassic, 214–217; Mesozoic, 213–219; Northern, invade South America, 271; origin, 211, 212; Paleocene, 221–227; Pleistocene, 264–273; Southern, invade North America, 266, 268, 269; subclasses, 210, 211; teeth, 210
Mammonteus, 265, Fig. 141
Mammuthus, Fig. 141

Man, ancestry, 274–291; characteristics of skeleton, 274–276; driven from the trees, 310; "expectation of life," 300–306
Mann, Albert, 30
Mantell, G. A., 128
Marsh, O. C., 16, 174, 253
Marsupials, 210, 215, 216, 218, 219, 222, 223
Martens, 229
Martynov, A. V., 200
Mastodons, 264, 265, 271, Fig. 141
Matthew, William Diller, 232, 236, 253, 267
Mayflies, 205
Mecoptera, 204, 207
Megalobatrachus, 116
Megalonyx, 269
Megaloptera, 204
Megalosaurs, 145, 146
Meganura, 203
Megasecoptera, 203, 205
Megatherium, 269
Merriam, J. C., 156, 236
Merychippus, 255, 311, Fig. 131
Merychyus, 247, Fig. 129
Merycodus, 250
Merycoidodon, 246, Fig. 128
Merycoidodonts, 227; carnivore-like characteristics, 248; described, 246–248
Mesohippus, 254, 255, 257, Figs. 131, 132
Mesosaurus, 156, Fig. 80
Metazoa, 25
Miacidae, 229, 231, 233
Miacis, 231, 233, Fig. 119
Micrococcus, 30
Microsaurs, 112–115, 117
Mimoceracones, 198, Fig. 100
Miohippus, 255
Miomastodon, Fig. 141
Miotapirus, Figs. 125, 125E
Mixosaurs, 156
Moas, 182, 183
Mocritherium, 264, Fig. 141
Molds, 6, Fig. 3
Moles, 221, 223
Mollusca, 21, 23, 27, 28, 33, 40
Monitors, 159
Monoclonius, 142
Monocotyledons, 298
Monotremes, 210, 218
Moodie, Roy L., 151
Moody, Pliny, 128
Moose, 250
Moropus, 243, Fig. 123C

Morosaurus, 149
Morse, Edward S., 21
Mosasaurs, 160, 162; described, 158, 159, Fig. 83
Moths, 207
Multituberculates, 213, 214, 218, 219, 222, Fig. 108
Mylodon, 269, Fig. 143

Nanosaurus, 137, 145
Nautilicones, 194, Fig. 100
Nautilus, 189–191, 194; described, 192, 193
Neanderthal man, 280, 281, 284, 287–289, 291, Fig. 148F
Needham, J. G., 201
Neoceratodus, 99
Neolenus, Fig. 21
Neolimulus, 56
Neoliths, 282
Neuroptera, 204, 207
Nimravus, 234, Fig. 120B
Nipponites, 198
Nopcsa, Baron F., 150, 183, 184
Notharctus, 276, Fig. 145A
Nothocyon, 231
Nothosaurs, 158
Nothrotherium, Fig. 143
Notoungulata, 223

O'Connell, Marjorie, 58
Octopus, 189, 190
Odonata, 204
Okapi, 250
Old age, racial, 161
Ophiacodon, 126, Fig. 68
Ophiocones, 198
Ophiurans, 68, Fig. 36
Opossums, 215, 217, 219, Fig. 111
Orangoutangs, 288
Oreodons, see Merycoidodonts
Ornithischia, 130, 136, 146, Fig. 71
Ornitholestes, 134, 186
Ornithomimus, 150
Ornithopoda, 130, 150; described, 136–139
Ornithosuchus, 145
Orohippus, 254, 311
Orthoceracones, 193, 194, Fig. 100
Orthogenesis, 300
Orthoptera, 207
Osborn, Henry Fairfield, 30, 124, 305
Osteolepis, 98, 100, 104, 105
Ostracods, 40

Ostracoderms, 88, 94; described, 76-82
Ostrich, pelvis, Fig. 93
Owen, Sir Richard, 128, 168, 182, 256
Oxyclaenidae, 229
Oxydactylus, 262

Palaeanodon, 226
Palaeodictyoptera, 202, 203, 205, Fig. 102
Palaeoloxodon antiquus, 265
Palaeomeryx, 249, 250
Palaeoscincus, 141
Palaeosimia, 287
Palapteryx, 182
Paleanthropus, 280, 287, 288
Paleaspis, 76
Paleolimulus, Fig. 26
Paleoliths, 282
Paleomastodon, 264, Fig. 141
Paleopithecus, 287
Pantotheres, 216, 218, 219, 223, 228
Paradoxides, 48, Fig. 20
Parahippus, 255, 311, Fig. 133
Parapithecus, 279, 286-288, 290, Fig. 147A
Parapsids, 125, 126, Figs. 67A,B
Parasuchians, 144
Pareiasaurus, 123
Parelephas, Figs. 141, 142
Patten, William, 81, 87, 88
Peccaries, 227, 266; described, 244, 245
Pecora, described, 248-252
Pelecypods, 21, 40; range, 301, Fig. 154
Pelion, Fig. 59
Pelycosaurs, 126, 127, 211, Fig. 68
Pentremites, 70, Fig. 37
Perissodactyls, 226; characteristics, 239, 240, Figs. 123B,C; described, 239-243, 253-258
Perlaria, 204, 205
Peterson, Olaf A., 247
Phareodus, Fig. 41
Phenacodus, Fig. 114
Phiomia, 264, Fig. 141
Phororhacos, 181
Phrynosoma, 123, Fig. 65B
Phyllospondyli, 112
Pigs, 244, 245, Fig. 126
Pinacoceras, 196, Fig. 98
Pinnipedia, 229
Pisces, 22
Pithecanthropus erectus, 279, 280, 287, 288
Placentals, 211, 215-216; dentition, 211; oldest known, 218, Fig. 112; place of origin, 222, 223

Planipennia, 204
Plants, as food, 220, 238, 293, 294, 297-299; Cambrian, 24, 25; emerge from the water, 295; first fruit-bearing, 297; history, 292-299; late Paleozoic, 296, 297; Mid-Devonian, 296; oldest terrestrial, 294; Pre-Cambrian, 29, 30, 32
Plateosaurus, 131, 134, 145, 146, Fig. 75
Platyrrhines, 276
Plectoptera, 204, 205
Pleistocene glaciation, influence, 270, 271
Plesippus, 255
Plesiosaurs, 125, 160; described, 156, 157, Figs. 81, 82
Pliauchenia, 262
Pliohippus, 255, 311
Pliopithecus, 278, 287
Pocock, R. J., 60
Podokesaurus holyokensis, 131, 132, 134, 186
Poebrotherium, 261, Fig. 138
Polyprotodonta, 215
Porcupine, 223, 266
Porifera, see Sponges
Portheus, Fig. 40
Predentata, 130, 145, 146
Price, Llewellyn, 311
Primates, 223, 224, 275; subdivisions, 276
Primordial fauna, 22
Proavis, 188
Procamelus, 262, Fig. 139
Procompsognathus, 132, 186
Productids, 303, 304
Promerycochoerus, 247
Prongbuck, 251
Propliopithecus, 278, 286-288, 290, Fig. 148A
Protapirus, 241, Figs. 125A,C,D
Protelytron, Fig. 102
Protentomobrya, Fig. 103
Proteosaurus, 153
Protoceratops, 142, 146, Fig. 78
Protodonota, 203-205
Proelytroptera, 205, Fig. 102
Protohemiptera, 202, Fig. 105
Protohippus, 311
Protohymenoptera, 203
Protolabis, 262
Protolindenia, Fig. 106
Protomeryx, 263
Protoperlaria, 205, Figs. 104, 105
Protozoa, 19, 23-25, 27, 33, 36, Figs. 13A,B
Protylopus, 260, Fig. 138
Pseudocynodictis, 231, 233, Fig. 118

Pseudosuchians, as ancestors of birds, 186–188, Figs. 93, 94; as ancestors of dinosaurs, 145
Psilophyton, 295
Psocoptera, 204, 205
Pteranodon, 167, 168, 171, 172, Fig. 87
Pteraspis, 76, 77, 81, Fig. 42
Pteraspids, 76, 78, 88
Pterichthys, 80
Pteridosperms, 296, Fig. 152
Pterodactyls, 166, 167, Fig. 85
Pterodactylus, 167, 171
Pterosaurs, 125, 186; described, 164–168; habits of, 168–172
Pterygota, 201, 202, 206
Pterygotus, 31, 57, Fig. 29

Raasch, R. I., 56
Raccoons, 229
Radial evolution, 305
Radiates, 62–72
Radiolarians, 19, 24, 30, 33, Fig. 6
Reed, Bill, 17
Reindeer, 250
Relicts, 304
Reptiles, 22; characteristics of skeleton, 119, 120; classification, 124–126; extinction, 161–163; flying, 14, 162, 164–172; marine, 14, 153–163; phalangeal formula, 120; range, 301, Fig. 154; temporal fenestrae, 124–126
Rhachitomi, 111
Rhamphorhynchus, 166–168, Fig. 86
Rhinoceroses, evolution, 256–258
Rhynia, Fig. 150
Richthofenia, 304
Robergia, 52
Rodents, 223
Romer, Alfred S., 150
Ruedemann, Rudolf, 59, 192, 302
Ruminants, described, 248–252

Saber-toothed tigers, 233–235, 269, Figs. 121, 122
Salterella, 193
Sauripterus, 98, 100, 102, Fig. 54
Saurischia, 130, Fig. 71; described, 130–136
Saurolophus, 139
Sauropods, 130, 146, 161; described, 134–136
Sayles, Robert W., 65
Scelidosaurus, 140
Schuchert, Charles, 147
Scorpion flies, 204
Scorpions, 60, 88, Fig. 30

Scott, William B., 229, 250, 261
Scyphozoa, 24
Sea cucumbers, 20, 66, 72
Seals, 229
Sea urchins, 20, 40, 71
Seeley, H. G., 164, 165, 169–171
Selenodonts, 244, Fig. 129
Sertularia, 42
Sesamodon, 211
Seymouria, 121, Figs. 64B,C
Sharks, 74, 78, 97, 100, Fig. 39
Shimer, Hervey Woodburn, 312
Shrews, 221, 223
Sidneyia, 58, Fig. 10
Sigillaria, 296, Fig. 151
Simiidae, 276, 285, 286
Simpson, George Gaylord, 215, 218, 222
Sinanthropus pekingensis, 280, 287, Fig. 148E
Sinopa grangeri, Fig. 117
Skeleton, calcareous, formation of, 34–39, 41; in corals, 62; in echinoderms, 66, 94; in vertebrates, 94; siliceous, formation of: in animals, 39; in plants, 294
Skull, bones, 104; fenestrae, 124–126
Skunks, 229
Sloths, ground, 266; described, 268, 269; tree, 268
Smilodon, 234, 235, Figs. 121C, 122
Snakes, 119, 125, 158, 162, 164
Sollas, W. J., 42, 115
Sphaerocoryphe pseudohemicranium, 10
Sphenodon, 122, 125, 162
Spiders, 55, 61
Spirorbis, 302, 304
Spirula, 190, 191
Sponges, 19, 20, 23, 24, 26–28, 30, 32, 33, 36, 37, Figs. 2, 7, 8, 13C
Squamata, 187
Squids, 189
Starfish, 20, 40, 66–68
Stegocephalia, 117, 121; described, 110–115
Stegodon, 279, Fig. 141
Stegosaurs, 130, 140, 146, 151
Stegosaurus, 140, Fig. 74; sacral "brain," 149, 162; small brain, 148
Stenodictya, Fig. 102
Stenomylus, 263, Fig. 134
Stensiö, E. A., 79, 81
Stereospondyli, 112
Sternberg, Kaspar Maria, Graf zu, 53
Stetson, Henry Crosby, 13, 78, 80, 90
Stigmaria, 296

INDEX

Stoermer, Leif, 55
Stone flies, 205
Strongylocentrotus droebrachiensis, 10
Strophalosia, 303
Struthiomimus, 134, 150, 186
Stylonurus, 58, Fig. 27
Styracosaurus, 142, Fig. 73
Suioidea, 244
Sus scrofa, 245
Swim bladder, 99
Symmetrodonts, 216–219, Fig. 110
Symphysops, 48
Synapsids, 125, 126, 212, Fig. 66B
Synaptosauria, 125

Tabulata, 64
Taeniodonts, 223, 226
Taeniolabis, 213, Fig. 108
Tait, David, 78
Talbot, Mignon, 131
Tapirs, 266, 305; described, 240–242, Figs. 124, 125
Tapirus, 241, Fig. 125F
Tarsioids, 276; described, 277–279, 286, 288
Tarsius, 277
Taxonomy, 9–11
Teleoceras, 258, Fig. 135
Teleosts, 74, 75, Figs. 39, 40, 41
Temnospondyli, 111
Tetonius, 277, 278, Fig. 146
Tetrabranchiata, 189, 193, 194
Tetracorals, 63, 64, Fig. 31; range, 301, Fig. 154
Tetrapods, bones of limbs, 97; oldest, 99, 108–118
Thallatosuchia, 160
Theriodonts, 211, 212, 219
Thelodus, 77, 78, 80, 81
Therapsida, 211, 212
Theropoda, 139, 145; described, 130–134
Thorpe, Malcolm R., 246, 248
Thrinaxodon, 211, Fig. 107
Thrips, 205
Thysanoptera, 204, 205
Tillodontia, 223, 225
Tillyard, R. J., 32, 200, 202
Titanichthys, 81, 83, 106
Tomarctus, 231
Torticones, 194, 198, Fig. 100
Toxodontia, 223
Trachodon, 138, 139
Traquair, R. H., 78, 80

Trails, 7, Fig. 1
Triceratops, 18, 141, 142
Trichoptera, 207
Triconodonts, described, 214–216, 218, Fig. 109
Trigonias, 257
Trilobites, 23, 32, 33; habits, 50–53; crawling, Figs. 23, 24; feeding, Fig. 25; molting, Fig. 22; swimming, Fig. 24; relationships, 50; structure, 47, 48
Trinucleus, 52
Trioracodon, 214, Fig. 109
Trituberculates, 216, 219
Tritylodon longaeus, 213, Fig. 108
Turriliticones, 198
Turtles, 124, 125, 160, 162, 304, Fig. 84
Tylopoda, 252
Tyrannosaurus, 133, 134, 141, 148, 149

Uintacrinus, 70
Uintatheres, 226
Undina, 100
Ungulata, described, 239–252
Urodela, 108
Ursus, 231, 232

Varanus, 159
Vermes, 20, 23, 24
Vertebrae, embolomerous, 111, Figs. 59A–C; lepospondylus, 113; phyllospondylus, 113, Fig. 57; double-ringed reptilian, 122, Figs. 59E,F; rhachitimous, 111, Fig. 59D
Vertebrates, 22, 73; characteristics, 75; emerge from water, 96; formation of skeleton, 94; origin: anaspid theory, 88–93; amphioxus theory, 84; annelid theory, 86; arthropod theories, 87; chordate theory, 84
Volborthella, 193
Vulpes, 231

Walcott, Charles D., 24, 25, 29–32, 34, 35, 58, 66
Walruses, 229
Wapiti, 250
Waptia, Fig. 10
Wasps, 208
Watson, D. M. S., 92, 93, 100, 105, 110
Weasels, 229
Westoll, J. S., 104
Wetmore, Alexander, 174, 180
Whales, 227
Wheeler, William Morton, 208
Whittard, W. F., 110

Wieland, George R., 160
Wilder, H. H., 86
Willey, Arthur, 192
Williston, Samuel W., 16, 120, 124, 156, 166
Wiman, Carl, 42, 79
Wolves, 230, 231

Woodward, Sir Arthur Smith, 282, 289
Worms, *see* Vermes

Xiphosura, 55, 56

Youngina, 126

Bei Fragen zur Produktsicherheit wenden Sie sich bitte an:
If you have any questions regarding product safety,
please contact:

Walter de Gruyter GmbH
Genthiner Straße 13
10785 Berlin
productsafety@degruyterbrill.com